Vincent Voet, Jan Jager, Rudy Folkersma
Plastics in the Circular Economy

Also of interest

Circular Plastics Technologies.
Chemical Recycling
Knauer, 2024
ISBN 978-1-5015-2328-1, e-ISBN 978-1-5015-1561-3

Plastics: The Environmental Issue.
Türk, 2025
ISBN 978-3-11-064139-4, e-ISBN 978-3-11-064143-1

Sustainable Products.
Life Cycle Assessment, Risk Management, Supply Chains, Eco-Design
Has, 2024
ISBN 978-3-11-131482-2, e-ISBN 978-3-11-131546-1

BioProducts.
Green Materials for an Emerging Circular and Sustainable Economy
Vijayendran (Ed.), 2023
ISBN 978-3-11-079121-1, e-ISBN 978-3-11-079122-8

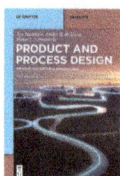

Product and Process Design.
Driving Sustainable Innovation
Harmsen, de Haan, Swinkels, 2024
ISBN 978-3-11-078206-6, e-ISBN 978-3-11-078212-7

Vincent Voet, Jan Jager, Rudy Folkersma

Plastics in the Circular Economy

2nd Edition

DE GRUYTER

Authors
Dr. Vincent Voet
NHL Stenden University of Applied Sciences
van Schaikweg 94
7811 KL Emmen
The Netherlands
vincent.voet@nhlstenden.com

Dr. Jan Jager
NHL Stenden University of Applied Sciences
van Schaikweg 94
7811 KL Emmen
The Netherlands
jan.jager@nhlstenden.com

Dr. Rudy Folkersma
NHL Stenden University of Applied Sciences
van Schaikweg 94
7811 KL Emmen
The Netherlands
rudy.folkersma@nhlstenden.com

ISBN 978-3-11-120029-3
e-ISBN (PDF) 978-3-11-120144-3
e-ISBN (EPUB) 978-3-11-120247-1

Library of Congress Control Number: 2024938440

Bibliographic information published by the Deutsche Nationalbibliothek
The Deutsche Nationalbibliothek lists this publication in the Deutsche Nationalbibliografie;
detailed bibliographic data are available on the Internet at http://dnb.dnb.de.

Contents

1   **Towards a circular economy —— 1**
1.1   History of plastics —— **1**
1.2   Introduction to the plastics economy —— **2**
1.3   The new plastics economy —— **5**
1.3.1   Challenges with regard to plastic pollution —— **5**
1.3.2   Transition to a circular economy —— **7**
1.4   The story continues —— **10**

2   **Introduction to polymer science —— 11**
2.1   Classification —— **11**
2.1.1   Origin —— **11**
2.1.2   Chemistry —— **12**
2.1.3   Properties —— **13**
2.1.4   Polymerization —— **13**
2.1.5   Application —— **13**
2.1.6   Nomenclature —— **14**
2.2   Structure of macromolecules —— **16**
2.2.1   Chain composition —— **16**
2.2.2   Chain architecture —— **17**
2.2.3   Chain regularity —— **19**
2.2.4   Chain length —— **20**
2.2.5   Characterization of macromolecules —— **23**
2.3   Synthesis of polymers —— **26**
2.3.1   Step-growth polymerization —— **26**
2.3.2   Chain-growth polymerization —— **35**
2.3.3   Copolymerization —— **46**
2.3.4   Polymerization techniques —— **51**
2.3.5   Post-polymerization reactions —— **54**
2.4   Polymer materials —— **57**
2.4.1   Polymer morphology —— **57**
2.4.2   Thermal transitions —— **58**
2.4.3   Characterization of T_g and T_m —— **61**
2.4.4   Viscoelasticity —— **62**
2.4.5   Crystallization of polymers —— **63**
2.4.6   Mechanical behavior of polymer solids —— **66**
2.4.7   Rheological behavior of polymer melts —— **68**
2.4.8   Electrical properties —— **69**
2.4.9   Liquid crystalline morphologies —— **69**
2.4.10   Block copolymer morphologies —— **69**
2.4.11   Vitrimers —— **70**

2.5   From polymer to product —— **72**
2.5.1   Polymer additives —— **72**
2.5.2   Processing of plastics —— **72**
2.5.3   Processing of fibers: Spinning —— **76**
2.5.4   Processing of foams —— **78**
2.5.5   Processing of fiber-reinforced plastics —— **78**
2.6   Exercises —— **80**
2.6.1   Classification —— **80**
2.6.2   Structure of macromolecules —— **81**
2.6.3   Synthesis of polymers —— **81**
2.6.4   Polymer materials —— **82**
2.6.5   From polymer to product —— **83**

3   **Bioplastics —— 84**
3.1   Biorefinery technologies —— **84**
3.1.1   Platforms —— **85**
3.1.2   Products —— **91**
3.1.3   Feedstocks —— **91**
3.1.4   Processes —— **92**
3.1.5   Process flow diagrams —— **94**
3.2   Biochemicals —— **95**
3.3   Chemistry of bioplastics —— **97**
3.3.1   Polyethylene (PE) —— **98**
3.3.2   Polypropylene (PP) —— **106**
3.3.3   Polyvinyl chloride (PVC) —— **113**
3.3.4   Polystyrene (PS) —— **115**
3.3.5   Polyvinyl acetate (PVAc) —— **121**
3.3.6   Polyacrylic acid (PAA) —— **123**
3.3.7   Polymethyl methacrylate (PMMA) —— **127**
3.3.8   Polyacrylonitrile (PAN) —— **130**
3.3.9   Polyethylene terephthalate (PET) —— **133**
3.3.10   Polyethylene furanoate (PEF) —— **143**
3.3.11   Polytrimethylene terephthalate (PTT) —— **144**
3.3.12   Polybutylene terephthalate (PBT) —— **146**
3.3.13   Thermoplastic polyetherester (TPEE) —— **150**
3.3.14   Isosorbide-based polymers —— **152**
3.3.15   Polyamide 6 (PA6) —— **155**
3.3.16   Polyamide 6,6 (PA6,6) —— **159**
3.3.17   Polyphthalamides (PPA) —— **168**
3.3.18   Polyether block amides (PEBA) —— **168**
3.3.19   Aramides (aromatic polyamides) —— **169**
3.3.20   Elastane —— **170**

3.3.21   Polybutylene succinate (PBS) —— **171**
3.3.22   Poly(butylene-*co*-succinate-*co*-adipate) (PBSA) —— **172**
3.3.23   Polyhydroxyalkanoates (PHAs) —— **173**
3.3.24   Polylactic acid (PLA) —— **179**
3.3.25   Polyglycolic acid (PGA) —— **183**
3.3.26   Poly(butylene-*co*-adipate-*co*-terephthalate) (PBAT) —— **186**
3.3.27   Poly(butylene-*co*-succinate-*co*-terephthalate) (PBST) —— **187**
3.3.28   Polycaprolactone (PCL) —— **188**
3.4   Bioplastics —— **189**
3.4.1   Bio-based plastics —— **191**
3.4.2   Biodegradable plastics —— **196**
3.4.3   Global production of bioplastics —— **204**
3.5   Exercises —— **207**
3.5.1   Biorefining technologies —— **207**
3.5.2   Biochemicals —— **207**
3.5.3   Chemistry of bioplastics —— **207**
3.5.4   Bioplastics —— **208**

4   **Recycling of plastics —— 209**
4.1   Circular plastics —— **209**
4.2   Plastic packaging —— **211**
4.2.1   Polymers used in packaging —— **211**
4.2.2   Multilayer packaging —— **215**
4.2.3   Decoration of packaging —— **218**
4.3   Mechanical recycling —— **219**
4.3.1   First sorting —— **220**
4.3.2   Second sorting —— **228**
4.3.3   Spectroscopic methods —— **233**
4.3.4   Marker systems —— **241**
4.3.5   Post-processing —— **243**
4.3.6   Applications of recyclates —— **250**
4.3.7   Challenges in mechanical recycling —— **252**
4.4   Chemical recycling —— **259**
4.4.1   Chemical recycling techniques —— **260**
4.4.2   Solvent-based purification —— **261**
4.4.3   Feedstock recycling —— **262**
4.4.4   Depolymerization —— **266**
4.5   Sustainable plastic product development —— **271**
4.5.1   Application of recyclates in products —— **272**
4.5.2   Design for recycling —— **276**
4.5.3   Recyclability incorporating in design —— **277**
4.5.4   Life-cycle assessment —— **281**

4.6 Exercises —— **284**
4.6.1 Circular plastics —— **284**
4.6.2 Plastic packaging —— **285**
4.6.3 Mechanical recycling —— **285**
4.6.4 Chemical recycling —— **286**
4.6.5 Sustainable plastic-product development —— **286**

Acknowledgment —— **289**

Bibliography —— **291**

Index —— **295**

1 Towards a circular economy

1.1 History of plastics

Many consumer goods that we use in our daily life are made from plastics. Today, we cannot imagine a world that exists without these versatile materials. Plastics help to keep us safe and healthy, and they improve the shelf life of food products. They make our everyday lives convenient in many ways. Halfway through the last century, however, we hardly used any plastics. The availability of plastic materials was fairly limited as was the number of possible applications. How did plastics transform from being so rare to becoming all around us?

In general, plastics consist of large molecules, named *polymers*, mixed with additives such as plasticizers, colorants, or flame retardants that improve the properties of the final product. Polymer science is considered a relatively new field of study. Nevertheless, polymeric materials have been applied for many centuries. In fact, more than 3,000 years ago, the Olmecs, an ancient Mesoamerican civilization in Mexico, played their "pok-ta-pok" game with a ball made from natural rubber. They are believed to be the first to actually process polymers from nature. It was not until the 1840s that Charles Goodyear in the United States patented the vulcanization of rubber using sulfur. After this discovery, rubber was soon adopted for multiple applications, including tires and footwear. The first synthetic polymer that was not derived from plants or animals but from fossil fuels was invented in 1907 by Leo Baekeland. His phenol formaldehyde *resin* was unique due to its hardness and heat-resistant properties, and it was named Bakelite®, after his inventor.

In 1920, Hermann Staudinger published his paper entitled "Über Polymerisation", in which he proposed that polymers are molecules with a high molar mass, composed of a large number of small building blocks. The fundamental understanding of this new class of materials, which Staudinger called *macromolecules*, lead to the development of a wide variety of new synthetic plastics and is therefore often considered to be the birth of polymer science. A few decades later, Staudinger was awarded the Nobel Prize in Chemistry for his pioneering work.

Today's commodity plastics, such as polyethylene (PE), polyvinylchloride (PVC), polystyrene (PS), and polyamide (PA), all became commercially available in the 1930s (Figure 1.1). The extensive needs of the military during the Second World War further increased the industrial growth of plastics, and other polymers were developed including polyester and Dupont's polytetrafluoroethylene (PTFE), better known as Teflon®.

Synthetic polymers, applied in plastics, rubbers, and fibers, became increasingly popular during the second half of the 20th century, when more and more products found their way to the market. Styrofoam™ was invented, and the PE bag made its first appearance, as well as Tupperware®, developed by Earl Silas Tupper who clev-

https://doi.org/10.1515/9783111201443-001

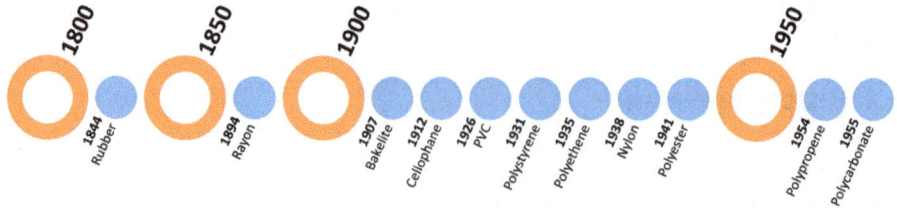

1800 *1850* *1900* *1950*

1844 Rubber — 1894 Rayon — 1907 Bakelite — 1912 Cellophane — 1926 PVC — 1931 Polystyrene — 1935 Polyethene — 1938 Nylon — 1941 Polyester — 1954 Polypropene — 1955 Polycarbonate

Figure 1.1: Timeline presenting the rise of the age of plastics.

erly promoted his air-tight PE food containers through a network of housewives. In the late 1950s, LEGO® patented its block coupling–decoupling system, and Mattel unveiled the first Barbie doll. Later on, acrylic paints were developed and polyethylene terephthalate (PET) beverage bottles were introduced. By the end of the 1970s, plastics were the most common type of material worldwide.

Nowadays, polymers can be tailored with high precision to meet the requirements of modern society. Annually, millions of tons of plastics are being produced worldwide. The applications are diverse: from low-cost products such as clothing, food packaging, and children toys to high-performance plastics in composites used in spacecraft, mobile phones, membranes, and medical implants such as artificial heart valves. *Plastics* seem to last forever. Unfortunately, this advantage turns out to be a serious disadvantage with respect to our environment. Since most plastics decompose very slowly, if at all, plastic waste can stay in landfills for thousands of years. Even more concerning is the part that leaks into our ecosystem, like the (micro)plastics that find their way to oceans forming floating junkyards better known as *plastic soup*. These problems need to be addressed, and current efforts focus on (micro)biologically degradable polymers, industrially compostable materials, and the process of recycling plastics [1].

1.2 Introduction to the plastics economy

The largest fraction of industrial polymers is derived from petroleum, or naphtha. *Crude oil* that comes out of the soil contains a multitude of components: various hydrocarbons, mostly alkanes, cycloalkanes, and various aromatic hydrocarbons. In an oil refinery, the crude oil is separated into different components, and large molecules are broken down into smaller molecules. This is a process referred to as *steam cracking*. The petrochemical industry receives the refined oil and creates monomers through chemical reactions and polymers through polymerization reactions. Next, the *plastics industry* produces all sorts of products from those polymeric materials. The production process is schematically depicted in Figure 1.2. Plastics are usually not unambiguous materials. There are many distinguishable types and numerous grades of plastics available on the market, each with its own characteristics, properties, and applications. Most often, plastics are designed for a specific application, by choosing the right combination of polymer and additive type, to meet the demands for that application in the most efficient manner.

Figure 1.2: Schematic production process for plastics from crude oil.

Plastics are produced in very large volumes. According to Plastics Europe, the association of plastic manufacturers in Europe, 400 million metric tons of plastics were produced worldwide in 2022 [2]. This enormous pile of materials equals 7,500 Titanic ships, or circa 40,000 Eiffel Towers. These plastics include thermoplastics, thermosets, elastomers, adhesives, coatings, sealants, and certain fibers. In past years, the global production capacities of plastics increased enormously. In contrast, only 1.5 million metric tons of plastics were produced worldwide in 1950.

The largest share of the plastics production volume is melt processable plastics, so-called *thermoplastics*. PE (26%), polypropylene (PP, 19%), polyvinyl chloride (PVC, 13%), PET (6%), and PS (5%) comprised 70% of the total global production volume of plastics in 2022. Assuming that the current world population is about 8.1 billion people (October 2023), which means that on average more than 49 kg of plastics are produced annually per capita of the world's population. We can expect that the production volumes of plastics will increase even further in upcoming years. Some recent studies even predict that the total production volume will double in the next 20 years. It has been estimated that in total ca. 10,000 million metric tons of virgin plastics have been produced globally to date.

China is the largest producer of plastics in the world, producing 128 million metric tons in 2022, which equals a 32% market share. As shown in Figure 1.3, more than half of the *production volume of global plastics* takes place in Asia, followed by North America (17%) and Europe (14%).

According to Plastics Europe, the total converter demand for plastics amounts to 50 million metric tons in Europe. The European *plastic converter demand* includes both thermoplastics and polyurethanes and some other plastics (thermosets, adhesives, coatings, and sealants). In terms of market segments, almost 39% of all plastics are used in the packaging industry. Other important sectors are building and construction, automotive, electronics, household, and agriculture. In addition, there are a multitude of other applications for which smaller quantities of plastics are used. Ex-

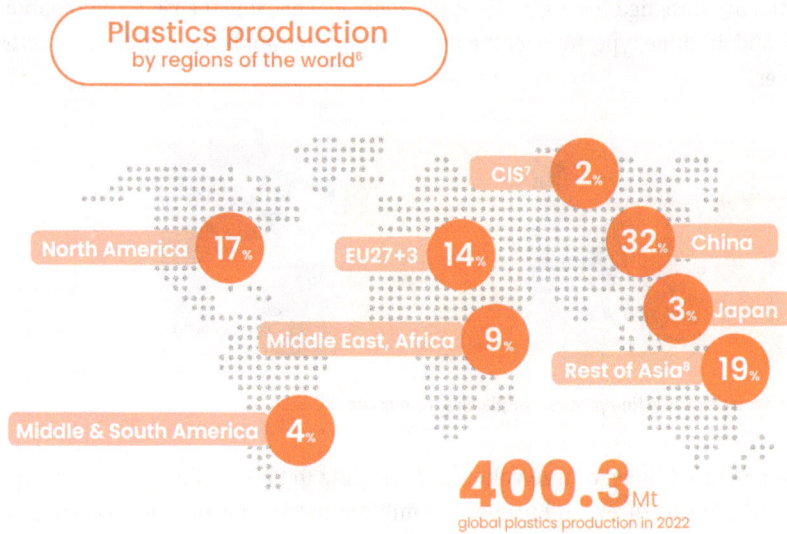

Figure 1.3: Distribution of global plastic production in the world in 2022, according to Plastics Europe [2].

Table 1.1: Plastic converter demand by market sector in 2021 [2].

Market sector	Total European converter demand by market sector in 2021 (million metric tons)	Percentage (%)
Packaging	19.7	39
Building and construction	10.7	21
Automotive	4.3	8.6
Electrical and electronics	3.3	6.5
Household, leisure, and sports	2.2	4.4
Agriculture	1.6	3.1
Others (e.g., medical)	8.6	17

amples are medical equipment, plastic furniture, and technical parts used for mechanical engineering. The total European converter demand by market sector in 2021 is presented in Table 1.1.

Only a few types of plastics are used in the *packaging industry*. Various grades of polyethylene (HDPE, MDPE, LDPE, LLDPE), PP, PET, PVC, and (expanded) PS ((E)PS) are the major players in the packaging industry (Table 1.2).

In 1988, the Society of the Plastics Industry (SPI) introduced the *Resin Identification Code* (RIC) system in an effort to develop consistency in plastics manufacturing and (recycled) plastics reprocessing. The Resin Identification Code assigns a number from 1 to 7, with a "chasing arrows" symbol around the number, to a piece of plastic to indicate its type. This coding system was developed for behind-the-scenes staff per-

Table 1.2: Plastic types used in the European packaging industry, including their Resin Identifications Codes (RIC) [2].

Plastic type	Packaging market share in Europe (%)	Resin Identification Code (RIC)
LDPE/LLDPE	31.1	4
PP	22.0	5
PET	18.9	1
HDPE/MDPE	18.3	2
(E)PS	5.7	6
PVC	2.1	3
Others	1.9	7

forming recycling of plastic household waste, to improve the sorting and recycling processes. Note that the coding system was never intended as a consumer communication tool. Consumers, however, often assume that this code indicates that a certain piece of plastic household waste is automatically *recyclable*, while that is not the case. A piece of plastic with the code may or may not be recyclable, as shown in Figure 1.4. The Resin Identification Code is currently under the control by ASTM International and is covered in a new international standard.

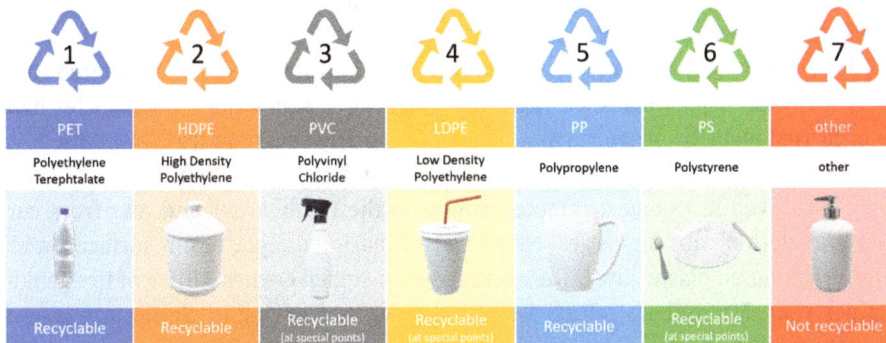

PET	HDPE	PVC	LDPE	PP	PS	other
Polyethylene Terephtalate	High Density Polyethylene	Polyvinyl Chloride	Low Density Polyethylene	Polypropylene	Polystyrene	other
Recyclable	Recyclable	Recyclable (at special points)	Recyclable (at special points)	Recyclable	Recyclable (at special points)	Not recyclable

Figure 1.4: Resin identification codes (RICs) according to the coding system of the Society of the Plastics Industry (SPI).

1.3 The new plastics economy

1.3.1 Challenges with regard to plastic pollution

Clearly, plastics are an integral part of our global economy. Besides all the benefits, the current plastics economy has led to serious (environmental) drawbacks as well. Millions of tons of plastics end up in the ocean each year. This equals the content of one garbage truck every minute. According to the Ellen MacArthur Foundation, our oceans will contain more plastics than fish (by weight) in 2050. The vast accumulation of plastic debris and microplastics in the world's oceans is typically referred to as *plastic soup*. The term effectively captures the idea of a plastic polluted ocean ecosystem, resembling a thick, soupy consistency rather than clear, clean water.

Larger pieces of plastic, such as packaging materials, fishing nets, or textiles, are typically called *macroplastics*. Under the influence of sunlight, oxygen, or friction, they become brittle and gradually break down into smaller and smaller plastic particles. While the rate of degradation depends on the circumstances, the process itself goes on forever. Fragments become *microplastics*, and microplastics finally become nanoplastics. The latter are so small that they can no longer be distinguished from grains of sand on the beach, and can hardly be observed even under the most advanced microscopes [3].

Surprisingly, a universally established definition for the term microplastics does not exist. In academic literature, however, microplastics are often defined as plastic particles up to 5 mm in dimensions with no defined lower size limit. Furthermore, nanoplastics are described as plastic particles in the submicron range, so below 1 µm. In the nanotechnology field, however, the term may refer to engineered particles below 100 nm, which is in line with the nanotechnology application size limit.

As an example, a single plastic bag made out of PE can break down into millions of pieces. Due to the process of ongoing fragmentation, the number of micro- and nanoplastics is increasing exponentially over time. Because of their lightweight, microplastics are easily transported over long distances. In 2014, it was discovered that polar ice is full of microplastics. At that time, it was assumed that the particles had been carried along by ocean currents and subsequently caught in ice. Nowadays, it is clear that microplastics are also carried by the wind. In fact, it "rains" microplastics every day, even in the most remote regions on the Earth. Wear and tear from car tires and clothing are the largest sources of primary microplastics in surface water after decomposed plastic litter. The average world citizen creates 0.81 kg of tire grinding per year. Microfibers released from washing and drying of clothes are also a major and difficult source of microplastics to combat. For each 5 kg of synthetic clothing, an average of 9 million microfibers is released per washing procedure.

Plastic pollution, and specifically the plastic soup, has far-reaching consequences, including harming marine ecosystems, endangering marine life through ingestion and entanglement, and potentially impacting human health. The latest research on the effects of microplastics on humans and other mammals shows significant behavioral changes due to microplastics added to the water and food supply. Researchers investigated the behavioral and inflammatory effects of plastic particles in the body, and found that microplastics pollute the body to the same extent as they do in nature and have reached the human blood stream [4].

1.3.2 Transition to a circular economy

More than 35 years after the launch of the RIC coding system (Figure 1.4), still only 14% of plastic packaging is collected for recycling [5]. Furthermore, at the current global production volume, about 5% of available crude oil is employed in the production of plastic products. If the strong growth of production volumes continues as expected, the consumption of crude oil by the entire plastics industry will account for a fifth of the total crude oil consumption by 2050.

At the beginning of 2016, a report was published with the title "The New Plastics Economy – Rethinking the Future of Plastics", issued by the Ellen MacArthur Foundation [6]. It envisions a time when plastic materials never become waste. Instead, plastics should re-enter the economy as valuable biological or technical feedstock, which aligns with the principles of a closed-loop economy, better known as the *circular economy*. In 2018, the Ellen MacArthur Foundation launched a new initiative, the New Plastics Economy Global Commitment. This initiative unites businesses, governments, and other organizations behind a common vision and targets to address plastic waste and pollution at its source. Mid-2019, over 400 organizations have signed the New Plastics Economy Global Commitment, like Apple, Phillips, Unilever, Coca-Cola, and Walmart. Applying the principles of a circular economy, the *New Plastics Economy* brings together key stakeholders to rethink and redesign the future of plastics, starting with packaging. The initiative advocates several ambitions, summarized in the following actions:

1. "Create an effective after-use plastics economy by improving the economics and uptake of recycling, reuse, and controlled biodegradation for targeted applications." (This is the cornerstone of the New Plastics Economy and its first priority and helps realize the two following ambitions.)
2. "Drastically reduce leakage of plastics into natural systems (in particular the ocean) and other negative externalities."
3. "Decouple plastics from fossil feedstocks by – in addition to reducing cycle losses and dematerializing – exploring and adopting renewably sourced feedstocks (biomass)."

It is rather difficult to attribute the concept of a circular economy to one specific initiator. In the 1970s, the principle of circularity gained more attention due to the groundbreaking report "The Limits to Growth" by the Club of Rome [7]. The report discusses the exponential growth of population and economy while having a finite supply of resources. In 1982, Walter Stahel published his paper "The Product-Life Factor," in which he first coined the term closed-loop economy [8]. Years later, Ellen MacArthur describes her vision of circular economy based on the above principles. In contrast to a *linear economy* that refers to a society that takes, makes, and disposes, the circular economy is both restorative and regenerative by design. In Figure 1.5, the butterfly model as proposed by the Ellen MacArthur Foundation is shown. A biological cycle regenerates living systems that provide renewable resources. On the other hand, the technical cycle aims to recover and restore materials and products. The general principle is to optimize resource yield by circulating products and materials at all times.

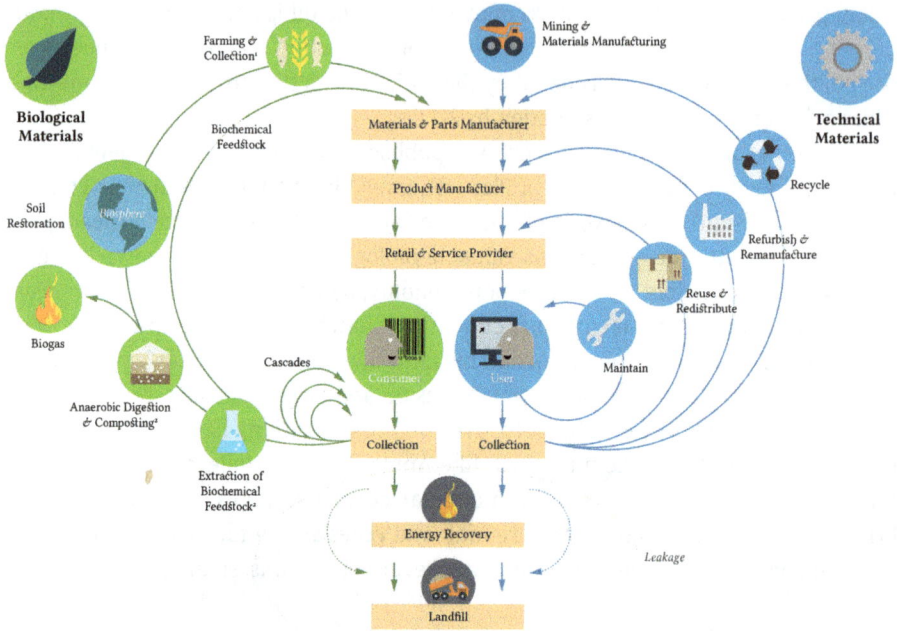

Figure 1.5: The circular-economy model, proposed by the Ellen MacArthur Foundation, representing a closed-loop system with biological and technical cycles of materials [6].

According to the United Nations (UN), the concept of circular economy, in which natural systems are regenerated, materials are kept in use, and waste does not exist, can contribute significantly to the implementation of 2030 Agenda for Sustainable Development (Figure 1.6), translated into the 17 *Sustainable Development Goals* (SDGs) [9]. A closed-loop economy holds promise for achieving SDG 7 (Affordable and Clean Energy), 8 (Decent Work and Economic Growth), 11 (Sustainable Cities and Communities), 12 (Sustainable Consumption and Production), 13 (Climate Action), and 14 and 15 (Life below Water and on Land). Clearly, the transition from a linear to a circular economy requires the joint effort of companies, policymakers, research institutes, and the society in general.

In June 2019, the World Economic Forum (WEF) published ten emerging technologies [10] that have the potential to provide major benefits to the global society and economy. Prior to this, an international steering committee of leading technology experts evaluated dozens of proposals from experts around the globe. In the end, the "Top 10 Emerging Technologies 2019" were identified. The topmost emerging technology was Bioplastics for a Circular Economy. According to the WEF, the development

Figure 1.6: The Sustainable Development Goals (SDGs), as proposed by the United Nations in the 2030 Agenda for Sustainable Development. Image used under license from Shutterstock.com, ID 1538140988.

of bio-based and/or biodegradable plastics can contribute to the goal of a *closed-loop plastic economy* in which plastics are derived from and converted back to biomass feedstock.

Later that year, the European Commission adopted the so-called *European Green Deal*. The European Green Deal is a set of policy initiatives with the overarching aim of making Europe climate neutral in 2050. In fact, it is the new roadmap for making the EU's economy sustainable. The Green Deal also includes a "circular-economy action plan" [11] that identifies initiatives along the entire life cycle of products, targeting, for example, their design, promoting circular-economy processes, fostering sustainable consumption, and aiming to ensure that the resources used are kept in the EU economy for as long as possible. This document states that "The Commission will develop requirements to ensure that all packaging in the EU market is reusable or recyclable in an economically viable manner by 2030, will develop a regulatory framework for biodegradable and bio-based plastics, and will implement measures on single use plastics." This led to the so-called single-use plastic (SUP) directive, which aims to significantly reduce the volume and environmental impact of disposable plastics [12]. A set of measures target plastic items found mostly on the European beaches, like cups, balloons, straws, bags, and bottles. In case sustainable alternatives are affordable and available, products made from SUP will no longer be sold on the markets of EU members.

1.4 The story continues

The closed-loop economy model presented in Figure 1.5 distinguishes between biological (or green) and technical (or blue) cycles. In the biological cycle, food and bio-based materials are designed to return through processes like anaerobic digestion and composting. A biological cycle regenerates living systems that provide renewable resources. On the other hand, the technical cycle aims to recover and restore materials and products. This is achieved through strategies such as reuse, repair, refurbish, and, eventually, recycling.

The structure of this book follows the circular economy model and its two cycles. First, an introduction to polymer science is given in Chapter 2 to communicate basic knowledge on the structure and behavior of plastic materials. It discusses various ways to make polymers and explores the unique structure–property relationship of macromolecules. Bio-based and biodegradable plastics are the main subject of Chapter 3. This chapter explores the processing of renewable feedstocks into biochemicals and ultimately bioplastics. In other words, the role of plastics in the biological cycle is discussed. Chapter 4 focuses on the mechanical and chemical recycling of plastics, as well as circular product development. The recycling of plastic products is part of the technological cycle. To summarize, in this book we will discover the role of "Plastics in the circular economy".

2 Introduction to polymer science

2.1 Classification

The success of plastics can be attributed to their low costs and wide range of properties that can be changed by tweaking the macromolecular structure. This unique structure–property relationship will receive specific attention in the remainder of this chapter. We will start by discussing various ways to categorize polymer materials.

The term polymer is derived from the Greek words πολύς (polus = "many") and μέρος (meros = "part"). *Polymers* can be defined as large molecules, sometimes referred to as macromolecules, composed of a large number of repeating units. The exact number required to meet the definition of a polymer depends on the size and chemical structure, but a good rule of thumb is at least 50 repeating units. Smaller molecules are named *oligomers*, which translates to "few parts".

The building blocks that form a polymer molecule are called *monomers*, literally meaning "one part". Monomers can undergo polymerization, in which they react together to form a polymer chain or network structure. Polymerizations are the subject of Section 2.3.

Considering the definition of a polymer, a tremendous amount of possibilities arise: flexible or stiff polymers, polymers derived from nature or those made in factories, polymers that absorb or repel water, polymers applied in airplanes or used in clothing, etc. To create some order, polymers are classified based on various categories, as displayed in Figure 2.1.

2.1.1 Origin

A wide variety of macromolecules is present in nature. Polysaccharides like starch and cellulose are formed in plants, while proteins such as silk and casein can be found in animals. Polynucleotides, known as DNA and RNA, play a crucial role in storing, coding, expressing, and regulating genetic information. While some *natural polymers* can be used directly as technical materials when harvested, other polymers need to be modified. For example, cellulose can be transformed into cellophane films or rayon and viscose fibers via the xanthate process. The hydroxyl groups in cellulose are modified in a basic environment, and cellulose is regenerated from viscose using acid. Also, technical rubber is produced by crosslinking latex from rubber trees via a process named vulcanization. Macromolecules made by chemically treating natural polymers are named *semi-synthetic polymers* [13].

Nevertheless, most polymers have a fully synthetic origin. They are created by the polymerization of monomers into macromolecules. One monomer polymerizes into a homopolymer, while two or more chemically different monomers make a copolymer

https://doi.org/10.1515/9783111201443-002

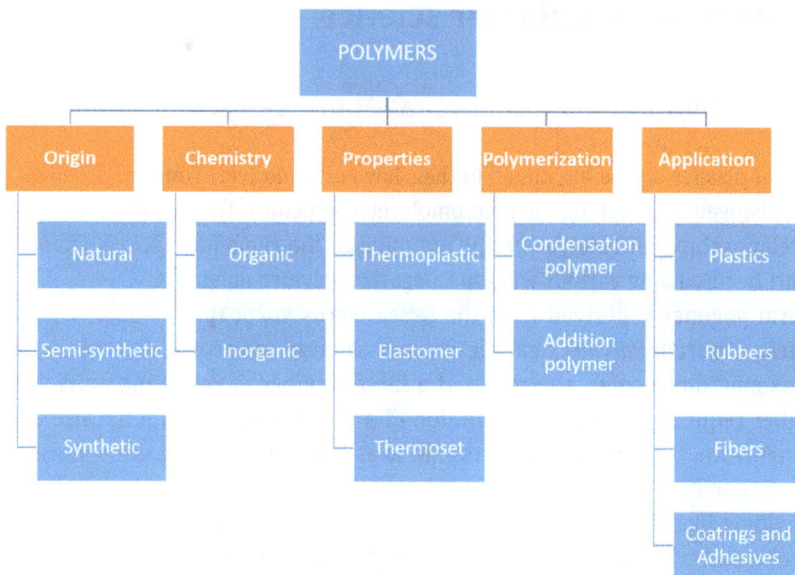

Figure 2.1: Classification of polymers based on origin, chemistry, (physical) properties, polymerization route, and application.

(Section 2.2.1). Polystyrene (PS), polyethylene (PE), polyethylene terephthalate (PET), and acrylonitrile-butadiene-styrene (ABS) are all examples of *synthetic polymers*.

2.1.2 Chemistry

Based on the chemical building blocks along the polymer chain, polymers can be subdivided into two classes: organic and inorganic macromolecules. In general, *organic polymers* are based on the element carbon (C). Their skeletal structure mainly consists of C, H, O, and N atoms. The vast majority of polymers is considered organic, including most synthetic polymers such as polyolefins, polyamides, polyesters, and acrylics.

In the most distinctive definition, *inorganic polymers* have a backbone that does not include carbon atoms at all. For instance, polydimethyl siloxane, better known as silicone rubber (when crosslinked), is composed of a –Si–O– backbone with methyl side groups. On the other hand, polyphosphazenes carry P and N atoms in the repeating unit. Polymers containing both inorganic and organic components, like DNA having organophosphate functionalities, are officially named *hybrid polymers* [14].

2.1.3 Properties

Related to the molecular structure, some polymer materials have the ability to flow at elevated temperatures. *Thermoplastics* are composed of separate linear or branched polymer chains. Approximately 90% of all plastics belong to this category, like polyethylene terephthalate (PET), polyethylene (PE), polymethyl methacrylate (PMMA), and polyamides (PAs). Upon heating, these materials can be processed in the molten state. Upon cooling, they return to their solid state. This cycle can be repeated, which enables the (mechanical) recycling of thermoplastic materials.

On the other hand, *thermosets* are crosslinked systems. Formation of the network, the so-called curing reaction, is typically initiated by heating or UV irradiation. Due to the crosslinked chains, a cured thermoset cannot be molten or dissolved. This type of materials is less suitable for recycling in a conventional way. The first synthetic polymer Bakelite® is a thermosetting phenol formaldehyde resin.

In between thermosets and thermoplastics, *elastomers* constitute weakly crosslinked systems. Elastomers, commonly known as rubbers, are recognized by their elastic behavior. Depending on the nature of the crosslinks, we can distinguish thermoplastic elastomers (TPEs) and thermoset rubbers, as will be explained in Section 2.2.2.

2.1.4 Polymerization

Chemists prefer to define macromolecules based on the manner in which the polymer chains are constructed during polymerization. *Step-growth polymerization* usually involves a condensation reaction. Classic examples are the reaction between difunctional acids with difunctional alcohols, or amines, resulting in the formation of polyesters and polyamides, respectively. The corresponding products are referred to as *condensation polymers*. Most *chain-growth polymerizations* occur via continuous addition of double bonds, for instance, during polymerization of ethylene (E), propylene (P), or styrene (S). Hence, polyethylene (PE), polypropylene (PP), and polystyrene (PS) are all *addition polymers*.

2.1.5 Application

Defined by the polymer industry, the most common areas of application are plastics, fibers, and rubbers. The *plastics* industry comprises packaging, consumer goods, coatings, and adhesives. Technical or textile *fibers* can originate from natural resources, like semisynthetic rayon (or viscose) fibers, or fossil resources, such as synthetic polyester yarn. The *rubber* industry comprises both thermoplastic as thermoset elastomers.

2.1.6 Nomenclature

The ultimate system of classification is, of course, naming. However, various terminologies have been developed in the polymer science community that can be rather confusing for outsiders. Table 2.1 provides an overview of the most common ways of *nomenclature*.

Table 2.1: Nomenclature of polymers, based on the building block(s), general structure, or industrial trade name.

Structural formula	Specific name	Abbreviation	General name	Trade name
	Polyethylene *or* Polyethene	PE		
	Polystyrene	PS		
	Polyethylene terephthalate	PET	Polyester	Terylene®
	Polydimethyl siloxane	PDMS	Silicone rubber	
	Polymethyl methacrylate	PMMA	Acrylic	Perspex® Plexiglass®
	Polycaprolactam	PA6	Polyamide 6	Nylon 6

Table 2.1 (continued)

Structural formula	Specific name	Abbreviation	General name	Trade name
	Poly(*para*-phenylene terephthalamide)	PPTA	Aramid	Kevlar® Twaron®
	Polytetrafluoroethylene	PTFE	Fluoropolymer	Teflon®

In general, macromolecules can be named for the typical functional groups within their repeating units, usually combined with the prefix "poly-". For instance, polyamide essentially refers to any polymer material with amide functionalities in the main molecular chain. In that sense, both proteins and nylons are designated as polyamides. Similar, the name fluoropolymer can refer to either polytetrafluoroethylene (PTFE), polyhexafluoropropylene (PHFP), or polyvinylidene fluoride (PVDF). In other words, naming polymers by their general structure is rather unspecific.

While (low molar mass) organic chemical compounds are systematically named by IUPAC nomenclature [15], another more practical terminology has been developed in polymer science. The basic principles of polymer nomenclature is founded on the original building blocks, for example, polymerization of ethylene leads to polyethylene, abbreviated as PE. In a similar way, methyl methacrylate (MMA) polymerizes into PMMA: polymethyl methacrylate. In certain cases, the repeating unit is used instead of the monomer. While the industrial synthesis of silicon rubber starts with dimethyldichlorosilane, PDMS stands for polydimethylsiloxane.

Occasionally, nomenclature based on building blocks is rather impractical for daily use. A good alternative in that case is the introduction of trade names. For instance, poly(*para*-phenylene terephthalamide) (PPTA) is better known worldwide as Kevlar® or Twaron®, the product names of PPTA given by DuPont and Teijin Aramid, respectively. Other examples include Perspex® (PMMA) from ICI and Teflon® (PTFE) from DuPont.

2.2 Structure of macromolecules

The previous paragraph already sketched the large diversity in the world of macromolecules. The chemical building blocks in the polymer backbone, the length of the chain, and the architecture of the macromolecules, all factors that influence the performance of polymers. It creates a valuable toolbox for polymer chemists and engineers to selectively change the properties by controlling the structure of the macromolecular chain.

2.2.1 Chain composition

A polymer is composed of building blocks that repeat along the (linear) chain. A chain that consists of only one single monomer type is called a *homopolymer*. The structural formula of homopolymers can be represented by depicting its repeating unit placed within square brackets (Figure 2.2). The number of repeating units is indicated by the subscript "*n*."

Figure 2.2: Schematic representation of homopolymer with "*n*" repeating units.

Macromolecules composed of at least two chemically distinct monomers are called *copolymers*. An infinite number of different copolymers can be synthesized by altering the types of monomers, as well as the chain sequence (Figure 2.3). For instance, if two monomers are polymerized in a statistically determined manner, a *random copolymer* (or statistical copolymer) is realized. Random copolymers are often denoted as poly(A-*co*-B), where A and B are two different monomers. The exact placement of monomers along the chain is determined by the A-to-B ratio and their relative reactivity, as will be explained in more detail in Section 2.3.3.

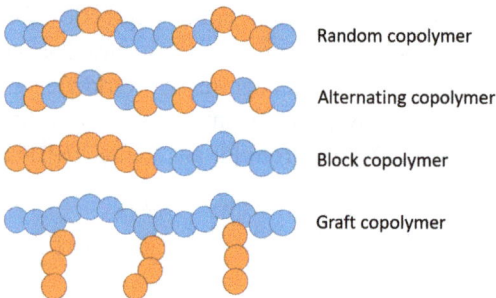

Random copolymer

Alternating copolymer

Block copolymer

Graft copolymer

Figure 2.3: Schematic representation of various copolymers with different monomer sequences.

If two monomers are built into the chain in an alternating fashion, the resulting macromolecule is an *alternating copolymer*, or poly(A-*alt*-B). Although commonly referred to as a homopolymer, polyethylene terephthalate (PET) is better described as an alternating copolymer resulting from the polycondensation of ethylene glycol and terephthalic acid. *Block copolymers* on the other hand are composed of clusters of identical repeating units along the backbone. Those clusters are referred to as blocks. Poly(A-*block*-B) is called a diblock copolymer, while poly(A-*block*-B-*block*-A) and poly(A-*block*-B-*block*-C) are triblock copolymers. Block copolymers can show intriguing phase behaviors at the nanoscale due to their well-defined molecular structure. Finally, if polymer blocks are attached as side chains to a different polymer, the resulting structure is called a *graft copolymer* or poly(A-*graft*-B). There are various ways to synthesize graft copolymers.

Copolymers synthesized from three chemically distinct monomers are named *terpolymers*. A famous example is acrylonitrile-butadiene-styrene (ABS), industrially made by polymerizing acrylonitrile (A) and styrene (S) in the presence of polybutadiene (B). The result is a graft copolymer having a polybutadiene backbone with side chains of a styrene-acrylonitrile copolymer. ABS combines the chemical inertness and heat resistance of acrylonitrile with the toughness of butadiene, while styrene adds to processability and rigidity. By varying the building blocks, the properties of ABS can be tuned as desired. Leaving out one of the comonomers leads to poly(styrene-*co*-acrylonitrile) (SAN), poly(styrene-*co*-butadiene) rubber (SBR), or poly(acrylonitrile-*co*-butadiene) rubber (NBR).

2.2.2 Chain architecture

Besides chemical diversity, the properties of polymers can be influenced by their (macro)molecular architecture. Thermoplastic polymers consist of either *linear* or *branched macromolecules*, or both (Figure 2.4). For instance, starch, a polysaccharide that is produced by plants as energy storage consists of both linear amylose and branched amylopectin. Synthetic polyethylene can be linear or branched depending on the polymerization method. While free radical polymerization leads to a highly branched structure referred to as low-density polyethylene (LDPE), coordination polymerization results in high-density polyethylene (HDPE) with a more linear chain architecture. Methods of polymerization are explained further in Section 2.3.

The molecular architecture of thermoset polymers is best described as a *network*. In principle, such a covalent network is in fact one large macromolecule. The network structure can be formed in various ways. One route is the covalent binding of separate linear chains (prepolymers) by reactive groups, so-called *crosslinking* agents (Figure 2.5(a)). The vulcanization of rubber using sulfur, invented by Charles Goodyear, follows this mechanism. The density of the crosslinked matrix depends on the concentration of crosslinking agents.

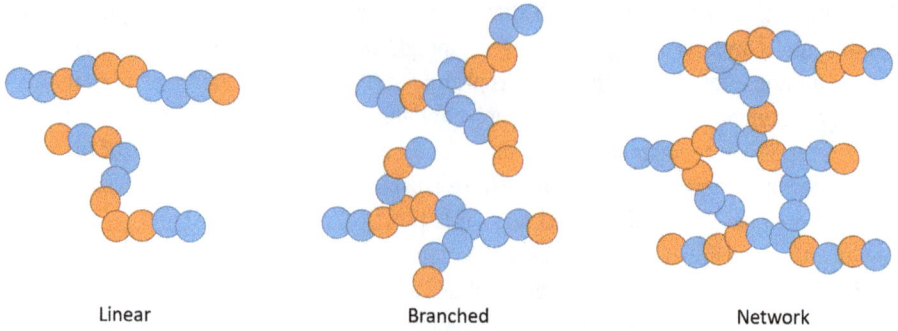

| Linear | Branched | Network |

Figure 2.4: Schematic representation of linear, branched, and network chain architectures.

Network formation can also be achieved via polymerization in the presence of mono-mers with a functionality of three or more (Figure 2.5(b)). For instance, polycondensa-tion of terephthalic acid with (trifunctional) glycerol instead of (difunctional) ethylene glycol leads to a macromolecular network structure.

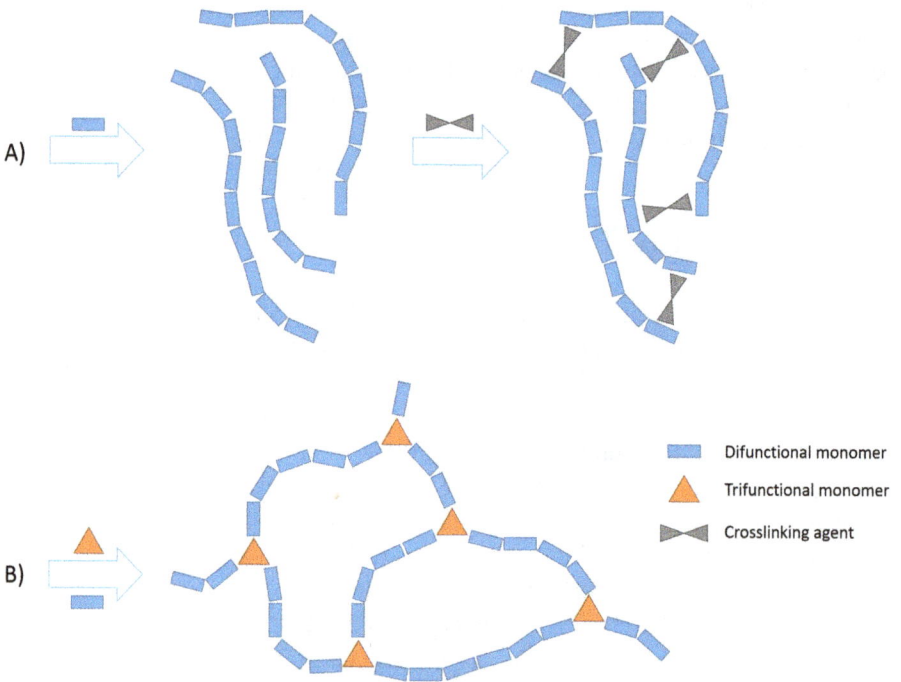

Difunctional monomer

Trifunctional monomer

Crosslinking agent

Figure 2.5: (a) Polymer network formation via (pre)polymerization followed by crosslinking reactions, and (b) polymer network formation via copolymerization in the presence of trifunctional monomers.

Not all network polymers are thermosetting materials. Depending on the type of crosslinks, a system may show thermoplastic behavior as well. While conventional rubbers rely on a covalently bonded network with so-called *chemical crosslinks*, elastomers such as thermoplastic polyurethane contain *physical crosslinks*. At lower temperatures, specific chain segments cluster in hard domains that behave as crosslinks. However, in contrast to covalent crosslinks, those domains are not stable at higher temperatures. In fact, the material becomes soft again, making a thermoplastic elastomer reprocessable. Upon cooling, the physical crosslinks return, i.e., the network structure of thermoplastic elastomers is reversible.

2.2.3 Chain regularity

The configuration along a polymer chain will influence the spatial (three-dimensional) arrangement and consequently the thermal and mechanical properties of a material. Like small molecules, macromolecules can exist as isomers. In other words, two polymer chain segments can be represented by the same molecular formula, but different structural formulas. Below we discuss a few examples.

2.2.3.1 Constitutional isomers

In constitutional isomerism, or structural isomerism, the atoms are connected in a different order. Considering the polymerization of a vinyl monomer like propylene, the monomer can attach to the growing chain in two manners. When it reacts in the same orientation as the previous monomer, a *head-to-tail* connection ($-CH_2$-CH**R**$-CH_2$-CH**R**$-$) is formed (Figure 2.6). Head-to-tail polymerization leads to a regular chain configuration, in which every second carbon atom along the chain possesses a methyl side group (**R**). *Head-to-head* ($-CH_2$-CH**R**$-$CH**R**-CH_2-$), or *tail-to-tail* ($-$CH**R**-$CH_2$-$CH_2$-CH**R**$-$) connections may also occur and result in structural isomers. Nevertheless, head-to-tail polymerization is usually favored, especially in case of bulky side groups [16].

2.2.3.2 Stereoisomers

In case of stereoisomerism, or spatial isomerism, the polymer segments have the same sequence of bonded atoms – that is, the same constitution – but differ in spatial form. If the polymer is partially unsaturated, the chain can adopt either a *cis*- or *trans configuration* at the double bonds (Figure 2.6). This seemingly small difference at the molecular level can have a surprisingly large effect on the macroscopic scale. For example, poly(*cis*-1,4-isoprene), better known as natural rubber, is soft, sticky, and elastic. In contrast, poly(*trans*-1,4-isoprene), or gutta percha, is a much harder and non-sticky material due to the more dense (crystalline) structure of the chains.

Enantiomers are a specific type of stereoisomers that originate from the presence of asymmetric carbon atoms. Let us examine the head-to-tail polymerization of propylene

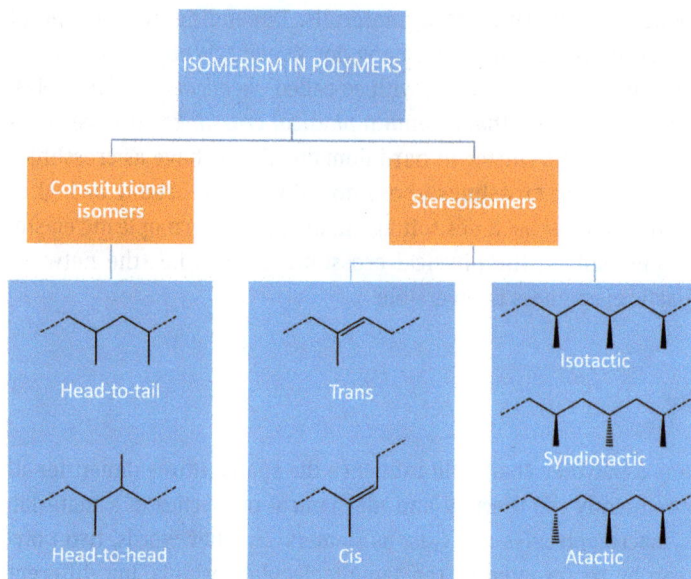

Figure 2.6: Examples of structural isomers and stereoisomers in macromolecular chains.

once more. Every second carbon atom along the chain is asymmetric. The methyl side group at such a *chiral center* can be oriented in two directions with respect to the polymer backbone. The two options are referred to as the *R*- and *S-configuration*. Repetition of an R/S configuration can result in different stereoisomers of polypropylene (PP): isotactic, syndiotactic, or atactic (Figure 2.6). In *isotactic* PP (PP-it), all asymmetric carbon atoms are configured in the same way. For instance, by this definition, a natural protein is isotactic with all side groups in the S configuration. *Syndiotactic* PP (PP-st) reveals an alternating orientation of methyl side groups, while in *atactic* PP (PP-at) the distribution of R and S configuration is completely random. Again, small differences at the molecular scale have important macroscopic consequences. While both PP-it and PP-st are (semi)crystalline materials with melting temperatures in the range of 150–170 °C, the irregular structure of PP-at cannot form a crystal lattice and is therefore an amorphous material without a melting point. The tacticity in a polymer chain can be directed by applying so-called stereospecific polymerization methods, as will be discussed in Section 2.3.2.3.

2.2.4 Chain length

The length of a polymer chain is quantified by the *degree of polymerization* (*P*). *P* is the number of repeating units in a macromolecule. By this definition, a monomer has a degree of polymerization of 1, while for a dimer *P* equals 2. In general, molecules are named macromolecules or polymers, when *P* is about 50 or higher.

Another parameter that is used to define the length, or to be more specific the size of a polymer molecule, is its *molar mass* (*M*), in g/mol. The molar mass, in polymer science traditionally referred to as *molecular weight*, equals the product of the degree of polymerization and the molar mass of one repeating unit (m_0), in g/mol, as depicted in equation (2.1).

$$M = m_0 \cdot P.$$ (2.1)

Clearly, molar mass increases with increasing degree of polymerization. The size of the macromolecule affects the thermophysical properties of the material. Consider the polymerization of ethylene as a model reaction (Table 2.2). The dimer is a gas, like the monomer itself. However, when increasing the molar mass, the polymerization products become liquid (*P* = 3–8) and eventually solid (*P* > 8).

Table 2.2: The length of a polymer chain can be expressed in degree of polymerization or molar mass (molecular weight). The molar mass is calculated directly from the degree of polymerization by equation (2.1).

Molecular formula	Degree of polymerization	Molar mass (g/mol)	Chemical name	State of matter at ambient conditions
$CH_2 = CH_2$	1	28	Ethylene	Gas
$CH_3\text{-}CH_2\text{-}CH_2\text{-}CH_3$	2	56	Butane	Gas
$CH_3\text{-}(CH_2\text{-}CH_2)_3\text{-}CH_3$	4	112	Octane	Liquid
$CH_3\text{-}(CH_2\text{-}CH_2)_7\text{-}CH_3$	8	224	Hexadecane	Liquid
$CH_3\text{-}(CH_2\text{-}CH_2)_{15}\text{-}CH_3$	16	448	Dotriacontane	Solid
$CH_3\text{-}(CH_2\text{-}CH_2)_{31}\text{-}CH_3$	32	896	Tetrahexacontane	Solid
$(CH_2\text{-}CH_2)_n$	>50	>1,400	Polyethylene (PE)	Solid

In practice, polymerization often leads to a mixture of macromolecules with different chain lengths and architecture [17]. In this case, the distribution of chains can be defined by an average value. Let us imagine a hypothetical polymer sample of five molecules (Figure 2.7): four chains with a molar mass of 250 g/mol (*P* = 5) and one chain with a molar mass of 1,000 g/mol (*P* = 20). The molar mass of the repeating unit is 50 g/mol. An intuitive manner of calculating the average is adding all molecular weights together and dividing by the total number of molecules. The answer of 400 g/mol is referred to as the *number average molar mass* (M_n), in g/mol.

$$M_n = \frac{\sum N_i M_i}{\sum N_i} = \sum n_i M_i.$$ (2.2)

The formula to calculate M_n is contained in equation (2.2). Here, N_i is the number of molecules having a molar mass M_i. The number average molar mass M_n can also be calculated using the mole fraction (n_i), like in the example of Figure 2.7(a).

Besides the number average, there are other valid methods to calculate averages. For instance, taking into account the mass of each chain provides another perspective

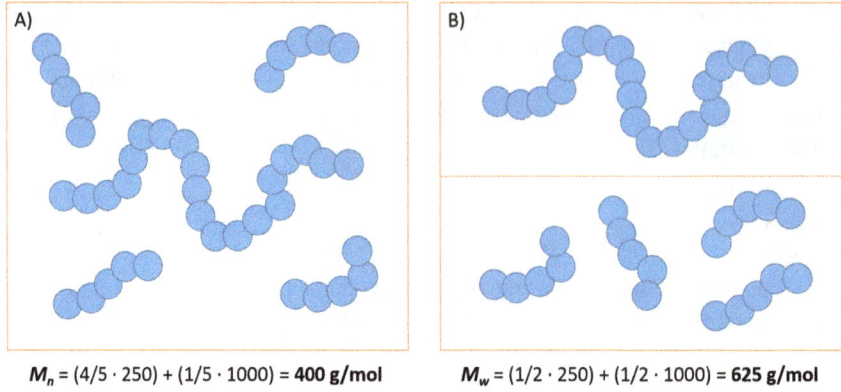

$M_n = (4/5 \cdot 250) + (1/5 \cdot 1000) = \textbf{400 g/mol}$ $M_w = (1/2 \cdot 250) + (1/2 \cdot 1000) = \textbf{625 g/mol}$

Figure 2.7: A mixture of chains can be defined by various averages: (a) number average molar mass and (b) weight average molar mass. The molar mass of the repeating unit (m0) is 50 g/mol.

for our polymer sample. Half of the mixture consists of macromolecules with a molar mass of 1,000 g/mol, and the other half consists of polymer chains with a molar mass of 250 g/mol (Figure 2.7(b)). The resulting average of 625 g/mol is referred to as the *weight average molar mass* (M_w), in g/mol.

The weight average molar mass M_w (equation (2.3)) equals the sum of the products of the mass fraction (w_i) and molar mass (M_i).

$$M_w = \frac{\sum N_i M_i^2}{\sum N_i M_i} = \sum w_i M_i. \tag{2.3}$$

By definition, $M_n \leq M_w$. Other averages can be calculated, but in practice M_n and M_w are most often used. As indicated, both of them provide unique information about the distribution of polymer chains in a sample. The (unitless) quotient of weight and number average molar mass, named *polydispersity* (D), is a measurement for the broadness of the molar mass distribution (equation (2.4)).

$$D = \frac{M_w}{M_n}. \tag{2.4}$$

A large polydispersity relates to a broad distribution of polymer chains. In the example in Figure 2.7, the polydispersity is 1.6. If all macromolecules in a sample have the exact same length (molar mass), M_w and M_n are equal. In that case, the polydispersity equals 1 and the sample is named *monodisperse*. For synthetic polymers, this is very unlikely to occur. However, natural polymers such as DNA and specific proteins can be exactly reproduced within organisms, making them monodisperse.

Similar to molar mass, an average for the degree of polymerization can be determined. Following from equation (2.1), the number average P_n (equation (2.5)) and weight average P_w (equation (2.6)) relate to the respective molar masses, in g/mol.

$$M_n = m_0 \cdot P_n, \tag{2.5}$$

$$M_w = m_0 \cdot P_w. \tag{2.6}$$

The molar mass and polydispersity of polymers is strongly influenced by the method of polymerization, as will be discussed in Section 2.3.

2.2.5 Characterization of macromolecules

Clearly, the structure and size of macromolecules greatly affect the polymer properties. Detailed characterization of composition, configuration, and molar mass (distribution) is of great importance to predict product performance.

2.2.5.1 Structure determination

Spectroscopic methods are a powerful tool to determine the chemical structure and chain regularity in polymer samples. Nuclear magnetic resonance (NMR) spectroscopy measurements can reveal the composition of chemical groups along the polymer chain in great detail. For instance, NMR is used to analyze the copolymer ratio in random co-polymers. Furthermore, the extent of head-to-tail coupling can be determined, as well as isotactic and syndiotactic sequences along the polymer chain. Additionally, Fourier transform infrared (FTIR) spectroscopy can be used to determine parameters such as degree of branching, crystal structure, *cis/trans* configuration, and tacticity.

2.2.5.2 Molar mass determination

Various techniques have been developed to calculate the molar mass of polymers. A pop-ular method to do so is end group determination. Quantitative analysis of end groups can be done either chemically, via titration, or by spectroscopic methods such as NMR. For the latter, the intensity of the signal(s) corresponding to the end group(s) is compared to the intensity of the signal(s) from the repeating units of the polymer chain. From the ratio between both signals, the number average degree of polymerization can be calcu-lated (Figure 2.8). The M_n follows from equation (2.5). Note that *end group analysis* is only applicable when the chemical structure and number of end groups per chain are known.

Dissolution of a polymer in a solvent results in changes of properties such as the freezing point, boiling point, vapor pressure, and osmosis. These are together referred to as *colligative properties* [18]. In a laboratory, colligative properties are used to de-termine the molar concentration of the solute, in this case, the polymer. If the mass of the polymer is also known, its molar mass can be determined. In fact, *membrane os-mometry* is an important method to provide M_n.

Other absolute methods to determine molar mass are (static) light scattering and ultracentrifugation. From light scattering experiments, the M_w can be calculated. Ultra-

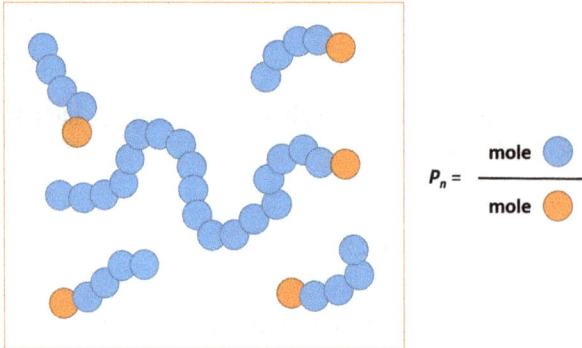

Figure 2.8: Quantitative analysis of end groups can be used to determine degree of polymerization (and molar mass) of a polymer sample.

centrifugation enables determination of molar mass distributions, and thus polydispersity, as well.

In contrast to the absolute methods just described, relative methods for molar mass determination need calibration by monodisperse samples with a known molar mass. Nevertheless, those characterization techniques are still very relevant and applied often. A famous example in polymer science is *viscometry*. When the concentration of the polymer in solution increases, it leads to a higher viscosity of that solution. However, the viscosity also increases for polymers with a higher molar mass. This effect is used in viscometry measurements. Using glass capillary viscometers, such as Ostwald or Ubbelohde, the *intrinsic viscosity* can be directly derived from the elution time of the polymer solution (measured in a concentration series). The viscosity average molecular weight (M_v), in g/mol, can be derived from the intrinsic viscosity ($[\eta]$), in cm^3/g, via the *Mark–Houwink equation* (equation (2.7)).

$$[\eta] = KM_v^a. \tag{2.7}$$

Here, K and a are the so-called Mark–Houwink parameters. Both are dependent on the polymer, solvent and temperature, which means M_v is not an absolute quantity as M_n and M_w. Clearly, during viscometry measurements, temperature needs to be controlled precisely.

2.2.5.3 Determination of molar mass distribution

Molar mass distributions can be obtained via techniques that separate the polymer sample based on size of the macromolecules. The most widely used method in polymer science is *size exclusion chromatography* (SEC), sometimes referred to as *gel permeation chromatography* (GPC). A polymer solution is introduced on a column packed with a porous gel. The smallest macromolecules within the sample will enter many pores when passing through the column and travel a relatively long distance. As a result, they will exit the column at a later stage than the larger macromolecules. Detection takes place at the end of the column, usually by refractive index (RI) or UV/VIS detectors. The resulting SEC curve (Figure 2.9(a)) depicts a size distribution. In order to relate elution time (or volume) to the molar mass, a calibration with narrow disperse polymers of known molar mass can be performed. This enables determination of both M_n and M_w and the polydispersity of the sample. Calibration is no longer needed when SEC is combined with static light scattering, an absolute method.

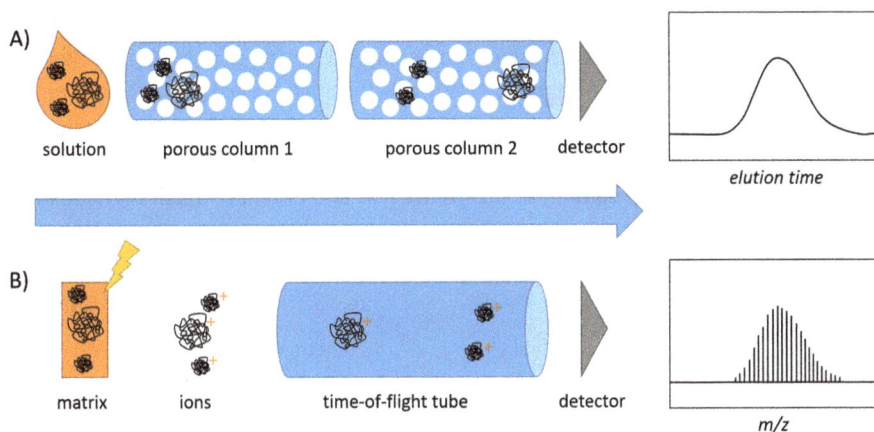

Figure 2.9: Schematic representation of (a) Size exclusion chromatography (SEC), and (b) MALDI-ToF. Both techniques can be carried out to determine the molar mass distribution of polymer samples.

In particular, a very sensitive method is *MALDI-ToF mass spectroscopy*. MALDI-ToF is short for "matrix assisted laser desorption ionization time-of-flight", which basically describes the technology involved. The method requires isolation of polymer molecules in a crystalline matrix.

The matrix can absorb energy from a laser pulse, leading to an explosive transition of matrix and embedded polymers into the gas phase. At the same time, the matrix also facilitates ionization, resulting in charged macromolecules. The ionized polymers are accelerated in an electrostatic field. After a certain time-of-flight (ToF), typically a few ms, the ions reach the detector. The relationship of mass to electric charge, denoted as m/z, is calculated from the flight time. Polymers with small molar mass reach the detector first, followed by larger ones. The resulting mass spectrum (Figure 2.9(b)) depicts a molar mass distribution, which enables absolute determination of M_n, M_w, and D. In contrast to SEC, which leads to a continuous distribution, in a MALDI-ToF mass spectrum individual signals can be identified. This means, in practice, $P_n = 50$ is distinguishable from $P_n = 51$. In fact, the distance between the two peaks equals the molar mass of the repeating unit. This high level of accuracy also allows determination of the molar mass of end groups.

2.3 Synthesis of polymers

Macromolecular architecture, whether it is a linear chain, branched molecule, or network structure, is constructed via polymer synthesis. The process in which monomers react together to form polymers is named *polymerization*. Polymerization can be classified by the mechanism of polymer-chain formation [19]. Two main polymerization mechanisms are distinguished: step-growth and chain-growth polymerization.

In step-growth polymerization (Figure 2.10(a)), all particles can react with one another during the reaction time, and their reactivity is independent on the length of the chain. Typically, a low molar mass byproduct is formed during the reaction. However, chain-growth polymerizations (Figure 2.10(b)), are initiated by specific reactive molecules. Monomers can only react with the initiated (growing) chains. Termination of chain-growth occurs after a short time period, while, at the same time, new chains are initiated and start growing. Both mechanisms are discussed in more detail in the following paragraphs.

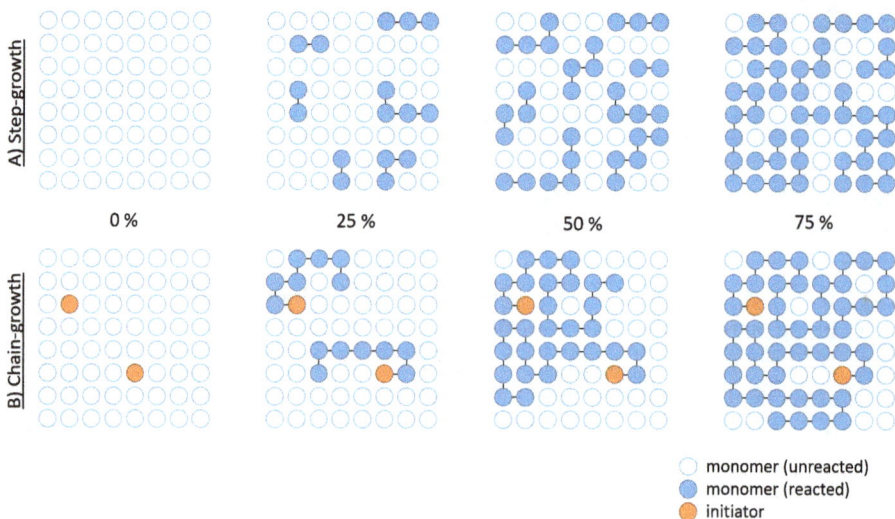

Figure 2.10: A schematic representation of the mechanisms for (a) step-growth and (b) chain-growth polymerization.

2.3.1 Step-growth polymerization

Polymerization requires monomers with at least two functional groups that can react with each other. Linear chains are formed via reaction of difunctional monomers, while polymerization in the presence of monomers with a functionality of three or higher results in branched or network structures (Figure 2.5(b)).

In case of *step-growth polymerization*, complementary functional groups can be united in the same monomer. Let us consider the polymerization of lactic acid. Lactic acid contains both a hydroxyl and carboxyl group in its chemical structure: a hydroxy acid. Those groups can react together to form ester moieties and water molecules (Figure 2.11). The term condensation polymerization, or *polycondensation*, refers to the formation of water as a byproduct. The esterification reaction is in equilibrium, indicated by the double arrow, since water can hydrolyze the ester group, forming the starting compound(s). The final product of the polycondensation is PLA, short for polylactic acid, part of the polyester family.

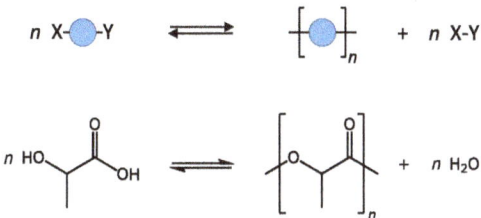

Figure 2.11: Polycondensation of lactic acid, containing both a hydroxyl (–OH) and carboxyl (–COOH) group, leads to polylactic acid (PLA).

Like every step-growth polymerization, the polycondensation of lactic acid takes place step by step. When two lactic acid monomers react, the resulting dimer contains again one hydroxyl and one carboxyl functionality. The dimer can react with another monomer, dimer, or oligomer to form a longer chain. The dimer has the same reactivity as the original monomer. The same holds for the oligomer. In other words, the reactivity of a functional groups seems independent on the length of the chain to which it is attached. This phenomenon is known as *Flory's principle of equal reactivity* [20].

Alternatively, two distinct difunctional monomers can undergo step-growth polymerization, as well. For instance in the polycondensation of ethylene glycol, a di-ol, and terephthalic acid, a di-acid (Figure 2.12). The monomers can polymerize in an alternate fashion to form polyethylene terephthalate (PET). A high degree of polymerization can only be obtained when starting with stoichiometric amounts, meaning both monomers are equally present in the reaction mixture. The esterification is an equilibrium reaction since hydrolysis can occur. Somewhat confusing, the common name of PET is simply "polyester". Technically speaking, however, PET is just one type of polyester, like PLA.

By definition, the *degree of polymerization* (P_n) is the average number of repeating units (n) per polymer molecule. This relates to the initial number of molecules (N_0) and the number of molecules at time t (N_t), as presented in equation (2.8).

$$P_n = \frac{N_0}{N_t}.$$

(2.8)

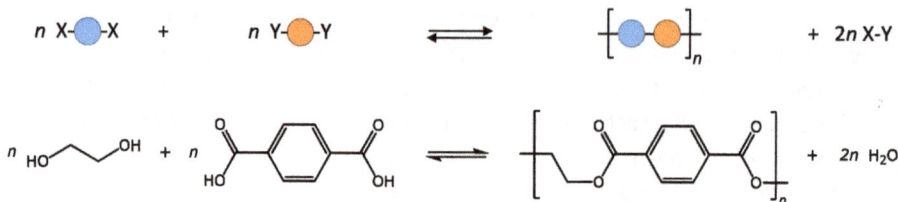

Figure 2.12: Polycondensation of ethylene glycol (di-ol) and terephthalic acid (di-acid) leads to polyethylene terephthalate (PET).

For instance, when examining the simplified polymer sample in Figure 2.7, initially 40 monomers were available ($N_0 = 40$). After polymerization, at time $= t$, five polymer chains have been formed ($N_t = 5$). As a result, the average degree of polymerization equals eight, according to the previous equation.

The conversion of the condensation polymerization (equation (2.9)) at t is related to N_0 en N_t as well, which results in an equation for N_t (equation (2.10)).

$$p = \frac{N_0 - N_t}{N_0} = \frac{N_0}{N_0} - \frac{N_t}{N_0} = 1 - \frac{N_t}{N_0}, \tag{2.9}$$

$$N_t = (1 - p)N_0. \tag{2.10}$$

Subsequently, P_n can be written as a function of conversion. Equation (2.11) is known as *Carothers' equation*.

$$P_n = \frac{N_0}{N_t} = \frac{N_0}{(1 - p)N_0} = \frac{1}{1 - p}. \tag{2.11}$$

Thus, the degree of polymerization can be controlled by conversion. Carothers' equation implies actual polymers are formed only at high conversion rates, above ca. 98%, as indicated by the graph in Figure 2.13. In practice, it means that water has to be removed from the reaction to shift the equilibrium in Figures 2.11 and 2.12 to the right, leading to a polymer product. Note that Carothers' equation holds solely for linear condensation polymers and that its simplicity is based on Flory's principle of equal reactivity.

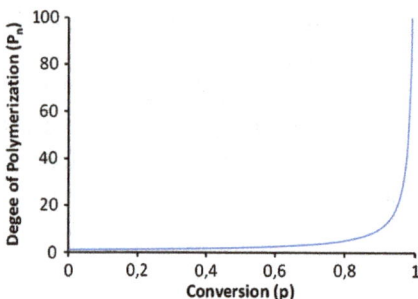

Figure 2.13: Degree of polymerization as a function of conversion: Carothers' equation. Polymers ($P_n \geq 50$) are formed only at very high conversion ($p \geq 0.98$).

In addition to the number average P_n, the weight average degree of polymerization (P_w) can be written as a function of p as well (equation (2.12)). Combining both relations with the definition of polydispersity (equation (2.4)), leads to equation (2.13). Since the conversion can reach up to 100% ($p_{max} = 1$), it follows that the polydispersity for a polycondensation has a maximum value of 2.0. The molar mass distribution resulting from condensation polymerization is commonly referred to as the *Schulz–Flory distribution*.

$$P_w = \frac{1+p}{1-p},$$

(2.12)

$$D = \frac{M_w}{M_n} = \frac{P_w}{P_n} = \frac{1+p/1-p}{1/1-p} = 1+p.$$

(2.13)

2.3.1.1 Linear polycondensates

Polyethylene terephthalate has the highest share of the *polyester* market. As presented in Figure 2.12, PET can be synthesized through polycondensation of ethylene glycol with terephthalic acid. Alternatively, the aromatic polyester is produced via transesterification with the dimethyl ester of terephthalic acid (Figure 2.14(a)). PET is known for its high stiffness and hardness and is used in textile fibers and in packaging, particularly in plastic bottles. The latter is due to the good O_2 and CO_2 barrier properties of PET. If ethylene glycol (1,2-ethanediol) is substituted for 1,4-butanediol, polymerization results in polybutylene terephthalate (PBT). PBT is applied in the automotive industry and household appliances.

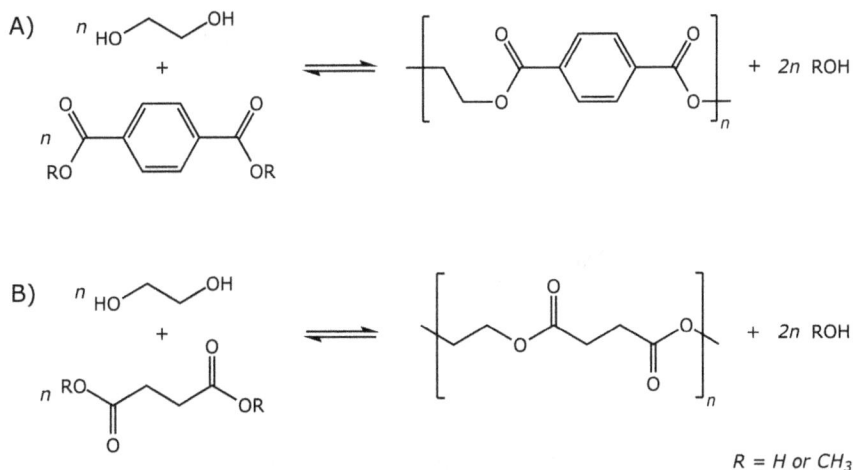

Figure 2.14: Synthesis of (a) partially aromatic PET and (b) aliphatic PES. Polymerization can proceed either via polycondensation of terephthalic and succinic acid (R = H) or transesterification of dimethyl terephthalate and succinate (R = CH₃). Starting from 1,4-butanediol rather than ethylene glycol (1,2-ethanediol) leads to PBT (instead of PET) and PBS (instead of PES).

Despite their excellent properties and recyclability, (partially) aromatic polyesters are not biologically degradable. Recently, aliphatic polybutylene succinate (PBS) and polyethylene succinate (PES) have gained renewed interest due to the increasing demand for bioplastics. Both polymers can be synthesized via polycondensation or transesterification (Figure 2.14(b)), similar to PET and PBT. PBS is both bio-based and biodegradable (Section 3.3.21) and has been marketed as the environmental friendly alternative to common plastics, such as polypropylene (PP) [21].

Among the family of polyesters, polylactic acid is one the most common *bioplastics*. The monomer can be produced from renewable resources like fermented plant starch. Figure 2.11 shows the polycondensation route towards PLA, however, it typically leads to low molar masses and competes with ring formation. The industrially more favored route is via ring opening polymerization of the cyclic diester of lactic acid, named lactide. Ring opening polymerization (ROP) follows a chain-growth mechanism, as will be discussed in Section 2.3.2.

The vast majority of the *polyamide* market is occupied by Nylon 6 and Nylon 6,6. Both polyamides have similar properties, can be applied in mechanical equipment, and are used extensively as textile fibers in clothing and carpets.

The numbering of polyamides refers to the number of carbon atoms in the monomer(s). Analogue to polyesters, polyamides can be synthesized from one monomer containing amine- and carboxyl functionalities, or by reacting diamines with diacids. A linear aliphatic polyamide originating from 6-aminohexanoic acid is named Nylon 6 (Figure 2.15(a)). Instead of the proposed reaction scheme, Nylon 6 however is rather produced by *ring opening polymerization* (ROP) of caprolactam (Section 2.3.2.4).

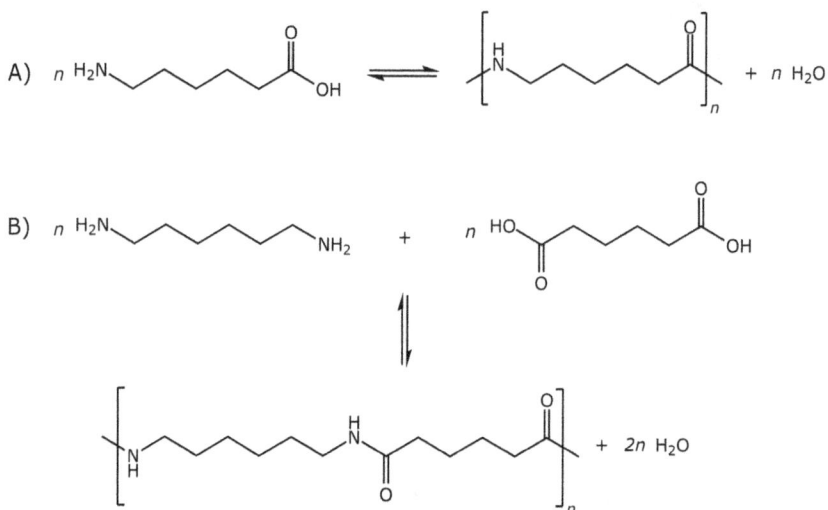

Figure 2.15: Possible synthesis routes of (a) Nylon 6 and (b) Nylon 6,6. The latter can be produced via an intermediate product called AH salt or by interfacial polymerization.

In contrast, Nylon 6,6 is synthesized from two monomers, hexamethylene diamine (HMD) and adipic acid (AA) (Figure 2.15(b)). The two digits refer to the presence of six carbon atoms in both starting compounds. Stoichiometric requirements are met by first crystallizing the monomers into an ammonium/carboxylate salt, a so-called AH salt (*nylon salt*). After purification, both constituents are present in an exact molar ratio of 1:1. Next, the nylon salt is successfully converted at high temperatures into Nylon 6,6. This procedure was developed in 1930s at DuPont by Carothers and his team.

Another preparation method of Nylon 6,6 involves the removal of polymer product during the reaction. First, hexamethylene diamine (HMD) is dissolved in water and adipoyl dichloride (ADC) in an organic solvent, such as cyclohexane. At the interface between the aqueous and organic phases, a polyamide film is formed via a so-called *interfacial polymerization*. A nylon fiber can be continuously drawn from the interface, and in the meantime new polymer is formed. The synthesis method is a popular demonstration experiment, known as the *nylon rope trick*, however, it is not commercially viable.

Polyaramides, or *aramids*, are the aromatic equivalent of nylons. The polymerization is more challenging with respect to aliphatic polyamides. Poly(*para*-phenylene terephthalamide) (PPTA), with trade names as Kevlar® (DuPont) and Twaron® (Teijin Aramid), is prepared by polycondensation of *para*-phenylene diamine (PPD) and terephthaloyl dichloride (TDC). Aramid fibers are spun from a PPTA solution in concentrated (fuming) sulfuric acid and show superior mechanical performance and heat resistance. The molecular structure in aramid fibers is highly ordered (Figure 2.16), which explains the directional dependence of their mechanical strength. Due to the high costs of Kevlar® and Twaron®, those fibers are only applied when strictly needed, for instance, for safety in bullet-proof vests and as reinforcing agents in sports car tires.

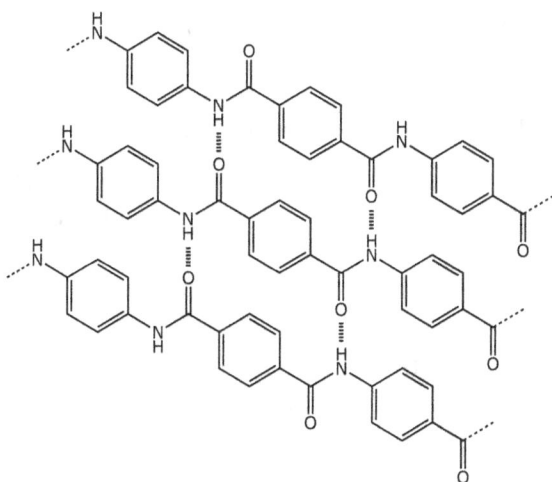

Figure 2.16: Highly ordered structure of PPTA macromolecules in aramid fibers. Dashed lines represent intermolecular hydrogen bonding between chains.

Polycarbonates (PCs) are basically polyesters of carbonic acid. The most important PC is prepared from diphenyl carbonate and bisphenol A (Figure 2.17). Alternatively, diphenyl carbonate can be substituted by phosgene. Its high transparency makes PC applicable in eye-protection and car windows.

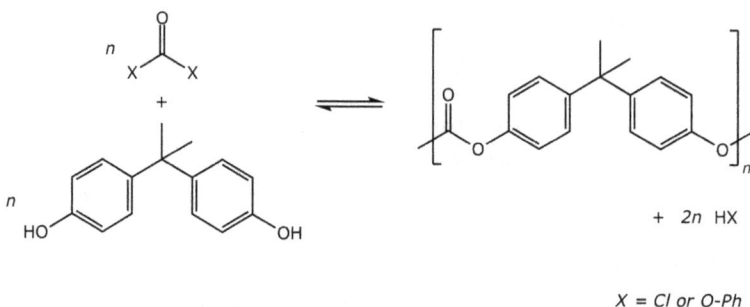

X = Cl or O-Ph

Figure 2.17: Synthesis of polycarbonate from phosgene (X = Cl) or diphenyl carbonate (X = O–Ph) and bisphenol A, giving hydrochloric acid (HCl) or phenol as a byproduct.

Inorganic *polysiloxanes* are obtained by hydrolysis of chlorosilanes and subsequent polycondensation. The polycondensation towards linear polydimethyl siloxane (PDMS), displayed in Figure 2.18, competes with the formation of cyclic siloxanes. PDMS is a viscous liquid and is used in hydraulic oils and lubricants, due to its excellent heat resistance. When crosslinked, polysiloxanes are referred to as *silicone rubber*. Silicon rubber is applied in seals and medical implants.

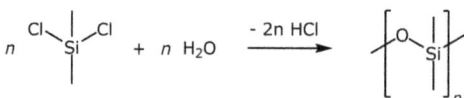

Figure 2.18: Synthesis of PDMS: hydrolysis of dimethyl dichlorosilane, followed by polycondensation.

Table 2.3 depicts an overview of the most important linear condensation polymers, discussed in this paragraph.

2.3.1.2 Polycondensate networks

As described in Section 2.2.2, polymer networks can be formed via various mechanisms. For example, *epoxy resins* are produced in a similar fashion as shown in Figure 2.5(a). First, low molar mass polymers with epoxy end-groups are synthesized. In the next step, these epoxy *prepolymers* are transformed into thermosets by reaction with crosslinking agents such as anhydrides or amines (Figure 2.19(a)). Conversion from a resin to a thermoset is called *curing*. Epoxy resins are often applied in coatings

Table 2.3: Overview of most important linear polycondensates and their chemical structures.

Polymer structure	Condensate polymer	Abbreviation	Chemical structure
Polyesters	Polyethylene terephthalate	PET	
	Polybutylene succinate	PBS	
	Polylactic acid* (Polylactide)	PLA	
Polyamides	Nylon 6*	PA6	
	Nylon 6,6	PA6,6	
Polyaramides (Aramids)	Kevlar® Twaron®	PPTA	
Polycarbonates	Polycarbonate	PC	
Polysiloxanes	Polydimethyl siloxane	PDMS	

*PLA and Nylon 6, however, are typically produced by ring opening polymerization of lactide and caprolactam, respectively. Ring opening polymerization will be discussed in Section 2.3.2.4.

and adhesives, or combined with fibers to fabricate fiber-reinforced plastics, a composite material used in automotive and construction industries.

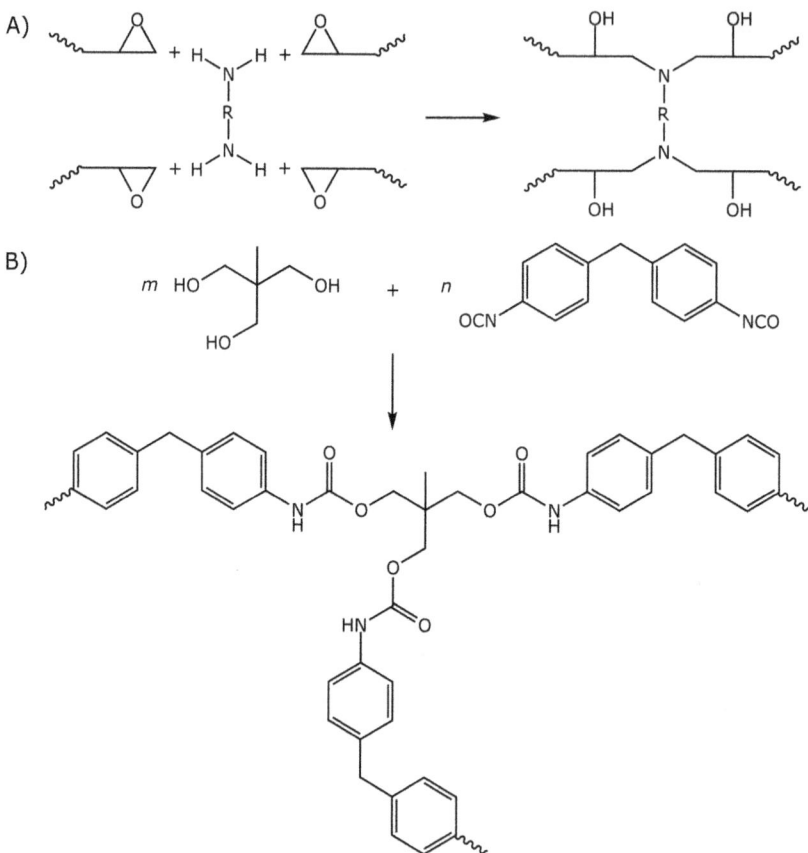

Figure 2.19: (a) Epoxy thermosets are formed by curing the epoxy resin prepolymers with amines (or anhydrides) as crosslinking agents. (b) Polyurethane networks can be formed via polymerization of diisocyanates and multifunctional alcohols.

Another successful application of network polymers is the family of *polyurethanes*, well known from their application in foams for insulation panels and mattresses. Polyurethanes result from the reaction of isocyanates with alcohols. The use of monomers with more than two functional groups leads to network formation. An example is the polymerization of diisocyanates with multifunctional alcohols (Figure 2.19(b)). The reaction follows network formation as presented in Figure 2.5(b). It has to be noted however that often polyols are used, which can be considered as prepolymers. The molar mass of the polyol can be varied, thereby controlling the density of the network. As a result, either rigid (with low-M_n polyol) or flexible (with high-M_n polyol) polyurethane foams are produced. Clearly, it is a powerful tool to vary the properties of the final product.

Other relevant crosslinking systems are phenolic and melamine resins, prepared by condensation of formaldehyde with phenol or melamine, respectively. Curing is normally achieved at elevated temperatures. Both have similar properties, but the use of melamine formaldehyde can be preferred since it is colorless. Phenol formaldehyde has been the first synthetic polymer commercially available, under the trade name Bakelite®.

2.3.2 Chain-growth polymerization

In the case of *chain-growth polymerization*, macromolecules are typically formed by addition to a reactive group, mostly a carbon–carbon double bond, or by the opening of a ring. Crucial is the formation of an active center that can be transferred along the chain at the moment a new monomer unit is added. Depending on the chemical nature of the monomer, different chain-growth mechanisms can occur.

2.3.2.1 Radical polymerization

Most chain-growth polymerization reactions proceed via continuous addition to double bonds. The resulting polymers are named addition polymers. When radicals are involved during initiation and chain growth, it concerns a *radical polymerization*. During conventional *free radical polymerization*, the chain growth process passes through several stages. Let us examine the polymerization of a typical vinyl monomer, such as vinyl chloride.

It all starts with the availability of a radical source, often an initiator molecule. The *initiator* dissociates into radical species under the influence of (UV) light or by heating. Common initiators are peroxides or azo compounds, like azo-*bis*-isobutyronitrile (AIBN). *Decomposition* of AIBN is depicted both schematically and by structural formulas in Figure 2.20(a).

Initiation of vinyl chloride takes place by addition of the radical species to the double bond of the monomer. As a result, a new radical center is formed (Figure 2.20(b)). The new radical can add additional monomers, a process referred to as *propagation*, thereby extending the PVC chain (Figure 2.21). During propagation, the polymer chain remains active. The radical center continuously shifts to the end of the chain when extended by a new monomer unit.

Chain *termination* ceases the formation of reactive intermediates. Depending on the reaction conditions, termination takes place by *radical combination* of growing chains (Figure 2.22(a)), or by *disproportionation* in the form of a hydrogen atom transfer (Figure 2.22(b)). Note that chain combination results in one chain per termination reaction, while disproportionation gives two chains, of which one has an unsaturated

A)

B)

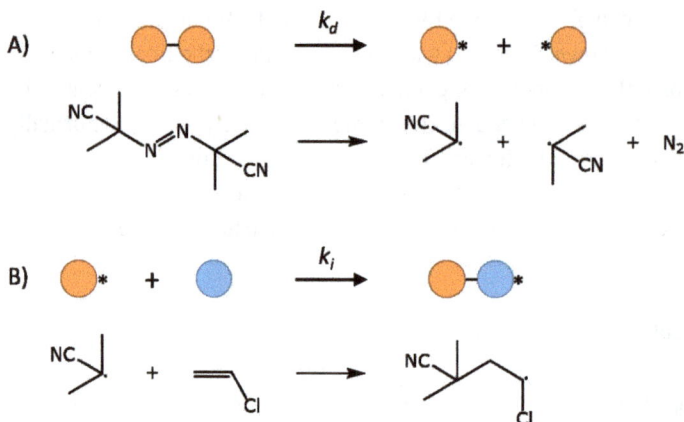

Figure 2.20: (a) Decomposition of AIBN, yielding two radical species and nitrogen gas. (b) Initiation of vinyl chloride. The rate constants of initiator decomposition and initiation are k_d and k_i, respectively.

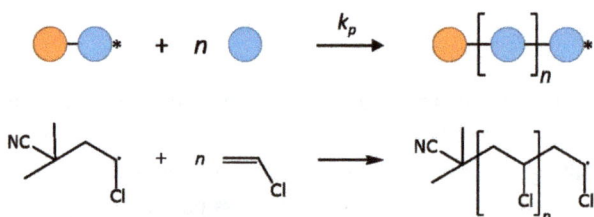

Figure 2.21: Propagation of vinyl chloride monomers. The rate constant of propagation is denoted as k_p.

end-group. Typically, both termination reactions occur, but disproportionation is favored at higher temperatures.

Another way to end the growth of a single polymer chain is by transferring the radical center to another species, a so-called chain transfer agent (CTA). In turn, the new radical species can initiate new chain growth. CTAs can be added intentionally to regulate chain growth. A higher concentration of chain transfer agent, in that case, lowers the average molar mass. *Chain transfer* can take place to a solvent or to the monomer itself. In our example, vinyl chloride is known to be an effective chain transfer agent (Figure 2.22(c)). In addition, transfer may occur to another polymer as well. In that case, it introduces branches on the polymer backbone, leading to a higher molar mass instead.

The *degree of polymerization* (P_n) of macromolecules prepared from radical polymerization is dependent on the kinetics of the possible reactions already described. The average chain length is related to the probability of chain growth (propagation) versus the probability of chain termination (by disproportionation). In other words, P_n equals the ratio of the rate of propagation (v_p) to the rate of termination (v_{td}), as stated in equation (2.14).

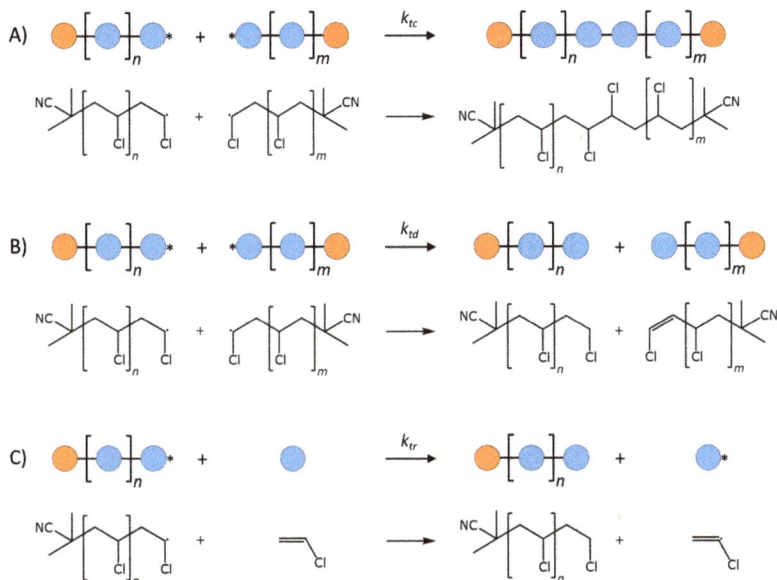

Figure 2.22: (a) Termination of chain growth by combination of radicals. (b) Termination of chain growth by disproportionation. (c) Chain transfer reaction to vinyl chloride monomer. The rate constants of termination by combination, termination by disproportionation, and chain transfer are k_{tc}, k_{td}, and k_{tr}, respectively.

$$P_n = \frac{v_p}{v_{td}}. \tag{2.14}$$

From Figure 2.21, we learn that the rate of chain growth is dependent on the rate constant (k_p), and the concentration of monomers ($[M]$) and concentration of radical chain ends ($[P\cdot]$) (equation (2.15)). In case of termination (Figure 2.22(b)), the rate of termination relates to the rate constant (k_{td}) and the concentration of radical chain ends (equation (2.16)). This leads to equation (2.17).

$$v_p = k_p[M][P\cdot], \tag{2.15}$$

$$v_{td} = k_{td}[P\cdot][P\cdot], \tag{2.16}$$

$$P_n = \frac{k_p[M][P\cdot]}{k_{td}[P\cdot][P\cdot]} = \frac{k_p[M]}{k_{td}[P\cdot]}. \tag{2.17}$$

The radical concentration ($[P\cdot]$) is dependent on the rate constant of initiator decomposition (k_d), the initiator concentration ($[I]$) and the rate constant of termination, according to equation (2.18).

$$[P\cdot] = \sqrt{\frac{2k_d[I]}{k_{td}}}. \tag{2.18}$$

Subsequently, P_n can be written as a function of monomer and initiator concentration (equation (2.19)). This equation only holds when termination occurs through disproportionation. When termination takes place via combination, two chains join, and the degree of polymerization is doubled (equation (2.20)).

$$P_n = \frac{k_p[M]}{\sqrt{2k_d[I]k_{td}}} = \frac{k_p}{\sqrt{2k_dk_{td}}} \cdot \frac{[M]}{\sqrt{[I]}}, \tag{2.19}$$

$$P_n = 2\frac{k_p}{\sqrt{2k_dk_{tc}}} \cdot \frac{[M]}{\sqrt{[I]}}. \tag{2.20}$$

It becomes clear that the average chain length obtained in radical polymerization is dependent on the concentration of monomer and initiator, at a specific time interval. During polymerization monomers are consumed, so the monomer concentration drops. Consequently, the P_n lowers during the reaction time. So, in the beginning of the polymerization reaction, longer chains are formed, while chains become shorter when the reaction proceeds. Therefore, free radical polymerization generally leads to higher values for polydispersity, in contrast to step-growth polymerization having a D_{max} of 2.0 (equation (2.13)).

A better control over polydispersity can be achieved by suppressing termination reactions. Let us compare the rate of propagation (equation (2.15)) with that of termination (equation (2.16)) once more. The former is a first-order reaction in radical concentration ($[P\cdot]$), while the termination rate is second-order ($[P\cdot]^2$). This means reducing the radical concentration has more influence on termination than on propagation (and thus polymerization). After all, termination only occurs when two radicals meet.

Controlled radical polymerization (CRP) uses this phenomenon by setting up an equilibrium between an active and inactive state (Figure 2.23). A stable radical ($X\cdot$) can react with a growing chain ($P_n\cdot$) and form an inactive macromolecule (P_n-X). This reaction is, however, reversible, and the "dormant species" can split into the active state. Chain growth can take place by reaction of $P_n\cdot$ with monomers (M). After a few propagating steps, $P_n\cdot$ deactivates into its dormant state by reacting with the stable radical, while other chains may activate. Since $k_{deact} > k_{act}$, the majority of chains are dormant during polymerization. In other words, the concentration of radical chain ends is significantly lowered. As a result, termination by combination or disproportionation is heavily suppressed and chain length can increase linearly with conversion.

In recent years, several CRP mechanisms have been developed that apply the concept described above. Most common are nitroxide-mediated polymerization (NMP), atom transfer radical polymerization (ATRP), and reversible addition fragmentation and transfer (RAFT) [22]. In contrast to free radical polymerization, controlled radical "active species"   "dormant species"

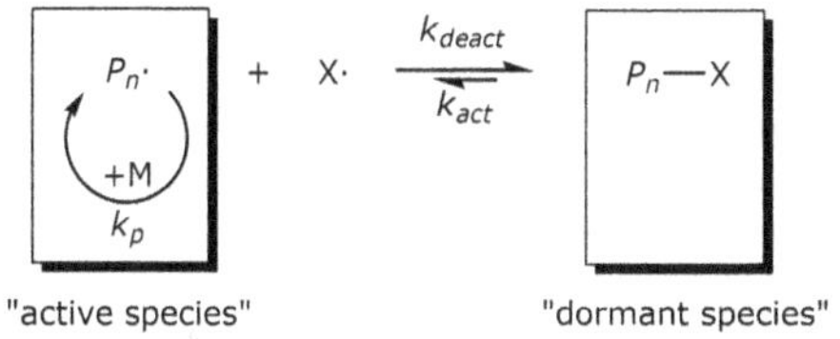

Figure 2.23: Controlled radical polymerization: equilibrium between active and dormant species lowers the concentration of radical chain ends, thereby suppressing termination reactions. Active species can continue propagation (polymerization) until they are deactivated once more.

polymerization leads to narrow molar mass distributions ($D \approx 1.1$). In addition, it enables the synthesis of well-defined polymer architectures such as block copolymers. Nevertheless, the technology is rather demanding and therefore expensive. Another method to synthesize block copolymers is via living anionic polymerization, described in the following paragraph.

2.3.2.2 Ionic polymerization

Ionic polymerization involves a chain-growth mechanism with cations or anions as active centers, rather than radicals. As stated before, most chain-growth polymerizations occur via continuous addition to double bonds. Whether a monomer can undergo cationic or anionic polymerization depends on its electron negativity, which is largely dependent on the side group(s). Monomers with electron donating groups favor *cationic polymerization*. The electron density on the double bond is increased, which facilitates electrophilic addition and stabilizes a cationic center. Examples are isobutylene and ethyl vinyl ether.

Figure 2.24(a) presents the cationic polymerization of isobutylene, initiated by a Brønsted acid (H^+X^-). Alternatively, Lewis acids such as BF_3, $TiCl_4$, and $AlBr_3$ can be employed as initiators. Chain growth occurs via electrophilic addition to isobutylene monomers. The counterion (X^-) can act as a free ion, or may exist as an ion pair with the active chain end. Since the growing chains are identically charged, they will not terminate by reacting with one another, in contrast to radical polymerizations. Nevertheless, termination of chain growth may still take place via deprotonation or, for example, chain transfer to the monomer. Those unwanted side reactions can be suppressed, but the reaction mixture needs to be very clean. Water and other protic compounds should be absolutely absent. In addition, polymerizations have to be carried out at low temperatures, typically below -100 °C, if a high degree of polymerization is desired. Those extreme requirements make cationic polymerization less suitable for industrial application.

Monomers with electron withdrawing groups favor *anionic polymerization*. The electron density on the double bond is decreased, which facilitates a nucleophilic attack and stabilizes the anionic center. Examples are acrylonitrile and MMA.

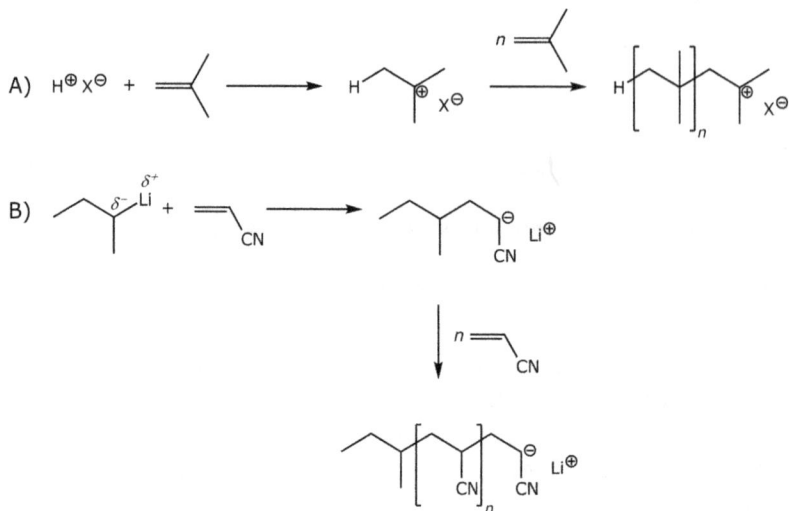

Figure 2.24: (a) Cationic polymerization of isobutylene, initiated by a Brønsted acid. Termination by combination or disproportionation is not possible due to positively charged chain ends. Nevertheless, alternative termination reactions can occur. (b) Anionic polymerization of acrylonitrile, initiated by an organometallic initiator. Termination by combination or disproportionation is not possible due to negatively charged chain ends. This enables living polymerization.

Figure 2.24(b) presents the anionic polymerization of acrylonitrile, initiated by *sec*-butyl lithium. Organic metal compounds are popular nucleophilic initiators for anionic polymerizations. They are commercially available and have good solubility in a wide range of organic solvents. Alternatively, electron transfer reactions can be used. Chain propagation takes place via nucleophilic addition to the acrylonitrile monomers. Again, termination via combination of active chain ends does not occur.

Anionic polymerizations are usually carried out below room temperature. As for cationic polymerization, traces of water and alcohols must be avoided to prevent unwanted termination reactions. By doing so, an ideal anionic polymerization can be achieved. When the rate of initiation is significantly larger than the propagation rate, all chains start growing at the same time. And, when termination and transfer reactions are absent, propagation continues until all monomers have been consumed. The growing chain remains active, and the polymerization is referred to as a *living polymerization* [23].

During living anionic polymerization, the chain length increases linearly with conversion. In the case of complete initiation, the average degree of polymerization (equation (2.21)) at full monomer conversion is determined by the ratio of initial monomer concentration ($[M]_0$) and initial initiator concentration ($[I]_0$). An increase in initiator concentration leads to lower chain lengths, and vice versa.

$$P_n = \frac{[M]_0}{[I]_0}.$$

(2.21)

Essentially, living polymerization leads to polymer products with a very narrow molar mass distribution, referred to as the *Poisson distribution*. The products are monodisperse, with values for polydispersity below 1.1. Higher molar mass leads to lower polydispersity (equation (2.22)).

$$D = 1 + \frac{1}{P_n}.$$

(2.22)

Since the chains remain active, even when all monomer has been consumed, adding a different type of monomer to the reaction mixture results in the formation of well-defined block copolymers (Figure 2.25). Eventually, living anionic polymerization can be stopped by intentionally adding a protic compound, such as methanol.

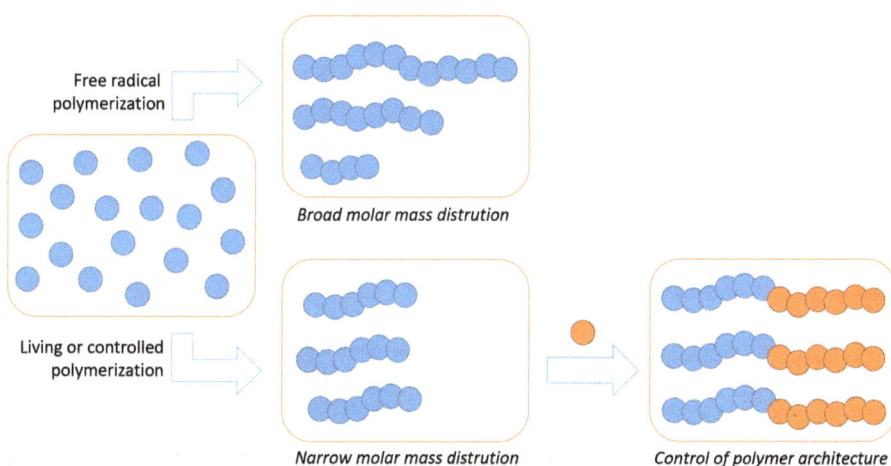

Free radical polymerization

Broad molar mass distrution

Living or controlled polymerization

Narrow molar mass distrution

Control of polymer architecture

Figure 2.25: Schematic overview of polymer products from free radical polymerization versus living/controlled polymerization. Free radical polymerization leads to polydisperse samples. Longer chains are formed at the start of the reaction, while shorter chains are formed when less monomer is available. Via controlled radical polymerization, or living anionic polymerization, narrow distributions can be obtained. Block copolymers can be synthesized by sequentially adding different monomers.

2.3.2.3 Coordination polymerization

Chain growth reactions that are catalyzed by transition metals are generally called *coordination polymerizations*. About half of all polymers worldwide are produced by this specific type of polymerization. Propylene, for example, cannot be polymerized via radical, cationic, or anionic polymerizations. It was only after the discovery of titanium-based catalysts by Ziegler and Natta in the 1950s that industrial production of polypropylene (PP) became possible [24].

Ziegler–Natta polymerization enabled the production of polyolefins at ambient conditions. Figure 2.26 depicts the polymerization of ethylene. First, ethylene coordinates with a transition metal catalyst to form a so-called π-complex. The ethylene molecule is inserted into the alkyl ligand, and a new coordination site becomes available. By repeating this process, a polyethylene (PE) chain is produced. Termination can take place through hydride elimination.

If chain migration occurs during the process (as in Figure 2.26), the next monomer will coordinate in the exact same way as before, due to the stereospecific catalyst. Polymerization with Ziegler–Natta catalysts can result in stereospecific polymers, which was a great achievement at that time and awarded the Noble Prize for Chemistry in 1963.

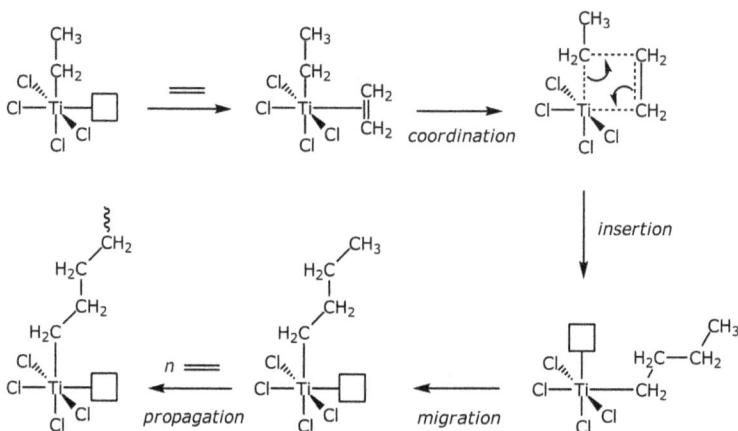

Figure 2.26: Chain growth via Ziegler–Natta polymerization: coordination of ethylene, followed by insertion and migration. By repeating the process, high-density polyethylene (HDPE) is obtained.

A few decades later, in the 1990s, Brintzinger discovered *metallocene catalysts* derived from zirconocenes [25]. Those homogenous catalysts introduced a higher level of microstructural control in comparison to the traditional systems from Ziegler and Natta. By catalytic polymerization of propylene in the presence of isoselective metallocene, isotactic PP (more than 98%) can be obtained. Alternatively, syndioselective metallocene results in highly syndiotactic PP (Section 3.3.2).

2.3.2.4 Ring opening polymerization

Polymerization that proceeds via ring opening is applied to produce several important commercial polymers. *Ring opening polymerization* (ROP) follows a chain-growth mechanism, and can be initiated via radical, ionic, or coordinative pathways, as described in the previous paragraphs. Suitable monomers for ROP are cyclic ethers, amides (lactams), esters (lactones), olefins, and siloxanes.

As pointed out before, the industrially favored route to produce PLA is not via step-growth polymerization of lactic acid, but through ring opening polymerization of cyclic lactide (Figure 2.27(a)). In a similar fashion, polycaprolactone (PCL) and Nylon 6 are produced by ionic ROP of caprolactone (Figure 2.27(b)) and caprolactam (Figure 2.27(c)), respectively.

Cyclic olefins can undergo a special type of ring opening polymerization. For instance, cyclooctene is polymerized via a metathesis reaction. *Ring opening metathesis polymerization* (ROMP) yields polycyclooctene (Figure 2.27(d)), marketed under the name Vestenamer®. The mechanism of ROMP follows the same principles as presented in Figure 2.26. During the insertion step, the ring is opened. ROMP of cyclic olefins has enabled the polymerization towards formally unknown architectures, such as polynorbornenes.

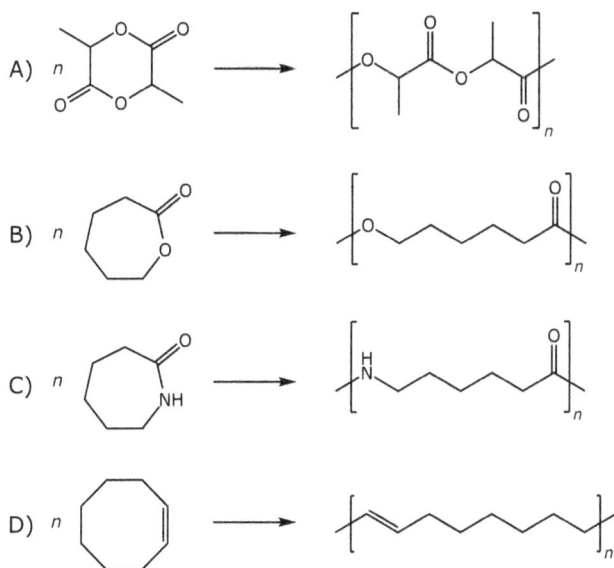

Figure 2.27: Ring opening polymerization of (a) lactide to PLA, (b) caprolactone to PCL, (c) caprolactam to Nylon 6 (PA6), (d) cyclooctene to Vestenamer®.

2.3.2.5 Important addition polymers

The majority of commercially available polymers are vinyl polymers, based on vinyl monomers ($CH_2 = CHX$). Polyethylene (PE), polypropylene (PP), polystyrene (PS), and polyvinylchloride (PVC) together share ca. 80% of the plastics market worldwide.

Free radical polymerization of ethylene leads to low-density *polyethylene* (LDPE). *LDPE* is a flexible material extensively applied in plastic films for packaging purposes. Ziegler–Natta polymerization of ethylene results in less branched high-density polyethylene (HDPE). *HDPE* has a better mechanical performance, leading to application in containers and pipes. A specialty PE is Dyneema®, an ultra-high molecular weight polyethylene (UHMWPE). Dyneema® fibers demonstrate excellent resistance to impact and wear and are used in high-end applications, such as ballistic protection and extreme sports.

Polypropylene (PP) is only produced by coordination polymerization. By carefully selecting the catalyst, chain tacticity can be varied, and isotactic, syndiotactic, or atactic PP is obtained. Depending on the specific grade, PP may be used in carpets, packaging, and molded parts for cars.

The term styrenics refers to *polystyrene* (PS) and its copolymers. The latter will be discussed in the next paragraph. PS is mostly produced via radical polymerization, although chain growth can take place via ionic propagation as well. Polystyrene is easily processable above its glass transition and used to fabricate various children's toys. Particularly important are polystyrene foams. Styrofoam™ (Section 2.5.4) is expanded polystyrene (EPS) used in food containers, coffee cups, and packaging.

Polyvinylchloride (PVC) is a very popular thermoplastic polymer, with a great number of applications. It can be processed via extrusion, injection molding, blow molding, pressing, and calendering. Credit cards, squeeze bottles, and transparent tapes are made from PVC. Recycled PVC is extensively applied in tubes and pipes. When produced with large amounts of plasticizers, soft PVC is obtained. This rubbery compound is used in flooring and cables.

While acrylonitrile is a crucial comonomer in ABS and SAN plastics, its homopolymer *polyacrylonitrile* (PAN) can be spun into synthetic fibers for rugs and blankets. Spun PAN is the number one precursor for high-quality carbon fibers as well. Carbon fibers can be applied in fiber-reinforced plastics or composites for aerospace and automotive industry.

In addition to vinyl polymers, polymers based on vinylidene monomers ($CH_2 = CXY$) are quite popular on the plastics market. *Polyisobutylene* (PIB) is produced on a large scale via a challenging cationic polymerization of isobutylene (Figure 2.24(a)). Commercial applications include chewing gum and skin-care products. One of the hardest plastics is *polymethyl methacrylate* (PMMA), well-known by its trade name Perspex®. PMMA has excellent transparency and is therefore used in spectacles, optical fibers, traffic signs, and solar panels.

Fully fluorinated ethylene can be radically polymerized to *polytetrafluoroethylene* (PTFE). PTFE shares 60% of the fluoropolymer market. It is better known under DuPont's trade name Teflon®. This fluoropolymer is chemically inert and anti-adhesive, resulting in its application in non-stick coatings for cookware.

Isoprene and butadiene can polymerize via 1,4-addition into *polyisoprene rubber* (PIR) and *polybutadiene rubber* (PBR). Alternatively, 1,2 linkages may also be formed. Both materials can be produced via radical, anionic, or coordination polymerization. Polybutadiene is crosslinked for application in car tires and dampers. Polyisoprene occurs in nature as natural rubber (*cis configuration*) or gutta percha (*trans configuration*). Like polybutadiene, it is crosslinked before practical use in shoes and gloves.

Table 2.4 depicts an overview of the technically most important addition polymers, discussed in this paragraph.

Table 2.4: Overview of most important (linear) addition polymers.

Addition polymer	Abbreviation	Polymerization route(s)	Chemical structure
Polyethylene	PE	– Radical (LDPE) – Coordination (HDPE)	
Polypropylene	PP	– Coordination	
Polystyrene	PS	– Radical – Cationic – Anionic – Coordination	
Polyvinyl chloride	PVC	– Radical – Coordination	
Polyacrylonitrile	PAN	– Radical – Anionic	
Polyisobutylene	PIB	– Cationic – Coordination	
Polymethyl methacrylate	PMMA	– Radical – Anionic – Coordination	
Polytetrafluoroethylene	PTFE	– Radical	
Polybutadiene *rubber*	PBR	– Radical – Anionic – Coordination	
Polyisoprene *rubber*	PIR	– Radical – Anionic – Coordination	

2.3.3 Copolymerization

Section 2.2.1 explored the large diversity of macromolecular structures, including various types of copolymers. An infinite number of copolymers can be synthesized by altering the type of monomers, as well as the chain sequence (Figure 2.3). The material properties are dependent on the chemical nature of building blocks and their ratio along the polymer chain. *Copolymerization* allows the synthesis of tailor-made macromolecules by tuning those variables.

Let us consider the radical copolymerization of acrylonitrile (A) and butadiene (B). During chain growth, several reactions may occur. When monomer A reacts with a growing chain, an active chain of A is formed (– a ·). This active chain may react with either A or B, as depicted in Figure 2.28. The rate constants of the possible chain-growth reactions are k_{11} and k_{12}, respectively. Analogously, an active chain end of B (– b ·) can react with either A (k_{21}) or B (k_{22}).

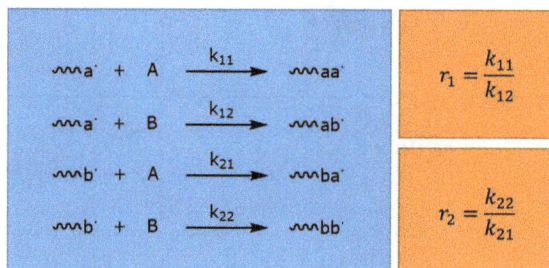

Figure 2.28: Radical chain growth for copolymerization of acrylonitrile (A) with butadiene (B). Four possible reactions with distinct rate constants lead to two reactivity ratios (r_1 and r_2).

The copolymerization parameter r_1 is defined as the *reactivity ratio* of k_{11} to k_{12} (Figure 2.28). In other words, a high value for r_1 means active chain ends of A favor reaction with monomer A. Low values for r_1 indicate a preference to react with the dissimilar monomer B. In a similar manner, r_2 describes the relative reactivity of active chain ends of B towards both type of monomers.

The final composition of the copolymer chain is described by the *Mayo–Lewis equation* for copolymerization (equation (2.23)). It can be determined simply from the concentration of both monomers, [A] and [B], and the reactivity ratios r_1 and r_2. The chemical composition of the copolymers is expressed in [a]/[b]. [a] and [b] are the concentrations of A and B incorporated into the copolymer that is formed.

$$\frac{[a]}{[b]} = \frac{r_1 \frac{[A]}{[B]} + 1}{r_2 \frac{[B]}{[A]} + 1}. \tag{2.23}$$

Note that the Mayo–Lewis equation describes the copolymer composition at a specific reaction time. During radical polymerization monomers are consumed. Therefore, depending on the reactivity ratios, the monomer composition in the reaction mixture may change significantly over time. As a result, the copolymers formed over time will also have a different composition. This phenomenon is typical for (radical) copolymerization reactions and is called *composition drift*. Composition drift may be prevented by continuously replacing incorporated monomers. A more practical solution is to stop the copolymerization at low conversions (<5%).

Using the Mayo–Lewis equation, we can determine the relationship between the monomer feed and composition along the copolymer chain. This relation can be visualized in a *copolymer composition diagram*, in which the mole fraction of monomer A in the polymer is plotted against the mole fraction of monomer A in the reaction mixture (Figure 2.29). The curve is dependent only on the reactivity ratio of the monomer pairs. For instance, in the case of $r_1 = r_2 = 1$, there is no preference at all for incorporation of A or B in the copolymer chain, and monomers are built in, in a random manner. Consequently, the copolymer composition corresponds to that of the initial monomer mixture (Figure 2.29(a)). No composition drift occurs. The scenario of $r_1 \cdot r_2 = 1$ is referred to as an *ideal copolymerization*. The radical copolymerization between MMA ($r_1 = 0.84$) and vinylidene chloride ($r_2 = 0.99$) is close to ideal [26].

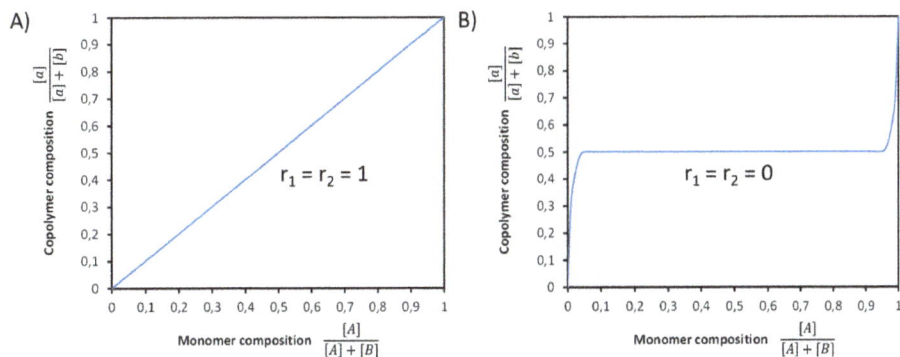

Figure 2.29: Copolymer composition diagram of (a) an ideal copolymerization with $r_1 = r_2 = 1$ and (b) an alternating copolymerization with $r_1 = r_2 = 0$.

When $r_1 = r_2 = 0$, it is impossible for a monomer to react with a growing chain end derived from the same monomer. In other words, both k_{11} and k_{22} (Figure 2.28) are zero. As a result, *alternating copolymerization* takes place. Independent on the monomer composition, the mole fraction of A (and B) in the copolymer chain is 0.5 (Figure 2.29(b)). The radical polymerization of acrylonitrile ($r_1 = 0.05$) with maleic anhydride ($r_2 = 0.005$) is an example of an alternating copolymerization [21].

It seems obvious that reactivity ratios are dependent on the specific pair of monomers. In addition, r-values also depend on the chain-growth mechanism [27]. A classic example is the copolymerization of styrene and MMA (Table 2.5). Radical polymerization leads to statistical copolymers, while polymerization in the presence of stereospecific Ziegler–Natta catalysts gives rise to alternating copolymers. For cationic polymerizations, $r_1 \gg r_2$. Addition of MMA monomers is unfavorable because it will result in a cationic center with an electron withdrawing side group, while, in contrast, $r_2 \gg r_1$ in case of anionic polymerization.

Table 2.5: Reactivity ratios for copolymerization of styrene (1) and MMA (2) via different chain-growth mechanisms.

Chain growth polymerization	r_1	r_2	Copolymer structure
Radical	0.52	0.46	Random
Cationic	40	0.01	Large sequences of styrene
Anionic	0.01	50	Large sequences of MMA
Coordination	0.01	0.05	Alternating

2.3.3.1 Graft- and block copolymerization

Block- and graft copolymers (Figure 2.3) are usually not obtained via simultaneous polymerization of two monomers. Special synthesis routes have been developed to prepare those well-defined copolymer structures (Figure 2.30).

Graft copolymers can be prepared via a coupling reaction between a functionalized polymer backbone and the end groups of another polymer or oligomer. This method, named *grafting (on)to*, has become more popular since the invention of so-called click chemistry [28], a class of reactions giving selective products in high yields by "clicking" molecular building blocks together. An elegant alternative is the *grafting from* strategy. Side groups along the polymer backbone function as initiation sites, from which side chains are polymerized. Finally, *grafting through* involves copolymerization in the presence of macromonomers. In this case, the ratio of monomer to macromonomer determines the graft density along the copolymer backbone.

Similar techniques are applied to produce *block copolymers* (Figure 2.30). Polymers with functional end groups can be joined together via coupling reactions, involving click chemistry for instance. Alternatively, polymer end groups may function as initiation sites, from which another chain is polymerized. This method is referred to as macroinitiation. As discussed previously, controlled radical polymerization or living anionic polymerization also enables the design of precise polymer architectures. In this way, block copolymers can be synthesized by sequentially adding different monomer types.

Poly(styrene-*block*-butadiene-*block*-styrene), also named SBS rubber, is a commercial block copolymer produced via anionic polymerization (Figure 2.31). First, the polymeriza-

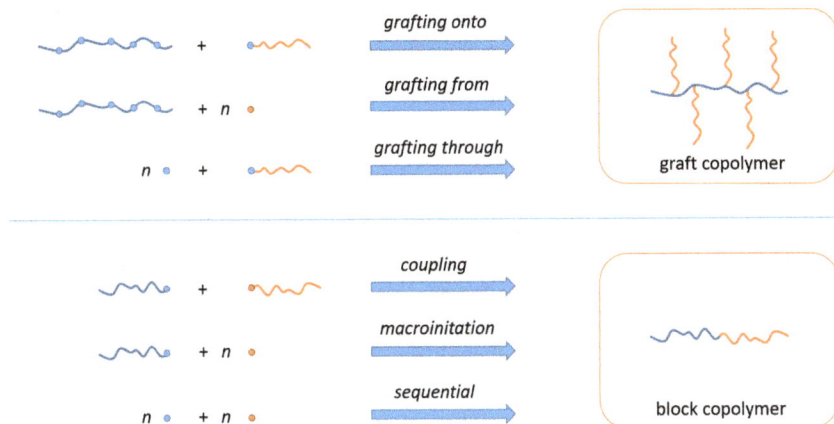

Figure 2.30: Schematic overview of various synthetic routes towards graft copolymers (*top*) and block copolymers (*bottom*).

tion of styrene is initiated by butyl lithium (BuLi). Butadiene is added to the living PS to yield a diblock copolymer. Then, dichlorodimethylsilane is added to link two growing chains. The results is SBS, a triblock copolymer known under the registered trademarks Kraton® and Styrolux®. Despite their covalent attachment, polystyrene and polybutadiene blocks are incompatible and therefore tend to separate on the nanoscale. As a result, they form a periodic nanostructure with soft and hard domains, a typical morphology for *thermoplastic elastomers* (TPEs). This will be discussed further in Section 2.4.10.

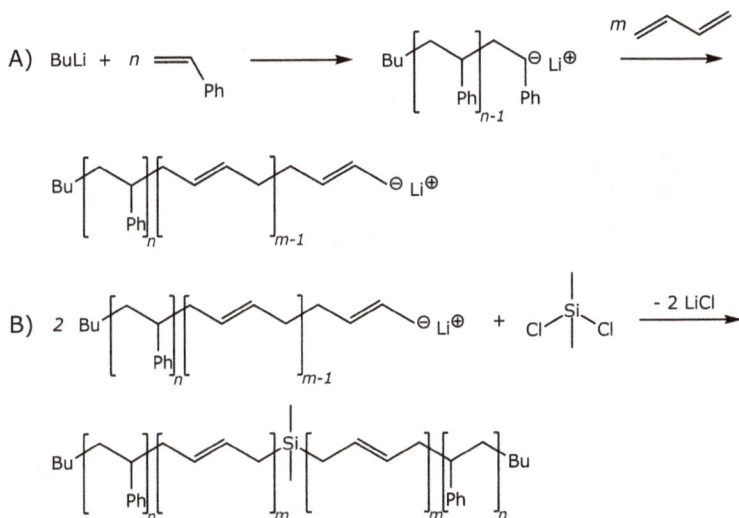

Figure 2.31: (a) Living anionic block copolymerization of styrene (S) and butadiene (B). (b) Termination proceeds with a bifunctional silane, resulting in SBS triblock copolymers.

2.3.3.2 Important copolymers

The advantage of copolymers is that the characteristics of homopolymers can be combined into a new material. By altering the chemical nature and ratio of different building blocks, the material properties can be tuned as desired. This strategy is called *copolymer engineering*.

Circa 60% of styrene produced ends up in thermoplastic copolymers rather than homopolymers. For example, polystyrene (PS) is quite brittle, and its resistance to solvents is moderate. The chemical resistance can be greatly improved by copolymerization of styrene with 10–40% acrylonitrile. Poly(styrene-*co*-acrylonitrile) (SAN) is extensively used in the automotive industry and household appliances. In great contrast, copolymerization with 65–80% butadiene yields a flexible elastomer, i. e., poly(styrene-*co*-butadiene) rubber (SBR). SBR is used to make tires, shoe soles, and backings of carpets. When elastomers specifically need resistance to oils and fat, e. g., in O-rings or hoses, poly(acrylonitrile-*co*-butadiene) rubber (NBR) is applied instead.

Acrylonitrile-butadiene-styrene, or ABS, combines the chemical inertness and heat resistance of acrylonitrile with the toughness of butadiene, while styrene adds to the processability and rigidity. By varying the [*A*] : [*B*] : [*S*] ratio, tailor-made terpolymers can be produced (Figure 2.32). The most imaginative application of ABS is in the colorful interlocking plastic bricks manufactured in Denmark by LEGO®.

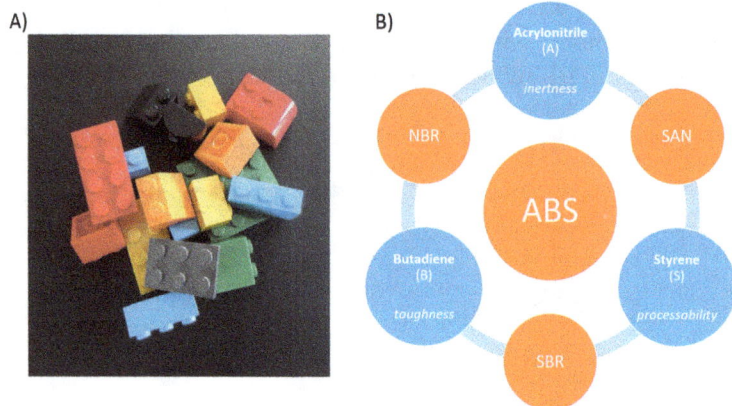

Figure 2.32: (a) Photograph of colorful LEGO® bricks, made from ABS copolymer. (b) Overview of variety of copolymers created from styrene, butadiene, and acrylonitrile. Copolymer properties can be varied by changing the monomer-to-monomer ratio.

Other commercially relevant copolymers are modified ethylene polymers. Ethylene-vinyl acetate copolymers, abbreviated as EVAc, can be produced in all possible ratios. Depending on the copolymer composition, applications can range from greenhouse sheeting to flip flops. Crosslinked ethylene-propylene-diene elastomer (EPDM), made via coordination polymerization, is applied as a membrane on flat roofs.

2.3.4 Polymerization techniques

The industrial production of polymers can take place by various methods. It depends on the chemical composition of the reactants, the required process conditions, and the desired product. Homogeneous polymerization techniques are mass and solution polymerization. Heterogeneous techniques include polymerizations in multi-phase systems, such as suspensions and emulsions.

Mass polymerization (or bulk polymerization) involves direct polymerization of monomers in the bulk (Figure 2.33). Due to the absence of other chemicals, like solvents or stabilizers, additional purification steps are unnecessary. In fact, bulk polymerization enables direct processing into the desired product form, without the need for additional post-processing. For example, Perspex® (PMMA) sheets are fabricated via this technique, starting from MMA and a peroxide initiator.

When addition polymerization is performed in bulk, sufficient heat removal becomes a great challenge. During the reaction, the viscosity of the undiluted system rapidly increases with increasing conversion. In a more viscous environment, polymer chains lose mobility. Therefore, the termination rate decreases. On the other hand, chain growth can continue since the mobility of the small monomers is hardly affected. In this particular case, Flory's principle of equal reactivity is no longer valid. The average chain length increases, which leads to an even higher viscosity, which leads to a higher P_n, etc. The rapid increase in viscosity causes problems in terms of heat removal since radical polymerizations are typically exothermic. As a result, local overheating can occur. This process is referred to as the *gel effect*, or *Trommsdorff effect* [29].

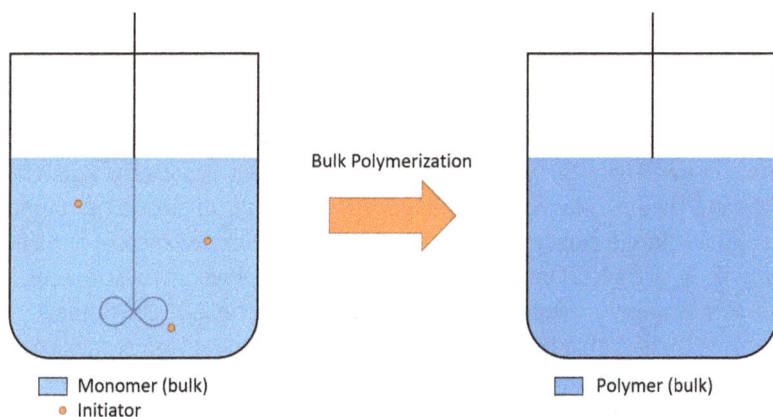

Bulk Polymerization

■ Monomer (bulk)
• Initiator

■ Polymer (bulk)

Figure 2.33: Bulk or mass polymerization: undiluted monomer is polymerized in bulk. These systems are susceptible to the Trommsdorff effect at higher conversions.

The Trommsdorff effect may be prevented by the addition of chain-transfer agents (Section 2.3.2.1), which lowers the molar mass of the product and therefore controls the viscosity of the system. Alternatively, the polymerization can be interrupted at low conversion.

As an alternative, monomers can be polymerized in solution (Figure 2.34). The diluted system improves heat transfer in comparison to bulk polymerization. *Solution polymerization* is popular on the laboratory scale. Industrial application is limited due to expensive product recovery that involves removal and recycling of solvents. Nevertheless, solution polymerization can be employed when the polymer product is commercialized as a polymer solution, which is the case for many coatings.

In certain cases, the polymer is not soluble in the reaction mixture. During polymerization, it will precipitate from solution, which allows filtration of polymer product afterwards.

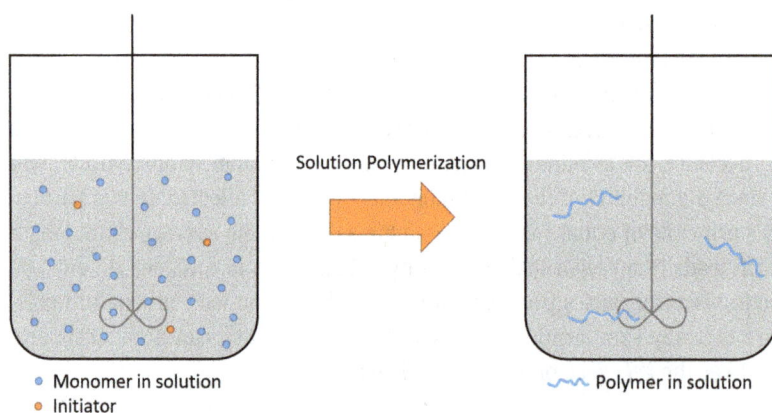

Solution Polymerization

● Monomer in solution
● Initiator

〜 Polymer in solution

Figure 2.34: Solution polymerization: diluted monomer is polymerized in solution. When the polymer is not soluble, it precipitates from solution.

Suspension polymerization involves polymerization in monomer droplets dispersed in a medium, typically water (Figure 2.35). This technique is particularly popular for (exothermic) radical polymerizations. Monomer droplets have a diameter of circa 0.1 to 1 mm. The initiator is soluble in the monomer. Upon heating, the initiator decomposes and initiates chain growth. In fact, a local bulk polymerization takes place within each monomer droplet, which acts as a tiny reactor. Temperature control is optimized due to the surrounding water that transfers the heat effectively. To stabilize the suspension, stabilizing agents are often added to prevent coalescence of droplets during polymerization. The reaction product, small polymer beads, can be separated from the medium. Clearly, suspension polymerization cannot be employed in the case of water-soluble monomers.

Surfactants, sometimes called "soap molecules", are amphiphilic molecules containing a hydrophilic and hydrophobic part. When dispersed in water, above a certain concentration, the so-called *critical micelle concentration* (CMC), surfactants aggregate

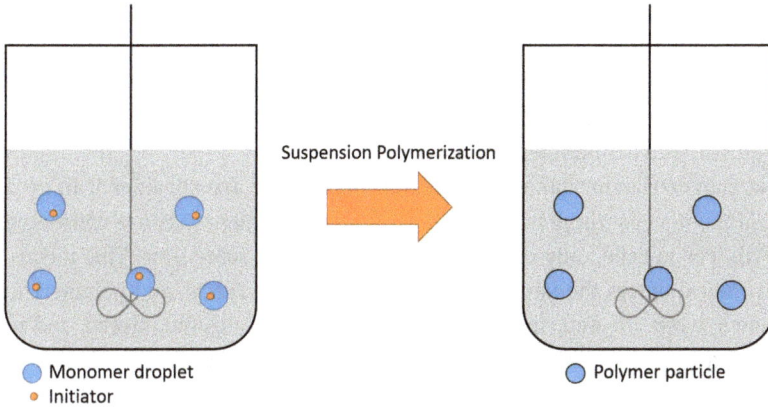

Figure 2.35: Suspension polymerization: monomer droplets are dispersed in a medium. The initiator is soluble in the monomer. Polymerization takes place in the monomer droplets.

into spherical micelles. In *emulsion polymerization,* a large concentration of surfactant is added, and polymerization takes place within these micelles (Figure 2.36) [30].

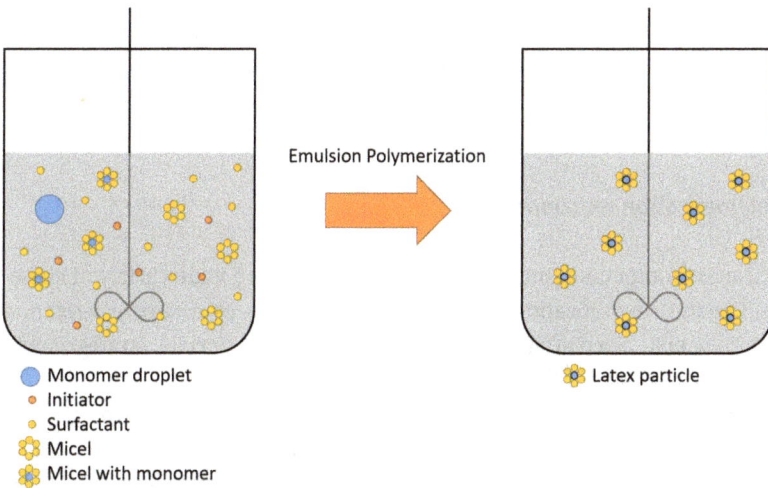

Figure 2.36: Emulsion polymerization: surfactants form micelles in a medium. Monomers diffuse from monomer droplets into the micelles. Initiators are soluble in the medium, where they initiate chain growth and diffuse into the micelles. Further polymerization takes place in the micelles that become latex particles.

The crucial ingredients for an emulsion polymerization are monomer, water-soluble initiator, surfactant, and water as a medium. After mixing, monomers diffuse from monomer droplets into the hydrophobic core of the micelles. The (radical) initiator is soluble in water, in contrast to suspension polymerization. When it decomposes, the active center attacks the diffusing monomers in the medium to form oligomers. In fact, the active oligomer is now a kind of (growing) surfactant itself. It can diffuse into a micelle, where it continues propagation. Since the surface area of micelles is considerably larger in comparison to that of the monomer droplets, the chance that oligomer radicals diffuse into a monomer droplet is close to zero.

As soon as polymerization takes place within the micelle, we consider it to be a *latex particle*. Monomers continue to diffuse into the micelle, where they are consumed by chain growth. Per micelle, only one radical can exist at the same time. This means, when another radical enters the latex micelle, the active chain reacts, and polymerization is terminated, while the entering of another radical reactivates the micelle, and so forth. The frequency of this "switch on/off" process influences the average degree of polymerization.

Latex particles grow in size as long as monomer is available. At the same time, the size of monomer droplets decreases. When all monomer has been consumed, the rate of polymerization decreases, and the process is completed. The latex particle size is typically 0.1 μm, which is about 1,000 times smaller than the polymer beads produced during suspension polymerization.

Emulsion polymerization can produce polymers with high molar mass, which is of great importance in industry. Many commercial vinyl polymers and copolymers are produced via this technique. The polymerized emulsion is often ready for use, in the case of paints, coatings, and glues. However, without further purification, surfactants and other additives remain in the product and may influence performance.

2.3.5 Post-polymerization reactions

Like low molar mass compounds, macromolecules can undergo a wide variety of chemical reactions. Chemical modifications on polymers are known as *post-polymerization reactions* since they take place after ("post") polymer synthesis. Various strategies will be discussed in the remainder of this section.

Reactions on functional groups of polymers without changing the degree of polymerization (P_n) are of great technical significance. An example of such a *polymer analogous reaction* is the hydrolysis of polyvinyl acetate (PVAc) to polyvinyl alcohol (PVA), as depicted in Figure 2.37. During the reaction, acceleration takes place since converted alcohol groups can undergo hydrogen bonding with incoming hydroxy ions (OH⁻). The conversion is crucial since PVA cannot be produced via direct polymerization from vinyl alcohol (VA), which is less stable in comparison to its *tautomer* (structural isomer) acetaldehyde. PVA is used in personal-care products and as a stabilizing agent in suspension polymerizations. In addition, PVA is the precursor to another commercial polymer, PVB, or polyvinyl butyral. PVB is obtained after reaction of PVA with butyraldehyde (Figure 2.37). Note that the irreversible reaction leaves unreacted alcohol groups. In fact, PVB should thus be considered as a copolymer. The industrial process has a maximum conversion of circa 80%. PVB is widely applied in laminated safety glass for the automotive industry. When glass windows break in a car accident, the glass splinters adhere to the PVB interlayer instead of falling free and causing injury.

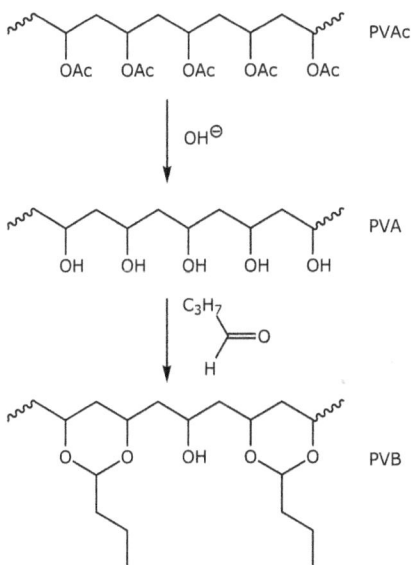

Figure 2.37: Two polymer analogous reactions: hydrolysis of polyvinyl acetate (PVAc) to polyvinyl alcohol (PVA), followed by conversion of PVA to polyvinyl butyral (PVB).

SBS block copolymers, depicted in Figure 2.31, can undergo hydrogenation to produce saturated triblock copolymers. Both saturated and unsaturated polymers are commercially available as high-performance elastomers (Kraton®).

Polysaccharides contain many functional hydroxyl groups suitable for polymer analogous reactions. For instance, cellulose can be converted into cellulose esters or cellulose ethers, semi-synthetic polymers named *cellulose derivates*. Many cellulose derivates with a broad range of properties are commercially available. Particular interesting is the conversion of cellulose in an alkaline environment into cellulose xanthate (Figure 2.38). Regenerated cellulose is produced from cellulose xanthate under acidic conditions, known as the xanthate or viscose process. The solubility of cellulose xanthate allows the fabrication of cellulose fibers (rayon or viscose) or sheets (cellophane).

In contrast to polymer analogous reactions, *crosslinking reactions* lead to an increase in the degree of polymerization (Figure 2.5). The degree of crosslinking influences both mechanical and thermal properties. Crosslinked systems are typically stronger, but hard to process and mostly insoluble. As pointed out in the introduction to this chapter, one of the oldest post-polymerization reactions is the vulcanization of rubber. The process converts sticky natural rubber into a valuable elastomer product, using sulfur as a crosslinking agent. Other popular crosslinkers are peroxide compounds, applied in the production of silicone rubbers and EPDM elastomers.

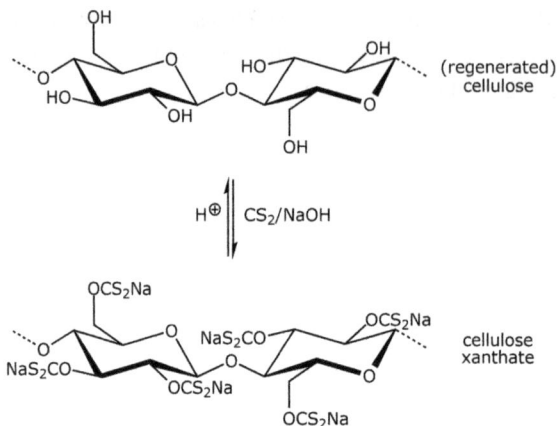

Figure 2.38: Representation of the viscose process: cellulose is converted into cellulose xanthate, followed by formation of regenerated cellulose in an acidic environment.

Unsaturated polyesters are extensively used as resins for the production of thermosets and fiber-reinforced plastics. The prepolymer contains carbon–carbon double bonds along the chain, originating from polycondensation in the presence of maleic anhydride. Styrene is commonly used to cure the unsaturated resin, a process initiated by peroxides (Figure 2.39). The network density can be tuned by either the copolymer composition of the polyester prepolymer or the concentration of styrene.

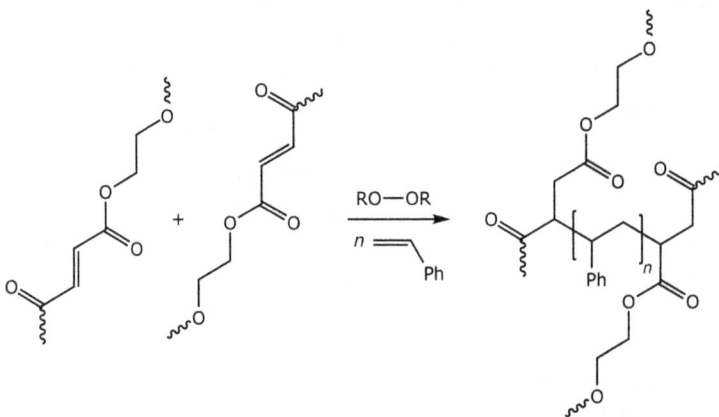

Figure 2.39: Peroxides initiate the crosslinking of unsaturated polyester resins with styrene.

Crosslinking reactions as previously discussed mostly involve heat curing. However, hardening can be initiated photochemically as well. Acrylate and methacrylate resins are often cured under influence of light. A photoinitiator decomposes into active species when irradiated with light and triggers network formation of multifunctional (meth)acrylates. Photopolymerization of acrylates is employed to produce clear coatings and dental fillings.

Degradation of macromolecules results in a decrease in the degree of polymerization. Depolymerization or chain scission may be caused by chemical reactions, (UV) light exposure, thermal or mechanical stress, or exposure to microorganisms. When degradation is undesired, polymer products can be stabilized with UV stabilizers or antioxidants. For certain applications, however, degradability is highly desired. In recent years, biologically degradable polymers have gained more interest due to environmental concerns. *Biodegradable polymers* can be decomposed under certain circumstances by nature (usually by microbial enzymes) into gases, water, and biomass. They will be further discussed in Chapter 3.

2.4 Polymer materials

So far, we have encountered a wide variety of macromolecules. Depending on the synthetic approach, polymers will differ in chain length, composition, configuration, degree of branching, and crosslink density. These factors drastically influence the performance of plastics. The unique characteristics of polymeric materials are the subject of the following section.

2.4.1 Polymer morphology

The term morphology is derived from the Greek words μορφή (morphe = "form") and λόγος (logos = "word" or "study"). In polymer science, *morphology* describes the structure of polymers on the (macro)molecular level. In the solid state, macromolecules can be arranged randomly or in an organized manner. Amorphous polymers present a random chain structure, while crystalline polymers are well-defined and show a highly organized morphology (Figure 2.40). The arrangement of macromolecules has a significant influence on the properties of the polymer material [31].

Crystalline materials have a highly ordered molecular structure. Common examples are sodium chloride (NaCl), or table salt, sugar, and diamonds. The transition from liquid to solid state is called crystallization. Due to their large molecular chain structure, the crystallization of polymers is usually imperfect. This results in a partially crystalline morphology with both amorphous and crystalline domains, i. e., a *semi-crystalline polymer*. *Crystallinity* refers to the degree of structural order in the macromolecular structure. Material properties like strength, stiffness, chemical resistance, melting temperature, and optical clarity are influenced by the degree of crystallinity. In practice, 100% crystalline polymers do not exist, and crystallinity ranges from 10 to 80%.

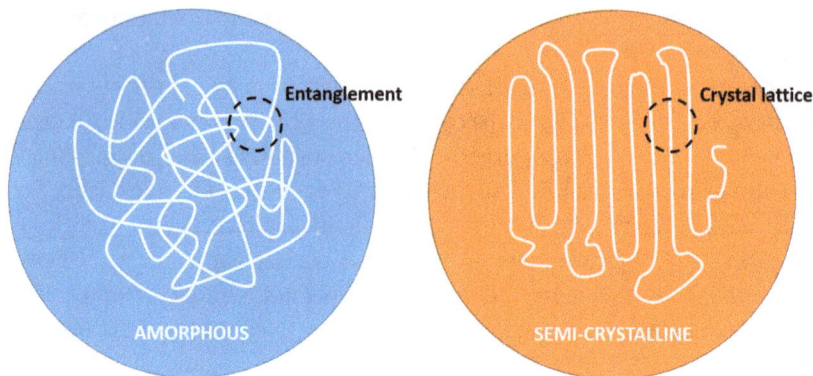

Figure 2.40: Schematic representation of molecular chain arrangement in solid state: amorphous and crystalline polymers.

Many polymers lack the ability to form ordered structures and so cannot crystallize. They form an amorphous phase, similar to window glass. In *amorphous polymers*, the molecular chains are randomly arranged and heavily entangled (Figure 2.40), much like a plate of cooked spaghetti. Intertwining polymer chain segments are therefore called *entanglements*.

2.4.2 Thermal transitions

Amorphous materials can be differentiated from crystalline solids by their thermal transitions. Let us first consider what happens to the mechanical properties of an amorphous thermoplastic when heated. The mechanical performance can be described in terms of *Young's modulus* (*E*), in MPa, also referred to as E-modulus. Young's modulus is defined in *Hooke's law* (equation (2.24)), where the applied stress (σ), in MPa, is proportional to elongation (ε). A high E-modulus resembles a stiff material, while a product that can easily be deformed has a low value for *E*.

$$\sigma = E \cdot \varepsilon. \tag{2.24}$$

Figure 2.41 displays the temperature dependency of the *E-modulus* for an amorphous polymer with a high degree of polymerization. At low temperature (*T*), polymer chains are immobile and represent a frozen state. The material is stiff and the modulus is high. The polymer is in its *glassy state* (I).

When the temperature is increased, chain segments become mobile. The tenacity of the material increases, and it starts to behave like leather (II). This process is called the glass transition, and the temperature related to this is the *glass transition temperature*, or T_g. During glass transition, the modulus drops to a certain extent.

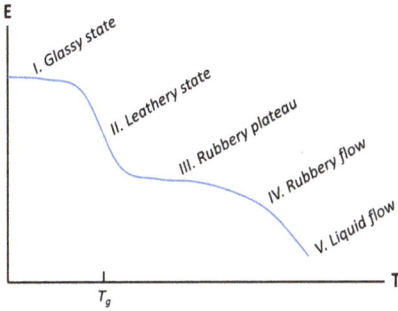

Figure 2.41: Elastic modulus as function of temperature: amorphous polymer with high molar mass. Elastic behavior dominates in the glassy state (I), while viscous behavior dominates in liquid flow (V). In between, the polymer material exhibits both elastic and viscous properties, so-called viscoelastic behavior.

Although polymer chains are mobile above T_g, the material does not behave as a liquid macroscopically due to the large amount of entangled chains. The material easily deforms when a small force is applied. Depending on the length of time that the stress is applied, the material may or may not return to its original shape. In this temperature range, referred to as the *rubbery plateau* (III), the polymer shows both elastic and viscous behavior. This property is very typical of polymers and is called *viscoelasticity*. It will be further discussed in Section 2.4.4.

When temperature is further increased, the mobility of the macromolecular chains also increases. As a result, they become disentangled. The viscous behavior becomes more dominant, and the material starts to flow macroscopically. This is a crucial temperature range for plastic processing via injection molding, extrusion, and melt spinning (Section 2.5). We distinguish rubbery flow (IV) and liquid flow (V), although in practice, there is no sharp transition from III to V.

Since the rubbery phase for amorphous polymers is dependent on the extent of entanglements, the "length" of the rubbery plateau is related to the polymer chain length. Smaller macromolecules will be less entangled. As a result, when degree of polymerization (P_n) decreases, liquid flow is achieved at lower temperatures (Figure 2.42).

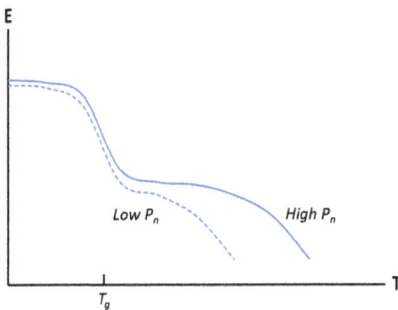

Figure 2.42: Elastic modulus as function of temperature: amorphous polymer with low molar mass (dashed curve) and amorphous polymer with high molar mass (solid curve). Liquid flow is reached at lower temperatures for low molar mass polymers.

When crystalline solids like salts are heated, a sharp melting point is observed. As just discussed, the crystalline structure of macromolecules in not perfect. Semi-crystalline polymers exist of both crystalline and amorphous domains. If the *glass transition* of a semi-crystalline polymer is reached, only the amorphous fraction transforms from the glassy to the rubbery state. The crystalline domains remain, and the E-modulus lowers to a lesser extent in comparison to a fully amorphous polymer (Figure 2.43). When the temperature is further increased, the highly ordered rearrangement within the crystalline domains is disrupted. This means the *melting temperature* (T_m) is reached. The macromolecular chains become mobile, no longer caught in their crystalline lattice, and the material starts to flow macroscopically. The T_g is always lower than T_m. In the hypothetical case of a 100% crystalline polymer, no glass transition is observed at all in the absence of amorphous domains.

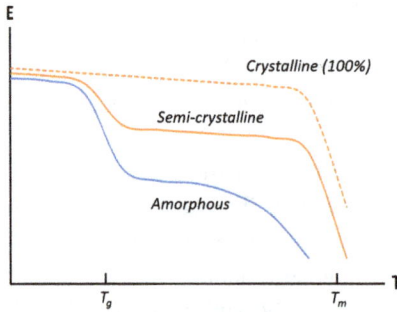

Figure 2.43: Elastic modulus as function of temperature: amorphous polymer (*blue curve*) in comparison to a semi-crystalline polymer (*orange curve, solid*) and hypothetical 100% crystalline polymer (*orange curve, dashed*). A semi-crystalline polymer exhibits both T_g and T_m.

In the case of crosslinked polymers (Figure 2.44), the network structure restricts the molecular mobility and therefore prevents macroscopic flow. In highly crosslinked networks, like thermosets, above the glass transition the Young's modulus is only slightly lower. In fact, thermosets can be used both below and above their T_g.

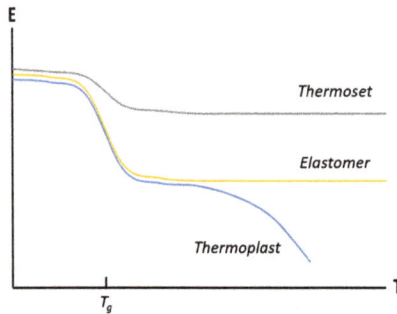

Figure 2.44: Elastic modulus as function of temperature: amorphous thermoplastic (blue curve) in comparison to an elastomer (*yellow curve*) and thermoset (*grey curve*). Network structures prevent liquid flow at elevated temperatures.

In slightly crosslinked networks, like elastomers, the molecular structure allows a larger extent of mobility. Therefore, the Young's modulus drops significantly above T_g. Nevertheless, the crosslinks still prevent liquid flow. As for thermosets, the modulus remains constant until thermal degradation occurs at elevated temperatures. Elastomers find applications above the glass transition, at which they demonstrate their elastic properties.

2.4.3 Characterization of T_g and T_m

The glass transition temperature (T_g) and melting temperature (T_m) are crucial material properties for a polymer product. They define an operating window for both processing and final application. Therefore, determination of T_g and T_m is of great importance. A method frequently used in laboratories is *differential scanning calorimetry* (DSC). DSC measures the difference in heat flow for a polymer sample with respect to a reference sample, while both are heated (or cooled) at a constant rate. The measured heat flow is related to the heat capacity (C_p), that is, the amount of heat needed to produce a defined change in temperature. A typical DSC curve for an amorphous thermoplastic is depicted in Figure 2.45. When the glass state transitions into the rubbery state, a sudden increase in heat flow is observed, visualizing the T_g.

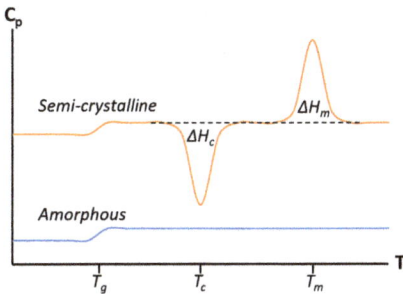

Figure 2.45: DSC curves: heat capacity (or heat flow) as function of temperature. Increase in heat capacity indicates a glass transition. The exothermic and endothermic peaks (visible for semi-crystalline polymers) represent crystallization and melting, respectively.

For semi-crystalline polymers, an exothermic (negative) peak above the glass transition indicates the *crystallization temperature* (T_c). Upon further heating, melting (T_m) occurs, visualized by the endothermic peak in the DSC curve (Figure 2.45). This is related to the fact that, during melting, the temperature of a polymer sample does not change, while heat is still absorbed. The area under the crystallization and melting peaks can be used to calculate the heat of crystallization (ΔH_c) and heat of melting (ΔH_m), respectively. From ΔH_c (or ΔH_m) the *degree of crystallization* can be determined [32].

Another technique to determine the glass transition is dynamic mechanical (thermal) analysis, abbreviated as DMA or DMTA. This measures the materials' modulus when subjected to a periodic (dynamic) stress. With DMA, it is possible to observe small thermal transitions that are difficult to visualize by DSC. An example of this is determination of T_g in crosslinked thermosets.

2.4.4 Viscoelasticity

As discussed in Section 2.4.2, amorphous polymers will not immediately behave as a liquid when heated just above T_g since the macromolecules are still heavily entangled at that point. In fact, they enter the rubbery state, in which the material can demonstrate both elastic and fluid behaviors. This unique characteristic of polymers is named *viscoelasticity*. When a polymeric material is applied to stress, the response is dependent on the duration of that stress.

If stress is only applied for a short time, the polymer chains have no time for reorientation and will adopt their original conformation afterwards. The material will return to its original shape. It behaves elastically. However, if the stress is applied for a longer period of time, the macromolecules have time to reorient and can move relatively to one another. Consequently, the material starts to flow. The mechanical behavior is not only dependent on time, but on temperature as well, as depicted in Figure 2.41. The mobility of polymer chains increases at higher temperatures, which induces fluid behavior.

Elastic properties can be visualized by a spring, an element of energy storage (Figure 2.46(a)). Applied stress at t_0 immediately results in elongation, according to Hooke's Law (equation (2.24)). When stress is removed at t_1, the spring instantaneously returns to its original shape. On the other hand, viscous behavior is exemplified by a dashpot, filled with a viscous liquid (Figure 2.46(b)). A dashpot is an element of energy dissipation. The deformation is permanent and related to the period of time that stress is applied ($t_0 - t_1$).

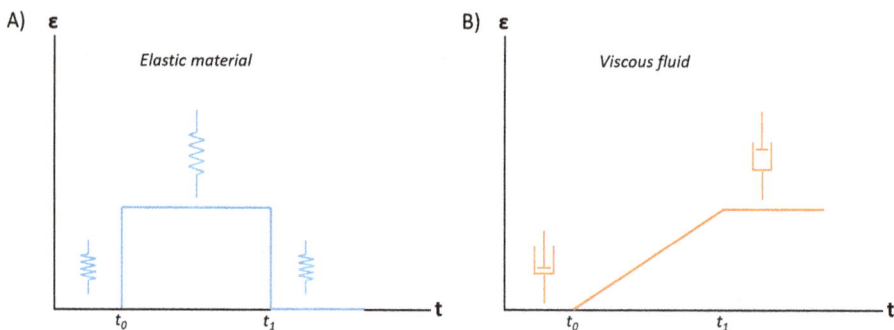

Figure 2.46: Elongation as function of time for (a) elastic material, represented by a spring (energy-storage element) and (b) viscous fluid, represented by a dashpot (energy-dissipation element). Stress is applied at t_0 and released at t_1.

The behavior of an elastic material and a viscous fluid can be combined, leading to various different models. A straightforward example is the *Maxwell model,* in which the spring and dashpot are connected in series (Figure 2.47(a)). The total elongation is therefore the sum of both elastic (spring) and viscous (dashpot) components. When the system is relieved of the stress (t_1), the spring retracts, while the deformation as a result from the dashpot remains. So, the model represents *irreversible creep*. Rather than predicting the mechanical performance under constant stress, the Maxwell model is used to describe the viscoelastic behavior under constant elongation, called *stress relaxation*.

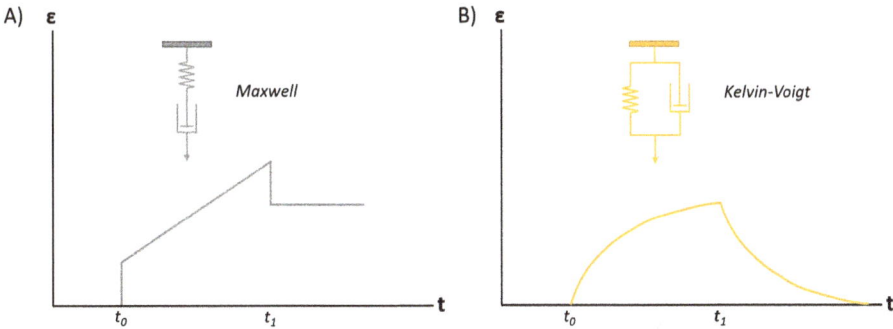

Figure 2.47: Elongation as function of time for (a) Maxwell model, and (b) Kelvin–Voigt model. Stress is applied at t_0 and released at t_1.

The *Kelvin–Voigt model* has a parallel arrangement of spring and dashpot elements (Figure 2.47(b)). In this case, the total elongation is equal to the elongation of one of the components. If stress is removed (t_1), retraction of the spring is delayed by the dashpot, but, eventually, the system returns to its original shape. In other words, this model represents *reversible creep*.

To describe viscoelasticity in a more accurate manner, we need to combine the Maxwell and Kelvin–Voigt models. This is achieved in the Burgers model, schematically depicted in Figure 2.48. When stress is applied at t_0, the "Maxwell spring" responds immediately. When stress is removed, this spring returns to its original position instantaneously. This relates to the spontaneous elastic deformation of a viscoelastic polymer (ε_a). If the applied stress is prolonged, both the "Voigt dashpot" and "Maxwell dashpot" respond. Removal of the stress results in recovery of the Kelvin–Voigt element. It corresponds to the retarded elastic deformation (ε_b). In contrast, the Maxwell dashpot remains elongated (ε_c), corresponding to the irreversible (viscous) flow of a viscoelastic polymer.

2.4.5 Crystallization of polymers

In practice, during plastic processing, *crystallization* often takes place when cooling from the melt. Upon crystallization, the entangled polymer chains prevent the formation of perfect crystals. The polymer chains have to order themselves, but a complete (macro)molecular reorientation is not possible in the viscous melt. Therefore, crystalline regimes will form alongside amorphous domains (Figure 2.49). This partially crystalline structure, with crystal lamellae interrupted by amorphous layers, is unique to polymers [33].

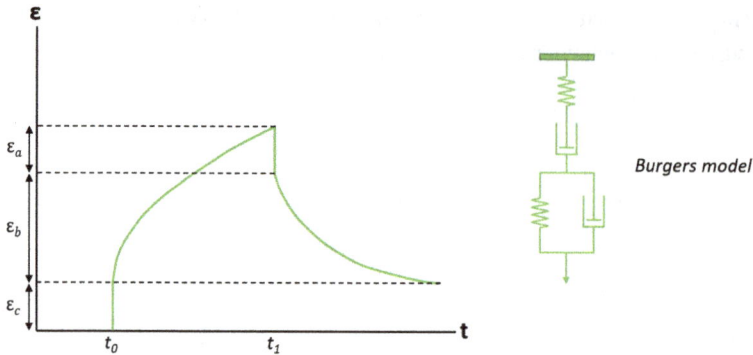

Figure 2.48: Elongation as function of time for Burgers model, used to predict viscoelastic behavior in polymer materials. Stress is applied at t_0 and released at t_1.

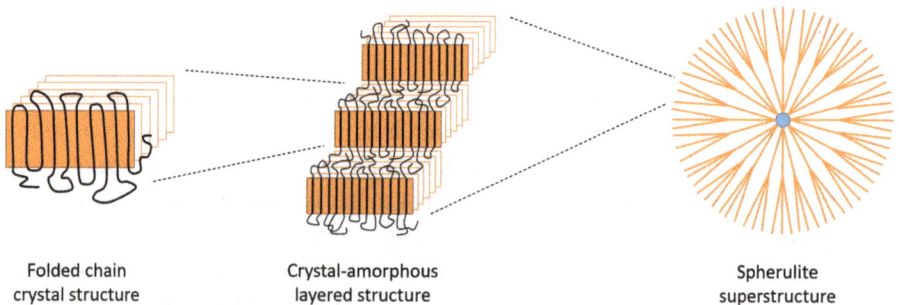

Folded chain crystal structure

Crystal-amorphous layered structure

Spherulite superstructure

Figure 2.49: Schematic representation of semi-crystalline morphology in polymers from nanoscale to microscale. The blue dot in the center of the spherulite superstructure represents the nucleus.

Imagine a semi-crystalline polymer above T_g and below T_m (Figure 2.43). In this temperature range, we find solid crystalline domains surrounded by a mobile amorphous phase. So, while both domains are composed of chains that are chemically identical, they exist in two different phases. Semi-crystalline polymers in between T_g and T_m typically show ductility. Macroscopic flow however is not possible because many polymer chains will be part of more than one crystal lamella, and movement of those *tie-molecules* is thus restricted.

Crystallization of polymers starts with *nucleation*, followed by crystal growth. Nucleation may occur from an ordered domain within the polymer itself, called homogenous nucleation. However, heterogeneous nucleation can also take place. Heterogeneous nuclei are – among others – impurities, additives, or specific nucleating agents. During *crystal growth*, the crystalline lamellae usually grow radially and form spherical superstructures (Figure 2.49), so-called *spherulites*, with a size in the order of 10–100 µm. Spherulites can be revealed by an optical microscope between crossed polarizers (Figure 2.50).

Besides cooling from the melt, polymers can crystallize from solution or upon mechanical stretching. The latter is applied in the drawing of polymer fibers, which will be discussed in Section 2.5.3.

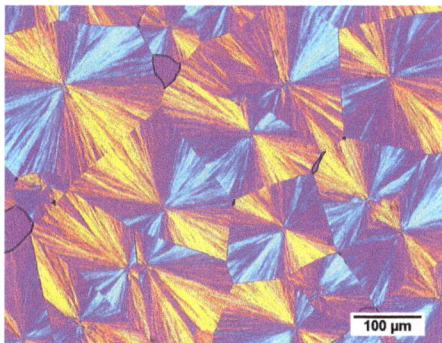

Figure 2.50: Spherulite superstructure revealed by a polarizing optical microscope [34].

In general, semi-crystalline polymers appear opaque. This is caused by the crystal domains that scatter visible light. Exceptions to this rule are crystalline polymers with crystallites smaller than the wavelength of light (as in transparent PET bottles). This can be achieved by adding a large amount of nucleating agents, which increases the number of nuclei and therefore reduces the size of the crystal domains significantly.

Amorphous polymers are usually optically transparent. For that reason, PMMA is used to produce Perspex® or Plexiglass® windows, while PC is commonly used in eye protection and as greenhouse glass.

2.4.5.1 Factors that influence crystallinity

Some polymers are able to crystallize from the melt, while others are not. Crystallization requires polymer chains to arrange themselves in a highly ordered manner. This may happen either via chain regularity or interaction. Let us discuss this based on a few examples.

The configuration along a polymer chain influences its crystallinity. For instance, atactic polystyrene (PS-at) is completely amorphous because of its irregular structure. Well-defined isotactic and syndiotactic PS, however, are both crystalline and have melting temperatures of 240–270 °C. *Crystallinity* greatly affects the properties of a polymer product. *Trans*-polybutadiene is a hard, semi-crystalline material. In contrast, *cis*-polybutadiene is a sticky, amorphous polymer. The *cis* configuration causes kinks in the polymer chains, preventing them from forming crystalline regions.

In addition, interactions between polymer chains can induce crystallization as well. An important member of the polyamide (PA) family, i. e., Nylon 6,6, forms an ordered structure due to intermolecular hydrogen bonding (Figure 2.51). Nylon 6,6 has a much higher melting temperature in comparison to high-density polyethylene (HDPE), which lacks the ability to form a hydrogen-bonded network. As a result, Nylon 6,6 is a very rigid polymer. Copolymerization supplies the opportunity to influence the properties of this material. For example, polycondensation in the presence of a larger diacid, such as dodecanedioic acid, disturbs the highly ordered network and leads to a decrease in T_m and E-modulus.

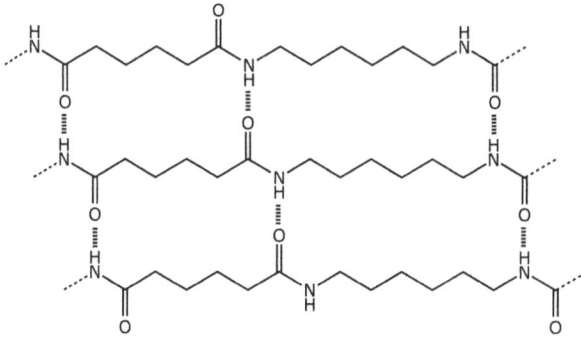

Figure 2.51: Ordered macromolecular structure of Nylon 6,6 due to hydrogen bonding leads to a high melting temperature and rigid performance.

2.4.6 Mechanical behavior of polymer solids

Important material properties of polymer solids can be obtained by stress-strain analysis. In a tensile tester, the polymer product is subjected to tensile force, and the response to stress is measured until failure. A typical stress-strain curve is displayed in Figure 2.52.

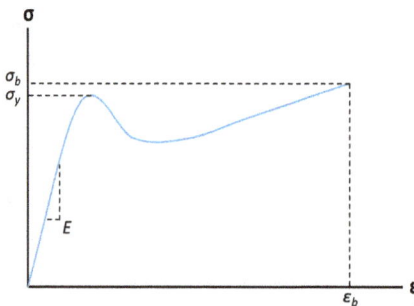

Figure 2.52: Stress-strain curve resulting from tensile testing.

The stress at which the material fractures is called the *tensile strength* (σ_b). The corresponding elongation is referred to as the *elongation at break* (ε_b). Young's modulus (E) is given by the initial slope of the curve, according to Hooke's law (equation (2.24)) and is a measure for stiffness. The area under the curve corresponds to the *fracture energy*, a measure for toughness. Clearly, tensile testing is a very powerful technique since material properties such the strength, stiffness, and toughness are obtained from one single measurement.

Let us consider three distinct stress-strain curves that relate to specific polymer materials (Figure 2.53(a)). For rigid, brittle polymers, a small elongation already leads to fracture. The steep curve indicates that minor elongation requires high stress. In other words, the tensile strength and *Young's modulus* are high. This behavior is typical of polymers below their glass transition, or highly crosslinked thermosets.

In contrast, elastomers give rise to flatter curves. Due to their rubbery behavior, elongation requires only a little stress. Those materials show a small modulus but a large elongation at break. This behavior is typical of polymers well above their glass transition.

In between, we find the curve for ductile, viscoelastic materials. At a certain elongation, those polymers start to flow. This is named the *yield point*, and the corresponding stress is referred to as *yield strength* (σ_y), as shown in Figure 2.52. The yield point marks the end of elastic behavior and the start of viscous (plastic) behavior. Once the yield point is passed, part of the deformation will be permanent. This typical viscoelastic behavior is observed near the glass transition. Compared to brittle and rubbery polymers, ductile materials demonstrate the highest fracture energy and thus the highest toughness.

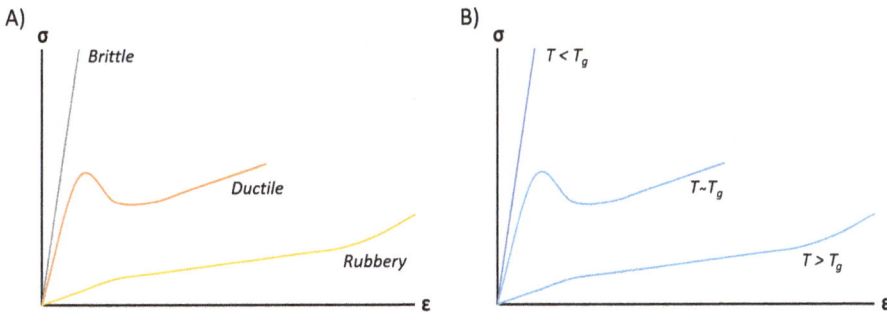

Figure 2.53: (a) Typical stress-strain curves for brittle, ductile, and rubbery polymer materials. (b) Mechanical performance is dependent on temperature, with a key role for the glass transition temperature.

Figure 2.53(b) demonstrates that the mechanical behavior of polymers is strongly dependent on temperature. With increasing temperature (T), the elastic modulus and strength will decrease, while the elongation at break increases. The materials become softer.

A similar dependence is observed for the speed of tensile testing. When subjected to a large tensile speed, brittle fracture is observed. Lower tensile speeds lead to ductile fracture. The fact that mechanical performance is dependent on both temperature and time has already been discussed when the viscoelastic models of Maxwell, Voigt–Kelvin, and Burgers were treated in Section 2.4.4.

Another important parameter for polymeric materials is impact strength. *Impact strength* is the capability to withstand force that is applied suddenly. Both Izod and Charpy impact tests have been developed to measure the amount of energy needed to fracture a material.

When polymers are subjected to mechanical stress in a harsh (chemical) environment, cracking may occur. This phenomenon is called *stress corrosion*. Typical examples are crack formation in rubber tubing under the influence of ozone and the failure of nylon fuel connectors due to (battery) acid attack.

2.4.7 Rheological behavior of polymer melts

Understanding the flow behavior of polymer melts is crucial to the successful processing of polymers. The study of material flow is called *rheology*. A shear rheometer measures the flow in response to applied shear stress. The rheological behavior of fluids is described by *Newton's law of viscosity* (equation (2.25)).

$$\tau = \eta \cdot \gamma. \tag{2.25}$$

In the case of a so-called *Newtonian fluid*, the shear rate (γ), in s^{-1}, is linearly proportional to the shear stress (τ) applied, in Pa. The gradient of the curve is defined as its *viscosity* (η), in Pa · s. Note the similarity between Newton's and Hooke's laws (equation (2.24)).

However, many fluids deviate from Newtonian behavior and are therefore referred to as *non-Newtonian fluids*. The rheological behavior of polymer melts is rather complex, due to the viscoelastic properties. At low shear rates, the entangled chains in the polymer resist flow. If the shear rate is increased, the macromolecules disentangle and flow is facilitated. Consequently, the viscosity of polymers melts is high, but typically decreases with increasing shear rate. This effect is called *shear thinning* (Figure 2.54).

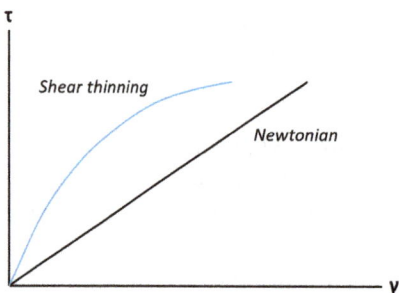

Figure 2.54: Shear stress as function of shear rate. Polymer melts typically exhibit non-Newtonian behavior called shear thinning.

2.4.8 Electrical properties

In general, polymers are electric insulation materials with a specific resistance exceeding ceramics and glass [16]. Examples of excellent *electrical insulators* are PE, PP, PS, and PTFE. Polyethylene (PE) has a very low loss factor, meaning only a minor fraction of electrical current is lost by dissipation in the material. As a result, PE enabled the use of radar in World War II.

The high electric resistance of most polymeric materials often leads to *electrostatic charging*, related to the inability to discharge. The resulting high electric tension may give rise to film sticking, dust attraction, or even spark formation. In order to decrease electrostatic charging, the conductivity should be increased. To do so, carbon black can be added to the polymer, or antistatic agents are used (Section 2.5.1).

Not all polymeric materials are electrical insulators. A special class of macromolecules, intrinsically *conducting polymers*, are organic polymers with the ability to conduct electricity. The main examples are polyacetylene (PAC), poly(*para*-phenylene vinylene) (PPV), and polythiophene (PT), in which the macromolecular backbone contains conjugated double bonds ($-C = C-C = C-C = C-$). Polythiophenes show potential application in organic solar cells.

2.4.9 Liquid crystalline morphologies

While most crystalline materials transition from an ordered solid state to a disordered liquid state, some substances form an intermediate state. This *liquid crystalline phase,* or mesophase, possesses characteristics of both a liquid and a crystal. In their mesophase(s), liquid crystalline materials demonstrate viscous flow behavior, but the molecular structure is still ordered, somewhat like a crystal [17]. Liquid crystals are widely used in liquid crystalline displays (LCDs).

Liquid crystalline polymers (LCPs) can be either lyotropic or thermotropic. Thermotropic mesophases are formed in bulk by changing temperature, while lyotropic mesophases can be formed in solution. A famous example of lyotropic LCPs are polyaramids. When aramid fibers are spun from solution in highly concentrated sulfuric acid, they form a liquid crystalline phase. The ordered structure is retained after processing (Figure 2.16). As a result, they exhibit excellent mechanical strength in the direction of the fiber.

2.4.10 Block copolymer morphologies

Block copolymers represent a special class of macromolecules, with intriguing phase behavior. Chemically distinct polymers are in general not miscible but tend to phase separate instead. In the case of block copolymers, however, two (or more) incompatible polymers are covalently bonded. Phase separation on the macroscale is therefore not possible. Instead, phase separation at the length scale of the macromolecules occurs. Although this happens at the nanoscale, this phase behavior is often referred to as *microphase separation* in the literature, which is somewhat confusing [35].

Depending on the volume ratio of the blocks, microphase separation may lead to various *block copolymer morphologies* (Figure 2.55). Low fractions of one component result in spherical aggregates in a matrix of the major component. Cylindrical morphologies are obtained when the fraction of the minor component is increased. If both blocks have similar lengths, a layered (or lamellar) morphology is obtained.

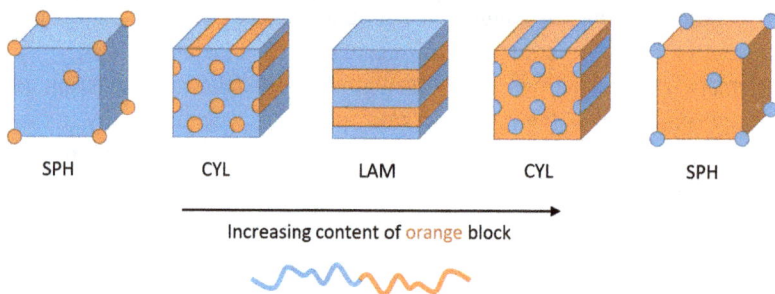

Figure 2.55: Schematic view of various block copolymer morphologies resulting from phase separation at the nanoscale: spherical (*SPH*), cylindrical (*CYL*), and lamellar (*LAM*) morphology.

Block copolymer phase separation is applied technologically in the case of *thermoplastic elastomers* (TPEs), such as the previously discussed SBS block copolymers (Section 2.3.3.1), known as Kraton®. The polystyrene (PS) outer blocks form hard spherical domains inside a soft and continuous matrix of polybutadiene (PB) inner blocks (Figure 2.56). The PS domains are glassy at room temperature and behave as physical crosslinks. The material acts as a chemically crosslinked rubber but is, in fact, thermoplastic. Above the T_g of PS, the hard domains become soft, and the material is malleable. Upon cooling below the glass transition, the phase separated nanostructure reforms. The morphology and properties of those TPEs can be tuned by altering the ratio of PS:PB blocks.

2.4.11 Vitrimers

The recent invention of a new class of polymeric materials blurred the classical distinction between thermoplastics and thermoset polymers. By introducing the so-called *dynamic covalent bonds* in a polymer network, crosslinks can be exchanged under specific conditions. Ludwig Leibler and coworkers were the first to point out the existence of such adaptable materials, capable of undergoing rearrangements in their molecular

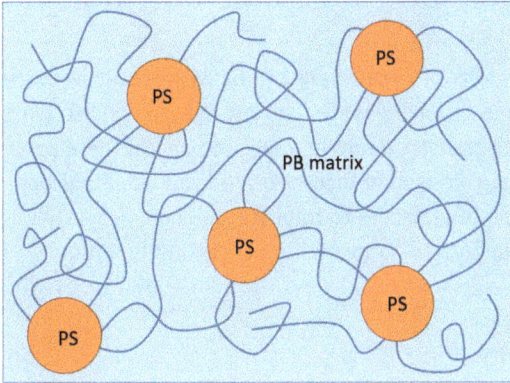

Figure 2.56: Microphase separation in SBS triblock copolymers, representing a spherical morphology. Glassy polystyrene domains (orange) physically crosslink the polybutadiene matrix (blue): an example of a thermoplastic elastomer (TPE).

network. The academic researchers named them *vitrimers*, referring to the resemblance to vitreous silica [36].

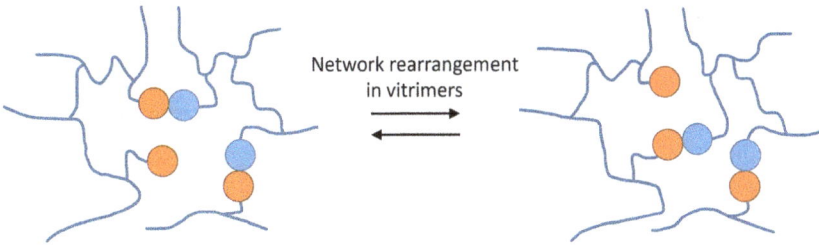

Figure 2.57: Schematic representation of network rearrangement in vitrimers. On the macroscopic scale, it enables (re)processing of crosslinked polymers.

When a vitrimer is heated, molecular arrangements can take place in the polymer network (Figure 2.57). In case that the exchange of dynamic bonds is fast enough, it can lead to macroscopic flow while processed. This unique behavior places vitrimers somewhere in between the traditional thermoplastic and thermoset materials. On the one hand, they possess the mechanical and chemical robustness of crosslinked polymers. When activated, however, their dynamic nature facilitates reshaping and reprocessing. This enables the recycling of (covalent) network polymers, opening a promising route toward a truly circular economy of plastics.

2.5 From polymer to product

Today, more than 400 million metric tons of plastics are produced worldwide every year. This enormous pile of plastics is used for the widest variety of applications: from packaging, textiles, and construction to transportation and electronics. All these products differ in shape, texture, and color. This section focuses on the conversion of polymers to a useful product. First, the use of additives to improve polymer properties is briefly discussed. The remainder is dedicated to available processing techniques to shape polymer products.

2.5.1 Polymer additives

During, or prior to, processing of polymers into a commercial plastic product, other substances are typically added to modify the characteristics of the material. These so-called *additives* may improve the mechanical performance, impart a desired color, optimize processing conditions, influence the chemical resistance, smoothen the surface, or simply reduce the overall cost of the material [37]. The exact formulation depends on the specific application and available processing equipment, but often multiple additives are included in one product. The process in which polymers are mixed with additives is referred to as *compounding*.

The importance of additives is reflected in the global market that is approaching €50 billion. The largest contribution is from *plasticizers*. Plasticizers reduce the T_g and modulus of a material, making it more flexible. Plasticizers like dialkyl phthalates are used to produce the soft PVC used in cables and flooring. *Fillers*, such as clay, talc, sand, or wood flour, serve to reduce costs. Reinforcement fillers are added to improve properties such as elastic modulus and tensile strength. Fibers may be also considered as reinforcement fillers. The processing of fiber-reinforced plastics, however, is discussed separately.

Other additives frequently employed are stabilizers, lubricants, flame retardants, colorants, biocides, blowing agents, antistatic agents, and compatibilizers. An overview of common polymer additives and their function is presented in Table 2.6.

2.5.2 Processing of plastics

In recent decades, a wide variety of processing techniques have been developed to create useful products from polymeric materials [38]. Most thermoplastics are processed in the molten state, where decreased viscosity allows flow of the material. After shaping, the material is cooled down to obtain a solid product. In contrast to cross-linked polymers (thermosets and elastomers), processing of thermoplastics can be repeated, allowing mechanical recycling of plastic products.

Table 2.6: Overview of polymer additives commonly used in plastic processing to enhance product performance.

Additive	Function
Plasticizers	Improve flexibility of material by lowering the glass transition temperature.
Fillers	Reduce material costs; improve processability or enhance mechanical properties (such as modulus, strength).
Stabilizers	Protect material against thermal or UV degradation during processing or actual application in environment.
Lubricants	Improve flow by reducing melt viscosity or adhesion between material and processing tool.
Flame retardants	Improve fire resistance of material by reducing the flammability.
Colorants	Impart (desired) color to material by dyes or pigments.
Biocides	Protect material against exposure to bacteria and fungi by controlling microbial growth.
Blowing agents	Transform the polymer into a foam by gas production via a physical or chemical process.
Antistatic agents	Reduce the electrostatic charge on material surface in order to prevent dust collection or sparking.
Compatibilizers	Promote miscibility between polymers; prevent phase separation.

2.5.2.1 Extrusion

Perhaps the most widely used technique to process polymers is *extrusion*. Extrusion takes place in an *extruder*, a cylindrical barrel containing a rotating screw (Figure 2.58). A polymer, typically in the form of granulate or powder, is introduced via the hopper. Three zones can be distinguished within the extruder: feeding, compression, and metering. The rotating screw transports the material, while the barrel is heated to ensure polymer melting. The screw is tapered so that the distance between the screw channel and barrel wall decreases along the extruder, resulting in an increased shear rate. At the end of the extruder, the polymer melt exits the system via an opening: the die. The die can shape the material into a desired form, such as a rod, pipe, or sheet. The molten shape is cooled using water or air and carried away mechanically. If this proceeds at a higher speed than the exiting speed of the material from the extruder, the shape is drawn into thinner dimensions. This can be applied to compensate for the "die swell" that occurs when the polymer melt leaves the extruder and tends to recover in an elastic manner.

Before the actual shaping process can take place as just described, a separate extrusion procedure may be needed in a so-called *compounder*. Compounding is performed if simple mixing of polymer and additives is not sufficient to obtain a

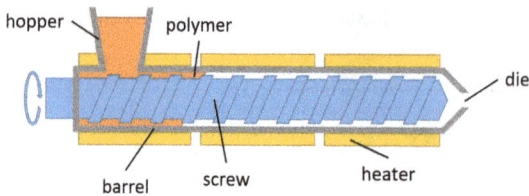

Figure 2.58: Schematic view of single-screw extruder.

homogenous compound. Typically, this is done in a twin-screw extruder. The compound that leaves the die in the form of a strand is granulated and dried, ready for further processing.

Besides shaping and mixing, extrusion can also be used for degassing. The geometry of the screw can be varied, depending on the required procedure. When chemical reactions are performed with polymers in an extruder, it is called *reactive extrusion*. An example is the chemical recycling (Section 4.4) of polyester. By addition of ethylene glycol, PET is depolymerized into its oligomers.

Blown film extrusion, or film blowing, is a specific type of extrusion used to produce packaging films and plastic bags. To achieve a very thin film, thin tubes are extruded through a ring die, and then immediately inflated by pressurized air. As a result, the tube is stretched, and the walls can reach a thickness of 10–100 µm. In an industrial setting, film blowing machines may exceed 15 m in height. After cooling, the tube is collapsed, and the film is rolled up. Subsequent sealing and cutting is performed to manufacture bags for trash bins.

2.5.2.2 Injection molding

In contrast to extrusion, *injection molding* is a batch process. It is the most popular technique to shape thermoplastics, although it is applied for rubbers as well. In an injection molding machine, a mold is filled with polymer melt by an extruder. This process is carried out under high pressure, which needs to be adjusted to optimize filling of the mold. The exact amount of material injected is called the shot weight. After the injection, polymer solidifies within the mold. The mold temperature is below T_g (for amorphous polymers) or T_m (for semi-crystalline polymers). Cooling is followed by demolding and ejection of the final product. Parts with precise geometry and high gloss can be produced on a large scale, using polymers such as PMMA, polypropylene (PP), polystyrene (PS), and polyamides (PAs). The process allows a large freedom of product design. Bottle caps, automotive parts, toys, musical instruments, chairs, LEGO® bricks, and storage containers are just a few examples of the many parts fabricated by injection molding. In the case of the processing of rubbers and thermosets, crosslinking occurs after injection into the mold. Consequently, this is referred to as *reaction injection molding* (RIM). Another popular molding technique is *injection blow molding*, widely used in the packaging sector to make plastic (PET) bottles.

2.5.2.3 Filament extrusion and 3D printing

Extrusion of thermoplastics through a cylindrical die leads to the fabrication of monofilaments. By controlling this process, filaments with a precise diameter can be produced. Those materials are used as plastic ink for *3D printers*. In *fused filament fabrication* (FFF), also known as fused deposition modeling (FDM), the monofilament is subjected to another extrusion process (Figure 2.59). Filament is fed into a hot extruder head that is able to move. Molten material leaves the extruder nozzle and is deposited as a layer on top of the previous layer, where it fuses and solidifies. Layer by layer, a 3D product is fabricated. The movements of the *3D printer* are controlled by a computer that translates 3D models into printable objects. 3D printing is also called *additive manufacturing* (AM), referring to the layer by layer addition of material [39]. FFF is one of the cheapest and most often used 3D printing technologies. It is applied for rapid prototyping, facilitating short production cycles and inexpensive testing. Other popular AM techniques rely on photopolymerization (stereolithography apparatus, SLA) and powder sintering (selective laser sintering, SLS) instead of extrusion. They produce more precise objects, but are more expensive as well.

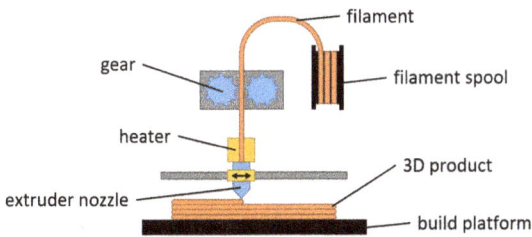

Figure 2.59: Schematic view of fused filament fabrication (FFF) 3D printer.

2.5.2.4 Pressing

While *pressing* can be used for the processing of all kinds of polymeric materials, it is in particular relevant for crosslinked systems. The uncured (pre)polymer is added to a mold. The mold can be part of the press itself or an independent piece that has to be placed in between the heating plates of the press. Regardless of the situation, the mold is closed and pressurized. As a result, the material is distributed evenly throughout the mold, before crosslinking occurs at elevated temperatures. Processing parameters have to be determined carefully to prevent premature crosslinking. Pressing is typically employed for elastomers and thermosets, such as epoxy- and unsaturated polyester resins.

2.5.2.5 Calendering

Calendering is employed for the production of foils and sheets. Polymeric material passes through a series of heated roller pairs that rotate in opposite direction at vari-

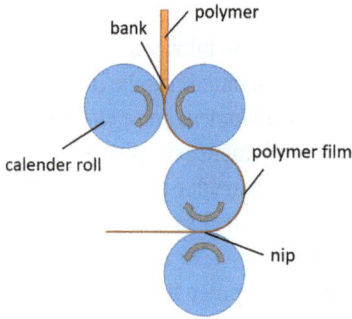

Figure 2.60: Schematic view of four-roll calender.

ous speed, and is rolled out into a broad film above T_g or T_m. A high melt viscosity is required for this process. For instance, polyamides cannot be processed via calendering. The thickness of the polymer film can be adjusted by the gap, or nip, between the rollers (Figure 2.60). A small gap leads initially to the accumulation of polymer melt, the so-called bank, where intensive mixing occurs. After the material passes the last roller pair, it cools down and can be collected by winding or cutting. This technique is applied in the fabrication of floors and shower curtains. If thinner films are desirable, blown film extrusion is the preferred processing technology.

2.5.3 Processing of fibers: Spinning

Fibers can be produced from polymers via a process called spinning. Polymer fibers are the synthetic equivalent of natural fibers such as cotton and wool. They find application in textiles, carpets, and in fiber-reinforced polymers (Section 2.5.5).

During the *spinning* process, polymer molecules are oriented in the direction of the fiber. The parallel alignment of the chains increases crystallinity. As a result, synthetic fibers behave anisotropically. In other words, they demonstrate a much larger tensile strength in the longitudinal direction of the fibers in comparison to the perpendicular direction. Interactions between the polymer chains stabilize the alignment. Therefore, symmetrical and unbranched polymers, such as isotactic PP, are suitable materials for fiber spinning. Even better is their ability to form hydrogen bonds, which is the case for PET, nylon, and high performance aramid fibers (Figure 2.16).

Several spinning processes have been developed in the past. In *melt spinning*, a polymer melt is forced through a perforated die plate, a *spinneret*, at high pressure using an extruder or pump (Figure 2.61(a)). A spinneret may contain more than thousand tiny holes with a diameter in the order of 1 mm. Beyond the spinneret, the fibers are pulled, cooled with air, and collected on a fast winding spool. The final thickness of the multifilament fibers is determined by the speed of the fiber take-up, rather than the diameter of the spinneret holes. In industry, extremely high spinning rates up to 8,000 m/min are reached. Melt spinning is an important technique for the production of nylon, PET, and PP fibers.

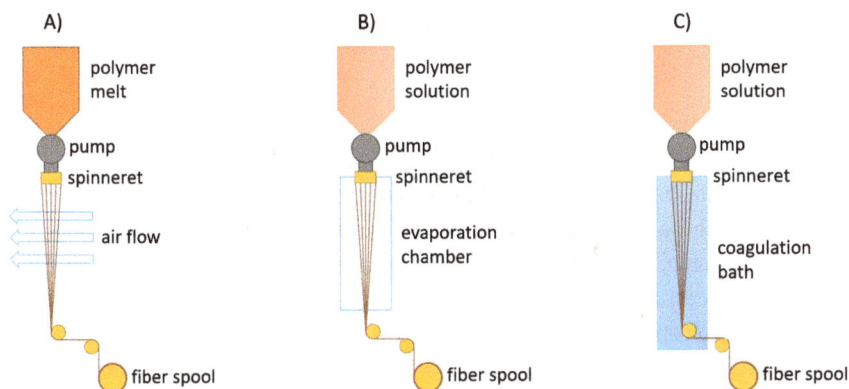

A) polymer melt — pump — spinneret — air flow — fiber spool

B) polymer solution — pump — spinneret — evaporation chamber — fiber spool

C) polymer solution — pump — spinneret — coagulation bath — fiber spool

Figure 2.61: Schematic view of (a) melt spinning, (b) dry spinning, and (c) wet spinning.

Alternatively, polymer fibers can be produced from solution. This technique can be applied at lower temperatures and is, somewhat confusingly, called *dry spinning*. A polymer solution is forced through a spinneret. After leaving the die plate, the fibers are formed by evaporation of solvent in warm air (Figure 2.61(b)). The solvent can be recovered and reused in the same process. Dry spinning is employed for the production of PAN and elastane (Lycra®) fibers.

Wet spinning also starts from a polymer solution (Figure 2.61(c)). However, in this case, the solution is pressed through a spinneret in a coagulation bath containing a non-solvent. As a result, the polymer fibers precipitate, while the solvent is washed out. Wet spinning is the oldest spinning technique, and it is applied in the viscose process for the production of rayon (or viscose) fibers. Cellulose cannot be processed in the melt. Therefore, it is converted into cellulose xanthate (Figure 2.38), which is soluble under basic conditions. The cellulose xanthate solution is spun into an acidic bath. In the acidic environment, cellulose (rayon) is regenerated. Aramid fibers are also produced via wet spinning. In contrast to the viscose process, an air gap is left between spinneret and precipitation bath. This technique leads to highly crystalline Kevlar® and Twaron® fibers with superior mechanical performance.

All spinning processes are usually followed by additional stretching of the spun fibers, a process called *drawing*. The fibers are stretched over (a sequence of) rollers rotating at increasing speeds. Drawing leads to an increase in alignment of the polymer chains. Typically, this results in an increased degree of crystallinity, stiffness, and strength, while elongation at break is reduced.

Other spinning techniques that are relevant are gel spinning, used for the production of UHMWPE fibers (Dyneema®), and electrospinning, applied in membrane fabrication.

2.5.4 Processing of foams

Foams are low-density materials that contain pores, also referred to as cells. The pores can vary in size and shape. In *closed-cell foams*, the pores are not connected and therefore not permeable. They cannot be compressed easily and find application in heat insulation and shock absorption, for instance, in car bumpers. *Open-cell foams* contain connected pores; liquid or gas is able to circulate. Those foams are usually soft and deformable and applied in cushions, cleaning sponges, and sound insulation.

To produce foam, gas bubbles have to be introduced into the material of choice. This can be achieved mechanically, chemically, or physically. A common process for foam production is extrusion, in which a melted polymer is extruded together with a blowing agent under high pressure. A popular example of a *physical blowing agent* used in this process is volatile solvent, like pentane. When leaving the extruder, the material expands into a foam due to the pressure drop. The expanded polymer strand can be chopped into granules and used for production of all sorts of molded parts. This technique is applied for the fabrication of expanded polystyrene (EPS), or Styrofoam™.

Alternatively, foams can be produced via an injection molding process. Pre-extruded granules with blowing agent can be used, but additional blowing agent can be introduced during injection molding, if needed. When the material is injected into the mold, it expands into a foam. Therefore, lower pressure is required compared to conventional injection molding of (non-porous) thermoplastics.

A very versatile class of foams is based on polyurethanes (PUs) formed by reaction of isocyanates and polyols (Figure 2.19(b)). By altering the functionality of both components, the density of the macromolecular network can be controlled. PU foams can be produced using either physical or *chemical blowing agents*. The latter is done by water that reacts with isocyanate groups to produce CO_2 gas (Figure 2.62). Both flexible (soft) and rigid foams are so fabricated, using techniques such as foam casting and reaction injection molding.

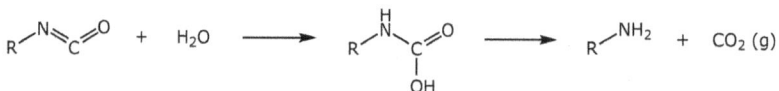

Figure 2.62: Foam formation in PU foams. Water acts as chemical blowing agent and reacts with isocyanate to form carbon dioxide.

2.5.5 Processing of fiber-reinforced plastics

Fiber-reinforced plastics (FRP) are composites of fibers fixed into a polymer matrix. To a great extent, the fibers determine the mechanical performance of the FRP. Typically, the modulus and strength of fibers is higher than the matrix. The matrix, however, plays a crucial role in stabilizing the fibers and protecting them from the environment. The matrix will distribute the loads evenly between the fibers, but only in the case of adequate adhesion between matrix and fibers.

Both inorganic fibers, like glass, and organic fibers, like carbon and aramid fibers, are extensively used. However, natural fibers (such as flax, hemp, and jute) have shown potential as sustainable alternatives. Fibers can be distributed randomly through the matrix, e. g., in the case of short (chopped) fibers or oriented in a unidirectional tape or multidirectional woven (Figure 2.63). The matrix material can be either thermoplastic or thermoset. The excellent mechanical properties of composites has encouraged application in the gripping arm of space shuttles and wings of airplanes.

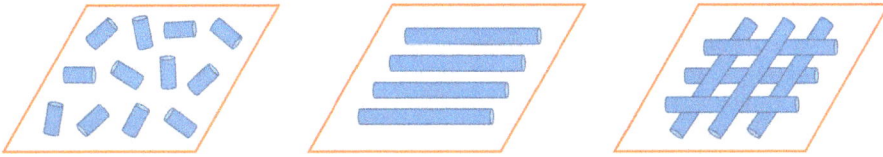

Figure 2.63: Various types of fiber orientation (blue) in composite matrix (orange): (a) random, (b) unidirectional, and (c) bidirectional.

A straightforward manufacturing process for composite materials is via *hand lay-up*. Layers of matrix and fibers are applied in an alternating fashion on a mold, covered with release agent. The process is quite labor intensive and the product is usually of lower quality compared to other processing techniques. Nevertheless, hand lay-up is used for the production of boats, containers, and prototypes. In the *spray-up* process, resin and chopped glass fibers are sprayed onto a mold, simultaneously with a spraying gun.

Alternatively, FRPs may be produced by *resin transfer molding* (RTM). The process usually takes place in a vacuum, so-called vacuum-assisted resin transfer molding, or *vacuum infusion*. Fiber mats are inserted into a mold, which is then closed and evacuated. Next, a (pre)polymer resin, for instance, unsaturated polyester or epoxy, is injected. The resin impregnates the fibers, assisted by the applied vacuum, and is cured. Composite parts, applicable in aerospace and wind turbines, of large scale and high quality can be manufactured. This is an attractive alternative to more expensive techniques using an industrial autoclave.

Glass mat-reinforced thermoplastics (GMT) or long fiber-reinforced thermoplastics (LFT) are semi-finished products that can be transformed into their final shape via *compression molding*. In the case of thermosets, pressing of sheet molding compounds (SMC) and bulk molding compounds (BMC) takes place. This technique is often used for the production of automotive parts.

Pultrusion is employed to produce mainly thermoset profiles reinforced with long and continuous fibers (such as carbon). During the process, fibers are first impregnated with resin. The materials get their final shape by being pulled through a die. Note the difference with extrusion, which involves pushing through a die instead. In the case of thermoset composites, crosslinking is induced by heating. Alternatively, thermoplastic pultrusion is employed to produce single profiles in which continuous fibers are combined with a thermoplastic polymer matrix. *Filament winding* is applied to manufacture hollow structures, such as pressure vessels and grain silos.

2.6 Exercises

2.6.1 Classification

1. Classify the following polymers based on origin:
   a. Casein
   b. PHA
   c. ABS
   d. Rayon
   e. Pectin
   f. mRNA
   g. Chitosan
2. Give the general name of the following polymers:
   a.

   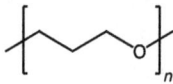

   b.

   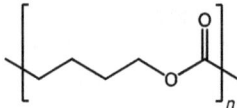

   c.

   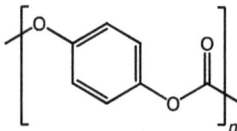

3. Write down the structural formula of the following polymers:
   a. Syndiotactic polypropylene
   b. Polyvinyl alcohol
   c. Polybutylene terephthalate
   d. Nylon 6,10

    e. Poly(1,1-difluoroethylene)

    f. Polyvinylidene fluoride

    g. Polycaprolactone

4. Discuss the difference between elastomers and rubbers.

5. Name the top three plastic industries and give an example of a product application for each market.

2.6.2 Structure of macromolecules

6. Explain why copolymers are sometimes referred to as "tailor-made" materials.

7. What is the minimum number of functional groups needed in a monomer to build a polymer network.

8. Polymerization of isoprene can lead to six different isomers. Write the structural formula for each of them.

9. Calculate the P_n and P_w of the polymer mixture presented in Figure 2.7.

10. 4 mole of a monodisperse sample of 20,000 g/mol is mixed with 6 mole of a monodisperse sample of 40,000 g/mol. Calculate the M_n, M_w and D of the product.

2.6.3 Synthesis of polymers

11. List the main differences between step-growth and chain-growth polymerization.

12. Figure 2.13 displays the relationship between conversion and degree of polymerization for a polycondensation reaction. Draw the curve for an anionic (living) polymerization.

13. An SEC measurement of a sample of poly(lactic acid) leads to the following data: Number average molar mass of 1,200 g/mol and a polydispersity of 1.94. We assume Flory's principle of equal reactivity is valid.

    a. Calculate the weight average molar mass of the product.

    b. Calculate the conversion of the polycondensate.

    c. Calculate the (number average) degree of polymerization.

14. PLA is an important bioplastic.

    a. Propose two different reaction schemes for the polymerization of PLA.

    b. Which synthesis route is preferred by industry? Explain why.

    c. Discuss the recyclability of PLA.

15. If proteins are considered as nylon derivatives, explain which number should be assigned to this natural polymer.

16. Explain the role of crystallinity on the mechanical performance of aramid fibers.

17. During bulk polymerization of MMA in a batch reactor, the reactor starts to over-heat after a monomer conversion of 50%.
    a. Write the reaction scheme for radical polymerization of MMA.
    b. Explain what happened above a conversion of 50%.
18. Reactivity ratios can predict the composition of a copolymer. What do you expect if both r_1 and r_2 are infinite?
19. For the cationic copolymerization of styrene (1) and MMA (2), $r_1 = 40$ and $r_2 = 0.01$.
    a. Calculate the (initial) copolymer composition when equimolar amounts of styrene and MMA are polymerized.
    b. Explain which termination mechanism does not occur during this copoly merization.
20. Explain why a narrow molar mass distribution is obtained in the case of emulsion polymerization.
21. Explain the large difference in the hydrolysis products of PMMA and PLA.
22. Discuss the role of the initiator in emulsion and suspension polymerization and explain the difference.
23. Which polymerization route is used to produce HDPE? Explain why.

2.6.4 Polymer materials

24. Sketch the macromolecular morphology of an amorphous (thermoplastic) material, if:
    a. $T < T_g$
    b. $T > T_g$
25. Explain the role that macromolecular entanglement plays in the rubbery state of an amorphous polymer.
26. Sketch the macromolecular morphology of a semicrystalline (thermoplastic) material, if:
    a. $T < T_g$
    b. $T_g < T < T_m$
    c. $T > T_m$
27. Which type of polymers find application above their T_g? Name three.
28. Draw the configuration of springs and dashpots in Burgers model for the crucial points in the ε–T diagram of Figure 2.48.
29. In case of shear thickening, the viscosity increases when the shear rate increases. An example is corn starch in water. This behavior is contrary to shear thinning, as depicted in Figure 2.54. Draw the curve that belongs to shear thickening.
30. A typical SBS block copolymer is used to produce a thermoplastic elastomer. However, the product has a very high elongation at break. Propose what changes can be made to the macromolecular structure to obtain a less elastic material.

2.6.5 From polymer to product

31. Nylon 6,6 and PPTA (aramid) are both examples of polyamides. Nylon fibers are typically produced via melt spinning, while aramid fibers are spun using a wet spinning process. Explain the different processing approach.
32. Propose the processing steps needed to fabricate the following products:
    a. PVC sewage pipe
    b. Custom-made ABS door handle
    c. Expanded polypropylene sheet
    d. Carbon fiber woven fabric
    e. Aramid fiber-reinforced epoxy car hood

3 Bioplastics

3.1 Biorefinery technologies

Bioplastics comprise a whole family of materials with different properties and applications. According to European Bioplastics, a plastic material is defined as a bioplastic if it is either bio-based, biodegradable, or features both properties [40]. Bio-based means that the polymer is (partly) derived from biomass such as corn, sugarcane, or cellulose. Biodegradation, however, is a process during which microorganisms that are available in the environment convert the polymer in natural substances such as water, carbon dioxide, and compost.

In 2009, the International Energy Agency (IEA) Bioenergy Task 42 [41] proposed a new classification approach [42] to describe distinguishable biorefinery processes. The International Energy Agency is a Paris-based autonomous intergovernmental organization. It facilitates the commercialization and market commitment of biorefinery technologies and advices policy and industrial decision makers in an appropriate way. They defined *biorefining* as the sustainable processing of renewable feedstocks (*biomass*), into a spectrum of different bioproducts, including biofuels, bioenergy, biochemicals, and bio-based building blocks to produce existing or newly developed bioplastics. Clearly, the use of biomass as renewable feedstock can provide a benefit by reducing the impacts on the environment. Industrial biorefineries have been identified as the most promising route to the creation of a new global bioeconomy. Therefore, biorefining is the bridge between agriculture and chemistry.

The IEA Bioenergy Task 42 biorefinery classification approach is based on four main features:
– Platforms
– Products
– Feedstocks
– Processes

In this classification approach, platforms are the most important feature because they are the main intermediate between a renewable feedstock and a final product, making use of one or more intermediate process steps. In *process flow diagrams*, the four main features are each represented by their own standard shape as indicated in Figure 3.1. Each of the mentioned main features can consist of several subgroups.

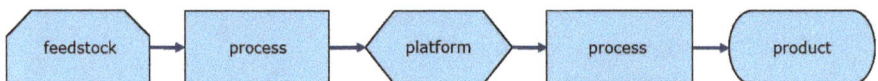

Figure 3.1: IEA process flow diagram.

https://doi.org/10.1515/9783111201443-003

3.1.1 Platforms

Platforms are intermediates which link renewable feedstocks (biomass) and final products, involving intermediate process steps. The platform concept is similar to that used in the petrochemical industry. In an oil refinery, crude oil is fractionated into a large number of intermediates that are further processed to fuels, materials, and chemical products (such as alkanes, cycloalkanes, and various aromatic hydrocarbons). The most important platforms that can be recognized in the development of bioplastics are listed below.

3.1.1.1 Biogas

Biogas is a mixture of primarily methane (CH_4) and carbon dioxide (CO_2) and may have small amounts of hydrogen sulphide (H_2S). Biogas can be obtained from an anaerobic *fermentation* (in the absence of oxygen) with methanogenic or anaerobic organisms of agricultural waste, manure, municipal waste, plant material, sewage, green waste, or food waste that digest in a closed system. This closed system is called an anaerobic digester, a biodigester, or a bioreactor. The composition of the biogas varies depending on the substrate, as well as the conditions within the anaerobic reactor (such as temperature, pH, and substrate concentration). Typical concentrations of methane in biogas range from 50% up to 75%. In general, the methane is scrubbed, compressed, and stored. Biogas is used for its energetic value as a renewable energy source.

Recent initiatives, however, use biogas as a *renewable feedstock* in a biogas fermentation process, together with specially developed biocatalysts, to produce bioplastics, such as *polyhydroxyalkanoates* (PHAs). In fact, bacteria in the bioreactor metabolize biogas into the biopolymer PHA. Both US-based biotechnology companies Newlight Technologies and Mango Materials are active in this field. In fact, Newlight is using their patented greenhouse gas-to-plastic bioconversion technology to produce plastics from air and methane-containing greenhouse gas emissions generated at a farm. Newlight's product is brand-named AirCarbon™ and is a poly(3-hydroxybutyrate) (PHB). Third-party organizations have certified that the carbon footprint of AirCarbon™ is significantly negative when produced using renewable sources of energy for electric power. Newlight Technologies started the operation of the world's first commercial-scale manufacturing plant for AirCarbon™ in 2020, and have been supplying the biomaterial to a wide variety of customers.

In early 2023, Mango Materials announced to seek a location for a facility that could produce up to 2,300 tons of PHB per year. VTT Technical Research Centre of Finland developed a process at which methanotrophic bacteria use methane-rich biogas to produce *polyhydroxyalkanoates*. VTT is a Finnish state-owned and controlled non-profit company, providing research and innovation services. The general chemical structure of PHAs is shown in Figure 3.2.

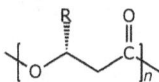

Figure 3.2: General chemical structure of a polyhydroxyalkanoate (PHA, R = alkyl functionality).

3.1.1.2 Syngas

Syngas or synthesis gas is a mixture of primarily carbon monoxide (CO) and hydrogen (H_2) and most often a small amount of carbon dioxide (CO_2) is present. Syngas is usually a product of *gasification* (over 430 °C) of biomass in the presence of oxygen (O_2) or air. After cleaning, the syngas can be used to produce power or can be converted into lower alcohols (e. g., methanol, CH_3OH), fuels (e.g., Fischer–Tropsch diesel), or chemical products. The Fischer–Tropsch process is a collection of chemical reactions that converts a mixture of carbon monoxide and hydrogen into liquid hydrocarbons, ideally having the formula C_nH_{2n+2}. These reactions occur in the presence of metal catalysts (most often cobalt, iron, or ruthenium), typically at temperatures of 150–300 °C and pressures of one to several tens of atmospheres. The more useful reactions produce alkanes as shown in Figure 3.3, where n is typically 10–20. Most of the alkanes produced tend to be straight-chain, suitable as diesel fuel. In addition to alkane formation, competing reactions give small amounts of alkenes, as well as alcohols and other oxygenated hydrocarbons. Historically, syngas has been used as a replacement for gasoline, when gasoline supply has been limited. For example, wood gas was used to power cars in Europe during World War II.

$$n\ CO\ +\ (2n+1)\ H_2\ \longrightarrow\ C_nH_{2n+2}\ +\ n\ H_2O$$

Figure 3.3: Fischer–Tropsch synthesis of hydrocarbons.

Syngas can also be fermented to give ethanol (CH_3CH_2OH), 1-propanol ($CH_3CH_2CH_2OH$), 1-butanol ($CH_3CH_2CH_2CH_2OH$), acetic acid (CH_3COOH), butyric acid ($CH_3CH_2CH_2COOH$), or other chemical building blocks. Syngas fermentation is a microbial process using microorganisms, which are mostly known as acetogens. In the past few years, US-based company LanzaTech developed processes for the production of bio-based 2,3-butanediol ($CH_3CHOHCHOHCH_3$) and ethanol (CH_3CH_2OH). Ethanol biosynthesis is probably the most researched process. After many years of preparation and testing, LanzaTech is now finally ready for three commercial plants on steelmaking companies' sites: in China with the Shougang Group, in Taiwan with China Steel, and in Belgium with the world's largest steelmaker, ArcelorMittal. In the past few years, LanzaTech announced several cooperations to develop new syngas-based chemicals, such as butadiene (Invista), 2-hydroxyisobutyric acid (Evonik), isobutene (Global Bioenergies), n-octanol (BASF), and isoprene (Sumitomo Riko). Also, US-based Synata Bio invested in pilot plant facilities to produce ethanol (CH_3CH_2OH) and 1-butanol ($CH_3CH_2CH_2CH_2OH$), starting from *syngas* feedstock.

3.1.1.3 Pyrolysis oil

Pyrolysis is the thermal decomposition of dried biomass at elevated temperatures in the absence of oxygen. In general, pyrolysis of organic substances, such as biomass, produces volatile products and leaves a solid residue enriched in carbon (bio-char). Extreme *pyrolysis* is called carbonization. Pyrolysis is the fundamental chemical reaction that is the precursor of both the combustion and gasification processes. The products of biomass pyrolysis include bio-char, bio-oil, and gases (like methane (CH_4), hydrogen (H_2), carbon monoxide (CO), and carbon dioxide (CO_2)). Depending on the thermal environment and the final temperature, pyrolysis will yield mainly biochar at low temperatures, less than 450 °C, when the heating rate is quite slow, and mainly gases at high temperatures, greater than 800 °C, with rapid heating rates. At an intermediate temperature and under relatively high heating rates, the main product is bio-oil or so-called pyrolysis-oil. Bio-oil can be used as a feedstock in a catalytic conversion process, using special types of zeolite catalysts. Zeolites are microporous aluminosilicate minerals (e. g., $Na_2Al_2Si_3O_{10}.2\,H_2O$) commonly used as catalysts in chemical processes. As a result of the catalytic conversion, *bioBTX* is produced. BioBTX, a basic feedstock for the petrochemical industry, is a mixture of aromatic compounds (in particular, benzene, toluene, and xylenes). Chemical structures of the BTX components are shown in Figure 3.4.

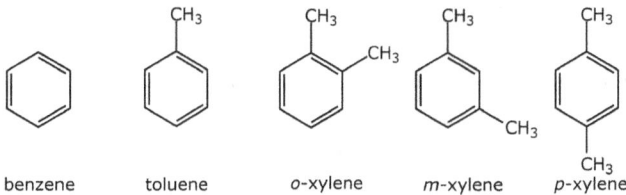

benzene   toluene   o-xylene   m-xylene   p-xylene

Figure 3.4: Chemical structures of BTX components.

Note that *p*-xylene is the basic raw material for the production of terephthalic acid (TA), one of the two chemical building blocks for polyethylene terephthalate (PET). Because biomass is the feedstock for pyrolysis oil, the BTX components are bio-based. So, bio-based terephthalic acid (bio-TA) is obtained after chemical oxidation of bio-based *p*-xylene.

3.1.1.4 Lignin

Lignin is an organic substance binding cells, fibres, and vessels that constitute wood and the lignified elements of plants, as in straw. After *cellulose*, it is the most abundant renewable carbon source on Earth. Lignin is a polymer of propyl phenol units (Figure 3.5), namely coniferyl alcohol and sinapyl alcohol, with a minor quantity of *p*-coumaryl alcohol (with the collective name monolignols, Figure 3.6). The complex of these components is crosslinked together through carbon–carbon, ester linkages,

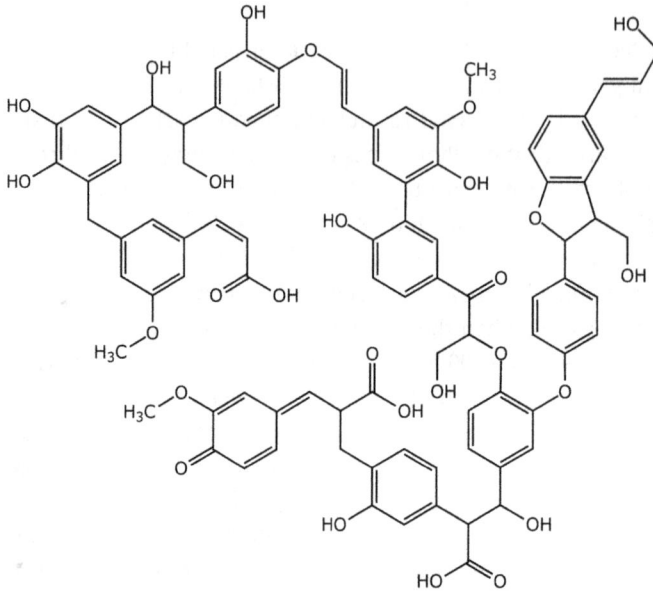

Figure 3.5: Proposed chemical structure of lignin.

and ether linkages. Via a chemical process, lignin can be converted into an intermediary product called lignin crude oil. Lignin crude oil can be fractionated and further converted into valuable products like phenols, resins, and octane-boosting fuel additives. The lignin crude oil itself is a suitable marine fuel.

Wood and other *lignocellulosic biomasses* have a complex chemical structure that consists mainly of cellulose, hemicellulose, and lignin. The composition depends on the biomass type. For wood it is around 35–45% cellulose, 25–35% hemicellulose, and 20–30% lignin.

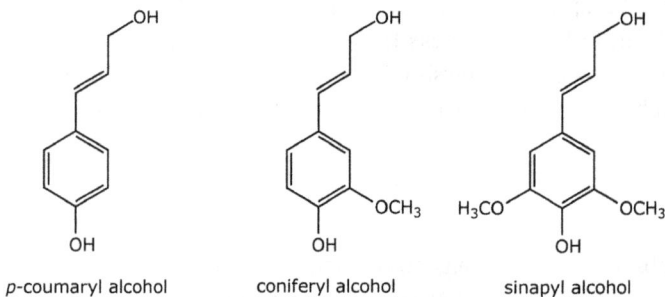

p-coumaryl alcohol   coniferyl alcohol   sinapyl alcohol

Figure 3.6: Chemical structures of lignin buildings blocks (monolignols).

3.1.1.5 C6 sugars

Carbon six (C6) sugar platforms are accessible from sucrose (a *disaccharide*) or through hydrolysis or enzymatic treatment of both starch or cellulose to give glucose. Glucose is a *monosaccharide* (Figure 3.7) and serves as a renewable feedstock for a broad range of biological fermentation processes providing access to a variety of chemical building blocks, including building blocks for polymer production. The number of chemical building blocks, accessible through fermentation, is considerable. Glucose can also be converted by (thermo)chemical processes to existing or new chemical building blocks.

Starch or amylum is a *polysaccharide*, a polymeric carbohydrate, consisting of a large number of glucose units joined by glycosidic linkages. Starch is made up of a mixture of amylose (15–20%) and amylopectin (80–85%), both depicted in Figure 3.8. Amylose consists of a linear chain of several hundreds of glucose repeating units, and amylopectin is a branched molecule made of several thousands of glucose units. In amylose, the glucose repeating units are joined together by α-(1,4)-glycosidic linkages. Amylopectin is a branched-chain polysaccharide composed of glucose units linked primarily by α-(1,4)-glycosidic bonds but with occasional α-(1,6)-glycosidic bonds, which are responsible for the branching.

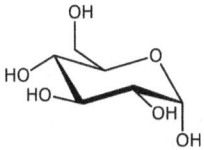

Figure 3.7: Chemical structure of glucose.

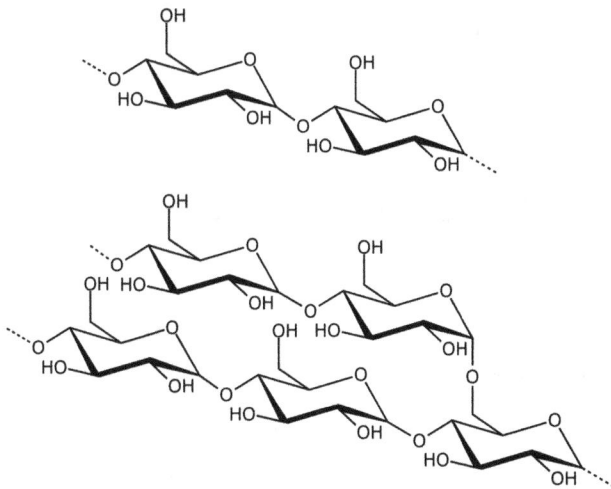

Figure 3.8: Chemical structures of amylose (top) and amylopectin (bottom).

Like amylose, cellulose is also a linear polymer of glucose repeating units. It differs, however, in the sense that glucose units are joined by β-1,4-glycosidic linkages, producing a more extended structure than amylose (Figure 3.9). Cellulose yields glucose after complete acid hydrolysis, yet humans are unable to metabolize cellulose as a source of glucose. Our digestive juices lack enzymes that can hydrolyze the β-glycosidic linkages found in cellulose, so although we can eat potatoes, we cannot eat grass. However, certain microorganisms can digest cellulose because they make the enzyme cellulase, which catalyzes the hydrolysis of cellulose.

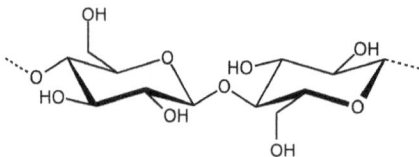

Figure 3.9: Chemical structure of cellulose.

3.1.1.6 C6/C5 sugars

A mix of carbon six (C6) and carbon five (C5) *sugars* can be easily obtained from hemicellulose after dilute acid or base hydrolysis or degradation with hemicellulase enzymes. Hemicelluloses are polysaccharides often associated with cellulose, but they have different compositions. Hemicelluloses contain many different sugar monomers (Figure 3.10), while cellulose only contains the glucose repeating unit. For instance, besides glucose, sugar monomers in hemicelluloses can include the C5 sugars xylose and arabinose, the C6 sugars mannose and galactose, and the C6 deoxy sugar rhamnose.

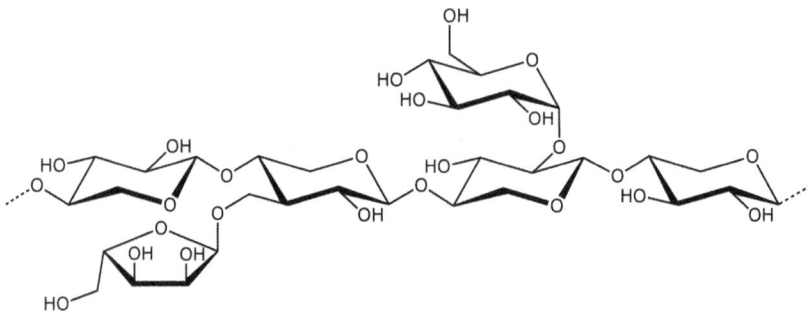

Figure 3.10: Chemical structure of hemicellulose.

3.1.1.7 Oil

Oil, a mixture of animal or vegetable triglycerides, can be derived from oilseed crops, algae, or oil-based residues. A triglyceride is an ester derived from glycerol

($CH_2OHCHOHCH_2OH$) and three long-chain *fatty acids*. An example is shown in Figure 3.11. The three fatty acids are usually different, the chain lengths of the fatty acids in naturally occurring triglycerides vary, but most often contain 16, 18, or 20 carbon atoms. Moreover, many fatty acids are unsaturated, some are polyunsaturated. As an example, palmitic (16:0) and stearic acid (18:0) are saturated fatty acids and oleic acid (18:1), linoleic acid (18:1), and *α*-linoleic acid (18:3) are unsaturated fatty acids. In the notation (C:D), C is the total amount of carbon atoms of the fatty acid, and D is the number of double (unsaturated) bonds in the chain. In polyunsaturated fatty acids, the double bonds are separated by one or more methylene (CH_2) groups.

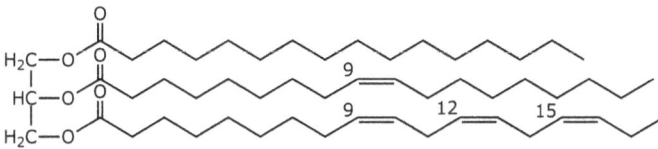

Figure 3.11: Chemical structure of a triglyceride with C16:0, C18:1, and C18:3 fatty acids.

3.1.2 Products

Various products can be obtained in biorefinery processes, depending on the type of feedstock used in the process. Those products can be roughly classified into two main classes:

Energy-related products
In this case, *biomass* is primarily used for the production of transportation biofuels such as bio-ethanol, biodiesel, or syngas.

Material-related products
Biomass is primarily used for the production of bio-based products such as biochemicals, biosurfactants, biosolvents, biolubricants, or bio-based chemical *building blocks* for existing or newly developed bioplastics.

Note that bio-ethanol for instance can be both an energy-related and a material-related product.

3.1.3 Feedstocks

Feedstock is the renewable raw material, or biomass, that is converted into marketable final products. Biomass can be subdivided into:

Primary feedstocks
Primary feedstocks include primary biomass that is harvested from forest or agricultural land.

Secondary feedstocks
Secondary feedstocks are process residues, such as sawmill residues or black liquor generated by the forests products industry.

Tertiary feedstocks
Tertiary feedstocks are postconsumer wastes or residues (such as waste waters or municipal solid waste).

Renewable feedstock may vary in composition with different shares of basic elements, such as cellulose, hemicellulose, lignin, starch, or triglycerides. In the classification approach two subgroups of renewable feedstocks are assumed:

Dedicated feedstocks
– Sugar crops (e. g., sugar beet or sugarcane).
– Starch crops (e. g., wheat or corn).
– Lignocellulosic crops (e. g., wood or miscanthus).
– Oil-based crops (e. g., rape seed oil, soya oil, palm oil, or jatropha).
– Grasses (e. g., green plant materials).
– Marine biomass (e. g., micro and macro algae).

Residues
– Oil-based residues (e. g., animal fat from food industries or used cooking oil from restaurants, households, and other sources).
– Lignocellulosic residues (e. g., crop residues or saw mill residues).
– Organic residues (e. g., organic urban waste or household waste).

3.1.4 Processes

To produce biofuels, biochemicals, lubricants, and chemical building blocks for bioplastics, the renewable feedstock has to be transformed into final products by applying different conversion processes. Four different basic types of processes can be distinguished, as displayed in Table 3.1.

In more detail, fermentation is the metabolic process that produces chemical changes in organic substances by the action of microorganisms or enzymes. Microorganisms such as yeast and bacteria usually play a role in the fermentation process. A well-known fermentation process is yeast fermentation, where yeast converts glucose into ethanol and

Table 3.1: Type of processes related to biorefineries.

Process	Description	Examples
Thermochemical	In thermochemical processes renewable feedstock undergoes extreme conditions (high temperature and/or pressure, with or without the use of a catalyst).	– Combustion – Gasification – Pyrolysis
Biochemical	Biochemical processes occur at mild conditions (low temperature and pressure) using yeasts, microorganisms, or enzymes.	– Aerobic fermentation – Anaerobic fermentation – Enzymatic processes
Chemical	In chemical processes, well-known chemical reactions take place.	– Esterification – Hydrogenation – Dehydrogenation – Hydration – Dehydration – Dimerization – Oxidation – Reduction – Hydrolysis – Polymerization
Mechanical/ physical	In mechanical/physical processes, mechanical or physical treatments of materials are carried out.	– Extraction – Separation – Fractionation – Filtration – Pre-treatment – Milling – Distillation – Crystallization – Purification

carbon dioxide. Aerobic fermentation is a metabolic process by which microorganisms metabolize sugars in the presence of oxygen. As an example, acetic acid bacteria (AAB) are capable of oxidizing ethanol to acetic acid. AAB are widespread in nature and play an important role in the production of food and beverages, such as vinegar. In anaerobic fermentation, microorganisms break down biodegradable materials in a sequence of processes in the absence of oxygen. Well-known examples are the ethanol fermentation and lactate fermentation. Over the years, the direct use of enzymes, instead of microorganisms, has grown. Enzymes are proteins that have high catalytic functions. Enzymes are classified depending on the compound they act on. Few of the common types of enzymes include amylases (break down starch into simple sugars) and cellulases (break down the cellulose). Industrial fermentation is a broader term used for the metabolic processes of applying microorganisms or enzymes for the large-scale production of biofuels and biochemicals. With the recent development of enzyme engineering, new possibilities were introduced for metabolic processes. Enzyme engineering is the modification of an enzyme's structure or modifying the catalytic activity of isolated enzymes. In this way, new pathways become available to produce new biofuels and biochemicals.

3.1.5 Process flow diagrams

The classification approach has been introduced for describing distinguishable biorefining pathways, but can also be applied to other chemical or biochemical processes. The approach is flexible as new subgroups can be added to the four main features.

The aforementioned four main features, together with their subgroups, are used for the classification of biorefinery pathways. By using a *process flow diagram*, biorefinery systems can be easily visualized. In Figure 3.12, the separate symbols are given for the four main features. Note that for the feature "process" a distinction is made between chemical processes, thermochemical processes, mechanical/physical processes, and biochemical processes. With respect to the feature "product," three separate products (energy product, material product, and (bio)plastic) are recognized.

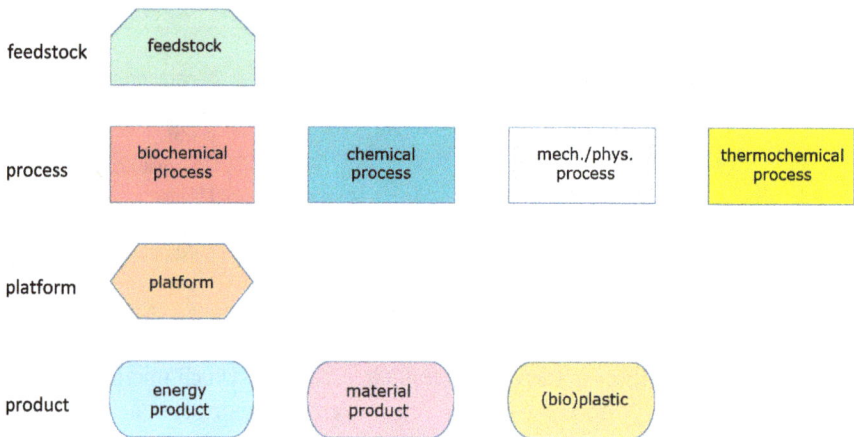

feedstock	feedstock

process	biochemical process	chemical process	mech./phys. process	thermochemical process

platform	platform

product	energy product	material product	(bio)plastic

Figure 3.12: Symbols used in IEA process flow diagrams.

Figure 3.13 shows an example of how these features are combined in a *biorefinery pathway*. In such a process flow diagram, the process starts with the feedstock, which is converted to one of the platforms, from which energy or materials or (bio)plastics can be produced.

Using these basic principles, the production of bio-ethanol (CH_3CH_2OH) is represented in the IEA process flow diagram shown in Figure 3.14. In this example, starch can easily be converted to the fermentable C6 sugar glucose. Glucose can then be converted, in a fermentation process, into bio-ethanol.

The production of biodiesel from oil crops (e. g. rape seed) is presented in the IEA process flow diagram in Figure 3.15.

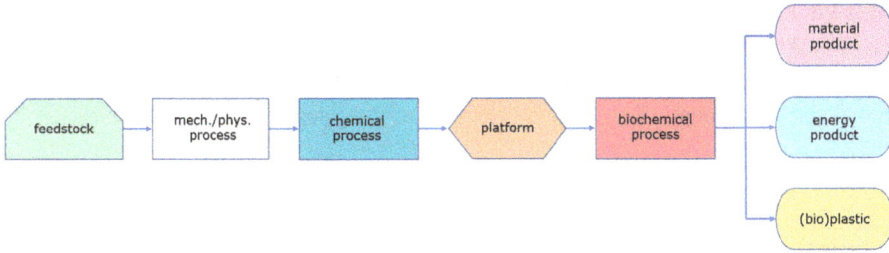

Figure 3.13: Example of a combination of main features for the classification of a general biorefinery pathway.

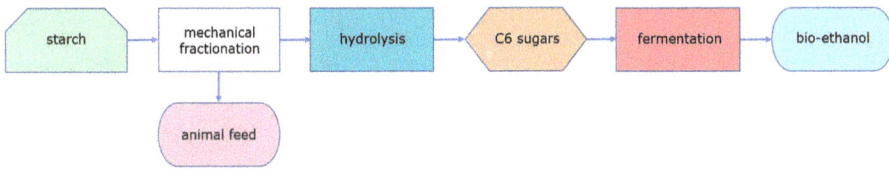

Figure 3.14: Fermentation of starch to bio-ethanol and animal feed.

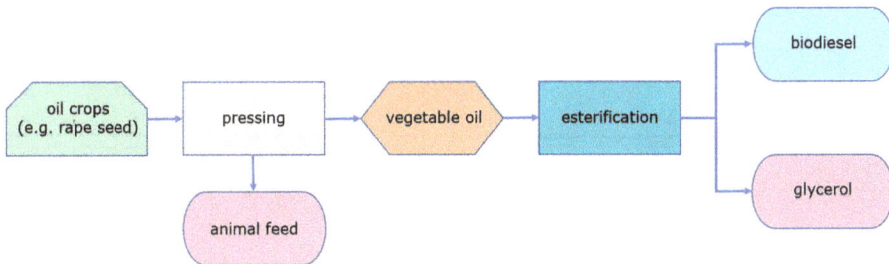

Figure 3.15: Production of biodiesel from oil crops (such as rape seed).

3.2 Biochemicals

Biofuels and *biochemicals* are main products from biorefinery processes. Biochemicals offer a sustainable alternative to chemicals based on fossil-based feedstocks. Developed from *biomass*, biochemicals can be structurally identical to existing chemicals (also referred to as drop-in chemicals) or have new structures that offer new opportunities for new derivatives or bioplastics. In 2004, the US Department of Energy (DOE) published a study titled "Top Value Added Chemicals from Biomass – Volume I – Results of Screening for Potential Candidates from Sugars and Synthesis Gas" [43]. This report identifies twelve building block chemicals that can be produced from sugars via both biological or (thermo)chemical conversions. Some of these building blocks are or can be easily converted into functional monomers for bioplastics. The twelve sugar-based building blocks are 1,4-diacids (succinic acid (SA), fumaric acid, and malic acid), 2,5-furandicarboxylic acid (FDCA), 3-hydroxypropionic acid (3-HPA), aspartic acid, glucaric acid, glutamic acid, itaconic acid, levulinic acid, 3-hydroxybutyrolactone, glycerol, sorbitol, and xylitol/arabinitol. Table 3.2 presents the chemical structures of these sugar-based *building blocks*.

Table 3.2: DOE's top value-added sugar-based building blocks.

DOE's building blocks		
glycerol	3-hydroxypropionic acid	succinic acid
aspartic acid	fumaric acid	malic acid
levulinic acid	3-hydroxybutyrolactone	itaconic acid
xylitol	arabinitol	glutamic acid
sorbitol	2,5-furandicarboxylic acid	glucaric acid

Biological transformations account for the majority of routes from renewable feedstocks (biomass) to these top building blocks, while (thermo)chemical transformations predominate in the conversion of those top building blocks to molecular derivatives and bio-based monomers for bioplastics. A few of the aforementioned top building blocks enjoy a huge interest these days: succinic acid (SA) and 2,5-furandicarboxylic acid (FDCA). Both building blocks can be used as bio-based monomers for the development and production of *bioplastics* (Table 3.3). Succinic acid (SA) is used as a mono-mer in the production of polybutylene succinate (PBS), and 2,5-furandicarboxylic acid (FDCA) is used as a monomer in the development of polyethylene furanoate (polyeth-ylene 2,5-furandicarboxylate, better known as PEF).

Table 3.3: Polybutylene succinate (PBS) and polyethylene furanoate (PEF).

Monomer	Chemical structure	Polymer	Chemical structure
SA		PBS	
FDCA		PEF	

Both PBS and polyethylene furanoate (PEF) are bioplastics. PBS is a bio-based and biode-gradable aliphatic polyester, prepared in a polycondensation from bio-based succinic acid (SA) and bio-based 1,4-butanediol (1,4-BDO). PEF is a bio-based, non-biodegradable aromatic polyester. PEF is also prepared in a polycondensation-type polymerization from bio-based 2,5-furandicarboxylic acid (FDCA) and bio-based ethylene glycol (EG). 2,5-Furandicarboxylic acid (FDCA) can be used as a replacement for TA, a petroleum-based monomer that is primarily used to produce PET. Both PBS and PEF are recyclable.

Recent findings show that environmental concerns regarding dangerous chemi-cals and fossil-fuel depletion is leading to increased production of biochemicals. Bio-chemicals (and bio-based monomers) are also of high interest for the development of bioplastics, but also markets such as biolubricants, biosolvents, biosurfactants, inks, and dyes are very interesting.

3.3 Chemistry of bioplastics

There are three groups of bioplastics, each with their own characteristics:
- *Bio-based* (or partly bio-based), non-biodegradable plastics, such as bio-based polyethylene (PE), bio-based polypropylene (bio-PP), (partly) bio-based PET (so-called drop-in solutions)
- Bio-based and *biodegradable plastics*, such as polylactic acid (PLA), PHAs, PBS, and starch blends
- Plastics that are based on fossil resources (*fossil-based*) and biodegradable, such as poly(butylene-*co*-adipate-*co*-terephthalate) (PBAT)

Figure 3.16 depicts these three common types of bioplastics and how they are classified according to their biodegradability and bio-based content. This graph is also
called the material coordinate system of bioplastics.

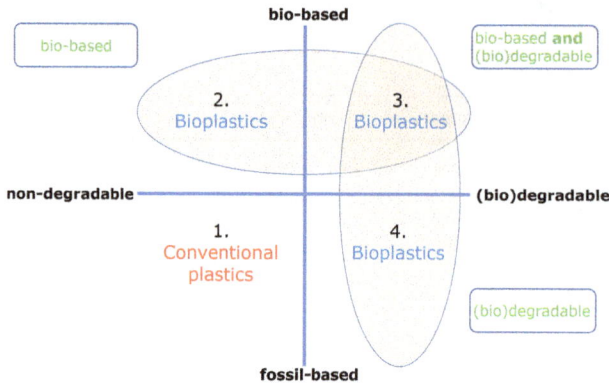

Figure 3.16: The material coordinate system of bioplastics.

In Table 3.4, a more detailed overview is given of known *bioplastics*. In the remainder
of this chapter, various types of bioplastics will be discussed in more detail.

3.3.1 Polyethylene (PE)

3.3.1.1 Fossil-based polyethylene (PE)

There are three main types of *polyethylene* (PE):
– High-density PE (HDPE)
– Low-density PE (LDPE)
– Linear low-density PE (LLDPE)

The major differences in the characteristics of these PE types arise from the changes
that exist during the polymerization reaction, starting from the fossil-based monomer
ethylene. Different polymerization conditions (e. g., temperatures, pressures, use of
catalysts, or use of additional comonomers) have drastic effects on the molecular
weight and the branching of the PE polymer backbone. Branching is the formation of
side chains on the polymer backbone. With a significant amount of branching, the
polymer morphology will become more amorphous. The different materials have
names associated with their respective densities (i.e., the degree of branching): HDPE,

Table 3.4: Overview of bioplastics.

Bioplastic type	Bioplastic properties	Bioplastic abbreviation	Bioplastic full name
1	**Conventional plastics** Fossil-based and non-biodegradable plastics		
2	**Bioplastics** Bio-based (or partly bio-based) and non-biodegradable plastics *Addition polymers*		
		bio-PE	Polyethylene (Section 3.3.1)
		bio-PP	Polypropylene (Section 3.3.2)
		bio-PVC	Polyvinyl chloride (Section 3.3.3)
		bio-PS	Polystyrene (Section 3.3.4)
		bio-PVAc	Polyvinyl acetate (Section 3.3.5)
		bio-PAA	Polyacrylic acid (Section 3.3.6)
		bio-PMMA	Polymethyl methacrylate (Section 3.3.7)
		bio-PAN	Polyacrylonitril (and copolymers ABS and SAN) (Section 3.3.8)
	Condensation polymers		
		bio-PET	Polyethylene terephthalate (Section 3.3.9)
		bio-PEF	Polyethylene furanoate (Section 3.3.10)
		bio-PTT	Polytrimethylene terephthalate (Section 3.3.11)
		bio-PBT	Polybutylene terephthalate (Section 3.3.12)
		bio-TPEE	Thermoplastic polyetherester (Section 3.3.13)
		bio-IT	Isosorbide-based polymers (Section 3.3.14)
		bio-PA6	Polyamide 6 (Section 3.3.15)
		bio-PA6,6	Polyamide 6,6 (Section 3.3.16)
		PPA	Polyphthalamides (Section 3.3.17)
		PEBA	Polyether block amide (Section 3.3.18)
		bio-PPTA	Aramides (aromatic polyamides) (Section 3.3.19) Elastane (Section 3.3.20)
3	**Bioplastics** Bio-based and biodegradable plastics		
		PBS	Polybutylene succinate (Section 3.3.21)
		PBSA	Poly(butylene-*co*-succinate-*co*-adipate) (Section 3.3.22)
		PHA	Polyhydroxyalkanoates (Section 3.3.23)
		PLA	Polylactic acid (Section 3.3.24)
		PGA	Polyglycolic acid (Section 3.3.25)
4	**Bioplastics** Fossil-based and biodegradable plastics		
		PBAT	Poly(butylene-*co*-adipate-*co*-terephthalate) (Section 3.3.26)
		PBST	Poly(butylene-*co*-succinate-*co*-terephthalate) (Section 3.3.27)
		PCL	Polycaprolactone (Section 3.3.28)

LDPE, and LLDPE. More information about the polymerization of ethylene to PE can be found in Chapter 2.

High-density polyethylene (HDPE)

HDPE is the most rigid type among the three common types of PE and has a density in the range 0.935–0.960 g/cm^3. HDPE is the product of limited branching that occurs when the polymerization is carried out at low temperatures and low pressures. Because of limited branching, HDPE is more crystalline, leading to an increased density. In practice, HDPE can be produced in three different ways: in a slurry particle reactor, in the gas phase, and by using a metallocene catalyst.

Low-density polyethylene (LDPE)

LDPE has a great flexibility, impact toughness, and stress cracking resistance. In general, LDPE has a density in the range 0.910–0.925 g/cm^3. LDPE is polymerized under conditions of high temperature and high pressures. Because LDPE is produced under these extreme conditions, branches are very high in quantity and length.

Linear low-density polyethylene (LLDPE)

LLDPE is created by a low pressure polymerization process much like HDPE but has more branches much like LDPE has. These branches are long enough to prevent the polymer molecules from being closely packed together. This results in a linear molecular structure like HDPE, but also a low density like LDPE. The density of LLDPE will typically in the range 0.918–0.940 g/cm^3. The LLDPE properties can be achieved by adding a comonomer (most often 8–10% of comonomer) during the polymerization process. This comonomer is usually 1-butene, 1-hexene, or 1-octene. The comonomer increases chain entanglements, which results in improved physical properties, as well as stronger secondary bonding. The downsides to LLDPE are higher melt processing temperatures, 8% greater shrinkage, less transparency (optically), and less flexibility.

The total global production capacity of PE in 2022 was 105.3 million metric tons [2]. In Table 3.5, the individual capacities are given.

Table 3.5: Production capacities of polyethylene (PE) grades.

Type of PE	Global production capacity in 2022 (in million metric tons)	Percentage (%)
HDPE	48.8	46.4
LDPE and LLDPE	56.4	53.6

The fossil-based ethylene monomer is derived from either natural gas, via an oxidative coupling of methane (OCM) using special developed nanowire OCM catalysts, or from the catalytic cracking of crude oil. In the OCM process, methane (CH_4) and oxygen (O_2) react over the catalyst exothermally to form the ethylene monomer (C_2H_4), water (H_2O), and heat. In addition, this technology can also co-feed ethane (CH_3CH_3) for conversion into ethylene. The OCM process, also called the Gemini technology, was developed by Siluria Technologies.

A fractional distillation of crude oil gives naphtha which after *steam cracking* and fractional distillation gives fossil-based ethylene ($CH_2 = CH_2$). Ethylene is subsequently polymerized into different types of PE. Both processes to produce fossil-based ethylene are shown in Figure 3.17.

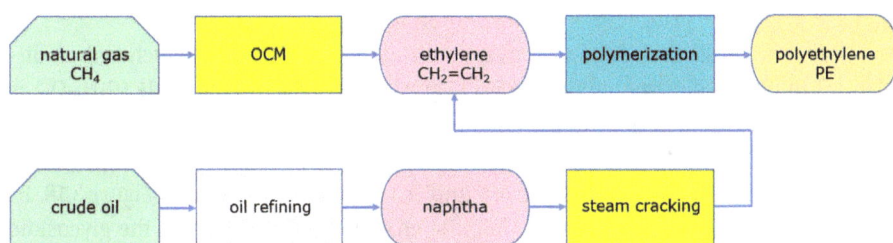

Figure 3.17: IEA process flow diagram for the production of PE.

3.3.1.2 Bio-based polyethylene (bio-PE)

Bio-based PE (bio-PE) from renewable raw materials can be produced by dehydrating bio-ethanol (bio-based ethanol, CH_3CH_2OH) to bio-based ethylene ($CH_2 = CH_2$) and subsequently polymerizing bio-based ethylene into *bio-based polyethylene* (bio-PE). Note that characteristics and properties of bio-based polyethylene (bio-PE) are equal to fossil-based PE: bio-PE is not compostable or biodegradable but is easily recyclable on existing plastic sorting and washing installations. Bio-PE can be easily introduced on existing production equipment, such as blow-molding or injection molding equipment, without major modifications of process settings.

Bio-ethanol is at present obtained from sugarcane (in Brazil) or corn starch (in USA). Bio-ethanol production in Brazil and the USA covers about 85% of the total world production of bio-ethanol. Both sugarcane and corn starch are first-generation (1G) feedstocks. Bio-ethanol is directly produced from food crops, such as sugarcane or corn starch. It is important to note that the structure and properties of the bio-ethanol itself does not change between generations, but rather the source from which the bio-ethanol is derived changes.

In the future, bio-ethanol could also possibly be produced using second-generation (2G) lignocellulose. Low-cost crop and forest residues, wood process waste, and the organic fraction of municipal solid waste can all be used as lignocellulosic feedstocks. The production of bio-ethanol from lignocellulosic feedstocks can be achieved through two different processing routes. In the biochemical route, enzymes and other micro-organisms are used to convert cellulose or hemicellulose components of the feed-stocks to complex sugars. A cellulase enzyme breaks down the cellulose molecule into monosaccharides such as glucose, or shorter polysaccharides and oligosaccharides. This process step is also termed saccharification. The specific reaction involved is the hydrolysis of 1,4-β-glycosidic linkages in cellulose. Because cellulose molecules bind strongly to each other, cellulose breakdown is relatively difficult compared to the breakdown of other polysaccharides such as starch. The sugars are further fermented to produce bio-ethanol.

In the acid hydrolysis route, lignocellulosic feedstock is pretreated and, subsequently, acid hydrolysis (using either dilute acid or concentrated acid) breaks down cellulose to glucose, prior to their fermentation to produce bio-ethanol.

After this, yeasts are added to convert the sugars to bio-ethanol, which is then distilled off to obtain bio-ethanol up to 96% in purity. One glucose molecule ($C_6H_{12}O_6$) is fermented into two molecules of bio-ethanol (CH_3CH_2OH) and two molecules of carbon dioxide (CO_2). Table sugar or sucrose (a disaccharide) is a dimer of glucose and fructose. Chemical structures of both glucose and fructose are presented in Figure 3.18. In the first step of the ethanol fermentation, the enzyme invertase cleaves the glycosidic linkage between glucose and fructose. Next, both fermentable sugars are fermented to bio-ethanol.

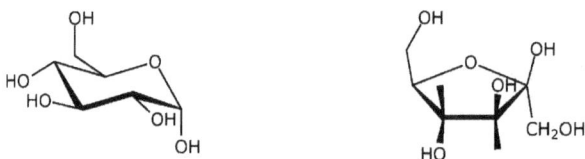

Figure 3.18: Chemical structures of glucose (left) and fructose (right).

Route 1

In the past years, POET-DSM Advanced Biofuels, Abengoa Bioenergy (now Synata Bio), GranBio, Clariant, and DuPont Industrial Biosciences developed novel 2G bio-ethanol production processes based on lignocellulosic feedstocks.

POET-DSM Advanced Biofuels, a joint venture between POET and DSM, started up their production project, called Project Liberty, in Emmetsburg, Iowa (USA) in September 2014. In 2017, the first stage pretreatment step was improved significantly. In this step, the feedstock (corn cobs, leaves, or husks) is processed so that enzymes and yeast can access the cellulosic sugars and ferment them into bio-ethanol. POET-DSM and other producers have identified this stage in the past as the major bottle-neck in commercial production of cellulosic 2G bio-ethanol. The commercial enzymes for this process are supplied by the Danish-based firm Novozymes. The initial annual production capacity of POET-DSM was about 75 million liters of 2G bio-ethanol. In July 2020, however, it was decided to shut down the 2G bio-ethanol factory due to lack of bio-ethanol demand. Note that POET itself operates 33 state-of-the-art bio-ethanol production facilities across the USA.

A subsidiary of the Spanish-based firm Abengoa, Abengoa Bioenergy, constructed a biomass-to-bio-ethanol facility in Hugoton, Kansas (USA), that produced cellulosic 2G bio-ethanol with a capacity of 95 million liters of bio-ethanol a year. The biorefinery utilizes cellulosic biomass, crop residue, and plant fibers as feedstock. Enzymatic hydrolysis converts the cellulose and hemicellulose of these agricultural residues to glucose (C6 sugar) and xylose (C5 sugar). Custom-made enzymes were developed with licence from Dyadic International, based on Dyadic's patented C1 Technology Platform®. The sugars are fermented to produce renewable 2G bio-ethanol. Lignin and animal feed are produced as by-products. Started in 2014, the Hugoton plant, however, never did reach this production level and was shuttered as a failure in 2015. By the end of 2016, Synata Bio bought the Hugoton assets. USA-based firm Synata Bio also owns patented cellulosic ethanol technology previously owned by Coskata. In February 2019, Seaboard Energy acquired the idled Synata Bio cellulosic ethanol plant in Hugoton. Since then Seaboard Energy has been evaluating the best potential opportunities and uses for the Hugoton biorefinery, including modifications to produce renewable diesel.

In Alagoas (Brazil), GranBio operates a cellulosic 2G bio-ethanol operation with an initial capacity of 82 million liters of bio-ethanol a year. The plant, called Bioflex 1, converts biomass from sugarcane residue, straw, and bagasse into 2G bio-ethanol and lignin. The cellulosic bio-ethanol plant was built adjacent to an already operational 1G plant that used sugarcane as a raw material. GranBio makes use of the so-called PRO-ESA™ technology, developed by Chemtex and Beta Renewables, subsidiaries of the former Italian PET producer Mossi & Ghisolfi (M&G). The PROESA™ process requires steam, yeasts, and enzymes to produce 2G bio-ethanol. The physical pretreatment of the biomass feedstock uses steam explosion to break the plant structures. Enzymes are introduced for enzymatic hydrolysis that breaks down the cellulosic fibers into simple sugar molecules. It then undergoes a fermentation process to turn sugars into bio-ethanol. The Danish-based firm Novozymes supplies the enzymes and the genetically modified yeasts for fermentation are supplied by DSM.

In September 2018, Clariant started construction of a new facility in Podari (Romania) with an annual capacity of 65 million liters of bio-ethanol. The plant will use Clariant's Sunliquid® technology to produce cellulosic 2G bio-ethanol from agricultural residues, including wheat straw and corn stover (the leftover stalks, and leaves of the corn plant) which will be sourced from local farmers.

In 2015, DuPont Industrial Biosciences opened the doors to the largest cellulosic ethanol plant in the world. The DuPont biorefinery, located in Nevada, Iowa (USA), is powered by corn stover and has a capacity to produce 115 million liters of bio-ethanol a year. Cellulosic 2G bio-ethanol is one of the cleanest-burning fuels on the planet, reducing carbon emissions by 90% over traditional fossil-based fuels. By the end of 2018, it was announced that the German-based firm VERBIO acquired DuPont's cellulosic 2G bio-ethanol plant and a portion of its corn stover inventory. VERBIO installed additional facilities to produce renewable natural gas (RNG) made from corn stover and other cellulosic crop residues at the site. VERBIO has two other cellulosic 2G bioethanol production facilities in Schwedt (Germany) and in Pinnow (Germany). DuPont Industrial Biosciences continues to participate in the overall biofuels market through specialty offerings, including both 1G and 2G biofuel enzymes and engineered yeast solutions that improve yield and productivity for biofuel producers.

Both 1G and 2G bio-ethanol production processes, starting from either sugarcane or cellulosic feedstock, respectively, are shown in Figure 3.19.

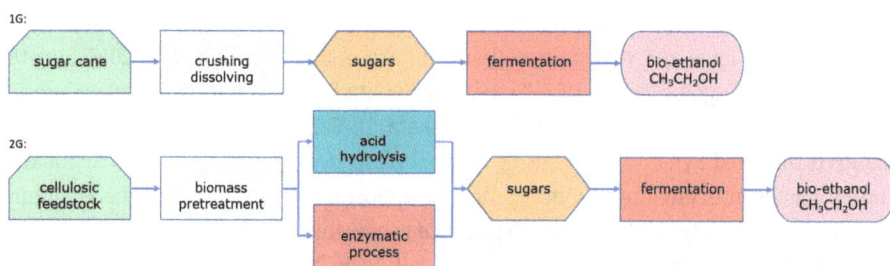

Figure 3.19: IEA process flow diagram for the production of bio-ethanol (route 1).

Route 2

USA-based firm Zea2 LLC (former ZeaChem) developed a totally different approach to produce 2G bio-ethanol using their patented biorefinery process. Unlike traditional processes, the Zea2 LLC technology is considered to be more economical and produces a higher yield (no CO_2 is generated). The hybrid technology is developed by combining a thermochemical process and a biochemical fermentation process. A cellulose carbon chain-digesting bacteria, acetogen, is used to convert biomass sugar molecules into acetic acid (CH_3COOH) in the core facility. An acetogen is a microorganism that generates acetic acid as an end product of anaerobic fermentation. The acetogen process produces 40% more bio-ethanol for a ton of biomass without any CO_2 of by-product. The acetic acid is then converted into an ester (ethyl acetate, $CH_3COOCH_2CH_3$), which is further reacted with hydrogen to produce bio-ethanol. Zea2 LLC generates the required hydrogen through gasification of lignin residue to create hydrogen-rich syngas. Hydrogen from syngas is used for ester hydrogenation, while the remaining gas is combusted for producing power and steam for the process. In Figure 3.20, the IEA process flow diagram for Zea2 LLC's process is shown.

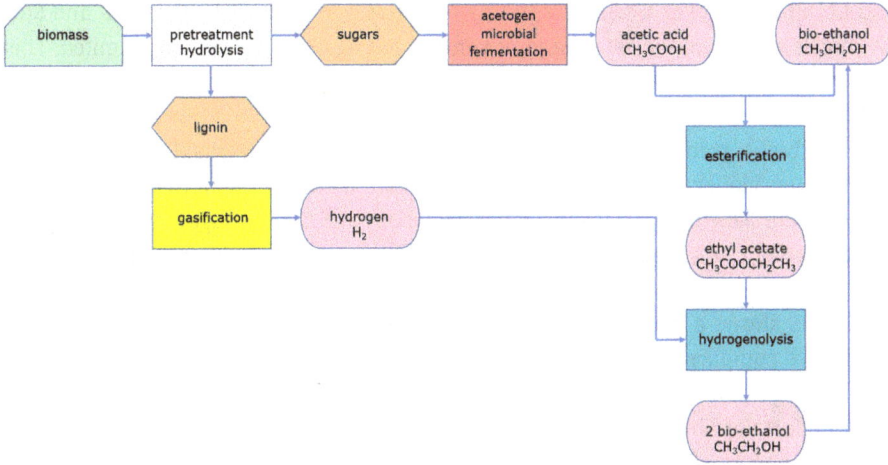

Figure 3.20: IEA process flow diagram for the production of bio-ethanol via Zea2 LLC technology (route 2).

Brazilian-based company Braskem produces bio-based PE (bio-PE) at commercial scale in Brazil (at the Triunfo Petrochemical Complex in Rio Grande do Sul, Brazil) since 2010 from sugarcane, under the "I'm Green"™ trademark.

Braskem's plant uses bio-ethanol produced from sugarcane as the feedstock, a process schematically displayed in Figure 3.21. The plant consumes about 462 million liters of bio-ethanol annually to produce the bio-PE. The bio-ethanol is initially being sourced from major Brazilian producers in the states Minas Gerais, Paraná, and São Paulo. It is transported by water and rail and a small portion by road. The state government of Rio Grande do Sul is taking the necessary steps to produce sufficient bio-ethanol feedstock entirely in the state.

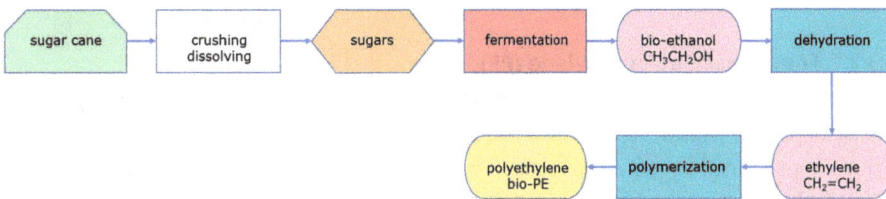

Figure 3.21: IEA process flow diagram for production of bio-PE.

The production capacity of bio-PE is about 260,000 metric tons a year. There are currently no other commercial scale plants for bio-PE; Braskem is the main commercial scale producer worldwide. Early 2014, Braskem announced the diversification of their bio-PE portfolio with plans to produce bio-based low-density polyethylene (bio-LDPE). Nowadays, Braskem produces approximately 30 grades in the bio-based HDPE, LDPE, and LLDPE, that cover a wide range of applications. In early 2023, Braskem and SCG Chemicals announced to set up a joint venture Braskem Siam Company Limited. This joint venture aims to produce bio-ethylene from bio-ethanol dehydration and to commercialize bio-PE in Thailand.

Bio-PE is a great example of a polymer made using a renewable feedstock, but bio-PE is not biodegradable. The first consideration is the source of the biomass starting materials. One hectare of land generates about 82.5 tons of sugarcane, 7,200 liters of bio-ethanol, three tons of bio-based ethylene, and subsequently about three tons of bio-PE.

Bio-PE shows a considerable reduction in *greenhouse gas* (GHG) emissions and energy use compared to the fossil-based equivalents. The current production of bio-PE from sugarcane realizes GHG emissions savings of more than 50%. According to a study undertaken by Braskem, one kg of bio-PE captures and stores about 2.15 kg of CO_2 while one kg of fossil-based PE releases almost two kg of CO_2. An early switch to renewable sources is important to the plastics industry to reduce oil dependency and greenhouse gas (GHG) emissions. Major packaging producers and A-brands in food, drinks, and cosmetics have already introduced bio-PE in their packaging products. Examples include the new Tetra Rex® cartons from Tetra Pak, Danone with Activia, and Proctor and Gamble with their Pantene bottles.

Recently, the LEGO Group from Denmark introduced the "botanical" elements range, such as trees, bushes, and leaves, prepared from Braskem's I'm Green™ bio-PE.

3.3.2 Polypropylene (PP)

3.3.2.1 Fossil-based polypropylene (PP)

Polypropylene (PP) is made from propylene monomer ($CH_2 = CHCH_3$). Propylene is the second most produced chemical building block in the petrochemical industry. Propylene is produced by ethylene steam crackers. It is a co-product of steam cracking of hydrocarbon fossil-based feedstock, and the quantity produced depends on the nature of the feedstock.

PP density remains between 0.895 g/cm^3 for amorphous and 0.92 g/cm^3 for crystalline material. However, this density may change with added fillers. Four main types of PP are commercially available.

PP homopolymer
This grade is considered as a general grade and can be used for a variety of general purpose applications. It is stronger and stiffer than PP copolymers. Main applications include packaging, textiles, healthcare, pipes, automotive, and electrical applications.

PP block copolymer
This grade contains a high ethylene content (between 5% and 15%). It has comonomer units arranged in a regular pattern (or blocks). The regular pattern makes the thermoplastic tougher and less brittle than the PP random copolymer. These polymers are suitable for applications requiring high strength, such as industrial usages.

PP random copolymer
This grade is produced by polymerizing together ethylene and propylene. It features ethylene units, usually up to 6% by mass, incorporated randomly in the PP main chain. These polymers are flexible and optically clear, making them suitable for applications requiring transparency and for products requiring an excellent appearance.

PP impact copolymer
This grade is a PP homopolymer containing a co-mixed PP random copolymer phase that has an ethylene content of 45–65%. PP impact copolymers are mainly used in packaging, houseware, film, and pipe applications, as well as in the automotive and electrical sectors.

Upon polymerization, PP can form three basic chain structures depending on the position of the methyl groups (-CH$_3$) along the polymer main chain. The term tacticity describes how the methyl groups are oriented. Commercially available PP is usually isotactic. Atactic propylene is amorphous and has therefore no crystal structure. Syndiotactic PP can only be prepared by using metallocene catalysts. Syndiotactic PP has a lower melting point compared with isotactic PP, depending on the degree of tacticity.

Production of PP, starting from crude oil, is shown in Figure 3.22.

Two main commercial technologies to produce PP from the propylene monomer are the Grace Unipol gas-phase PP technology® on a fluidized-bed reactor system (market share 16%) and the Lyondell Basell Spheripol Technology (market share 39%) for PP homopolymers and PP random copolymers.

3.3.2.2 Bio-based polypropylene (bio-PP)
Bio-PP is not commercially available yet. However, a number of different process routes are described for the purpose of producing the bio-based propylene monomer.

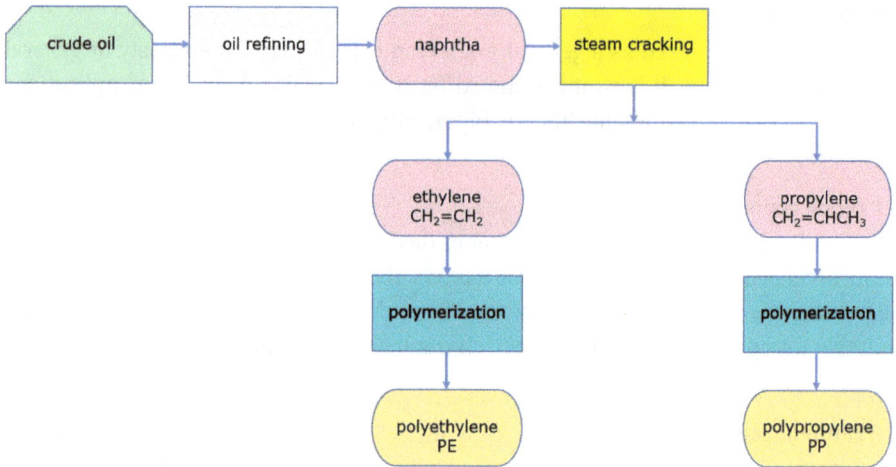

Figure 3.22: IEA process flow diagram for the production of PP.

Route 1

Mitsui Chemicals introduced their *bio-based polypropylene* project at the G20 Ministerial Meeting in Karuizawa (Japan, June 2019).

In Mitsui's process, raw biomass is fermented directly into bio-based isopropanol (bio-IPA). After *dehydration* of isopropanol, the resulting bio-based propylene monomer can be used for the production of bio-PP in existing production equipment. The chemical process is shown in Figure 3.23. The IEA process flow diagram for the production of bio-PP is shown in Figure 3.24.

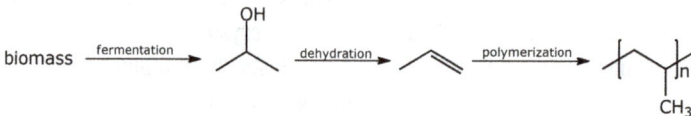

Figure 3.23: Production of bio-based polypropylene (bio-PP).

Mitsui Chemicals is considering the commercialization of their bio-PP. Production will start at the soonest in 2024.

Route 2

In general, biodiesel is produced from vegetable oil crops. Besides biodiesel, glycerol and animal feed are produced. The IEA process flow diagram for the biodiesel process is shown in Figure 3.25.

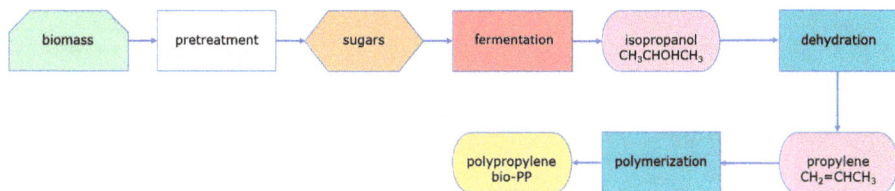

Figure 3.24: IEA process flow diagram for the production of bio-PP (route 1).

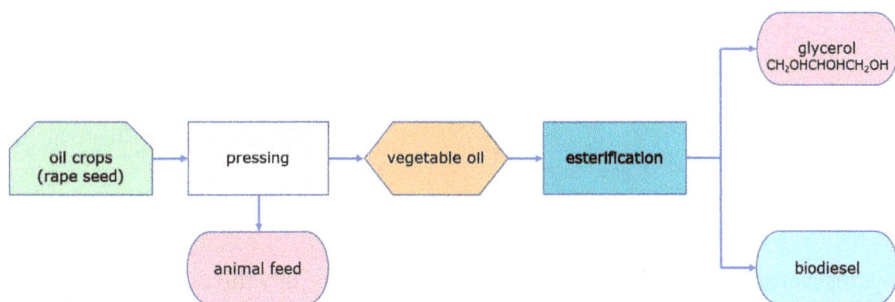

Figure 3.25: IEA process flow diagram for the production of biodiesel.

Note that, instead of vegetable oil crops (such as rape seed oil), also 2G oils can be used. The Dutch-based company Sunoil Biodiesel deploys used frying oil to produce biodiesel, also known as "2G biodiesel". Sunoil Biodiesel has a capacity of up to 100 million liters of biodiesel a year. As indicated, glycerol ($CH_2OHCHOHCH_2OH$) is a by-product in the biodiesel production process. Glycerol is widely used for its multiple properties, for example, in cosmetic products, food application, and pharmaceutical applications.

Glycerol, however, can also be transferred into propylene in a one-step catalytic process using molybdenum-based catalysts. The glycerol-to-propylene (GTP) process was first introduced by Quattor Petrochimica, headquartered in Rio de Janeiro (Brazil). In 2009, this company was incorporated by the Brazilian-based company Braskem. In a typical hydro-deoxygenation (HDO) reaction, using Fe-Mo catalysts (supported on black and activated carbon), a selective C-O bond cleavage is realized, thus converting glycerol to propylene with high yields and purities (Figure 3.26).

Route 3

In 2018, Borealis, a leading provider of innovative polyolefins, and Neste, a leading provider of renewable biodiesel, renewable jet fuel, and an expert in delivering drop-in renewable chemical solutions, entered into a strategic co-operation for the production of bio-PP. The co-operation will enable Borealis to start using 100% bio-based propane ($CH_3CH_2CH_3$) produced with Neste's proprietary NExBTL™ technology as renewable feedstock. Neste's NExBTL™ process entails a patented direct catalytic hydro-deoxygenation

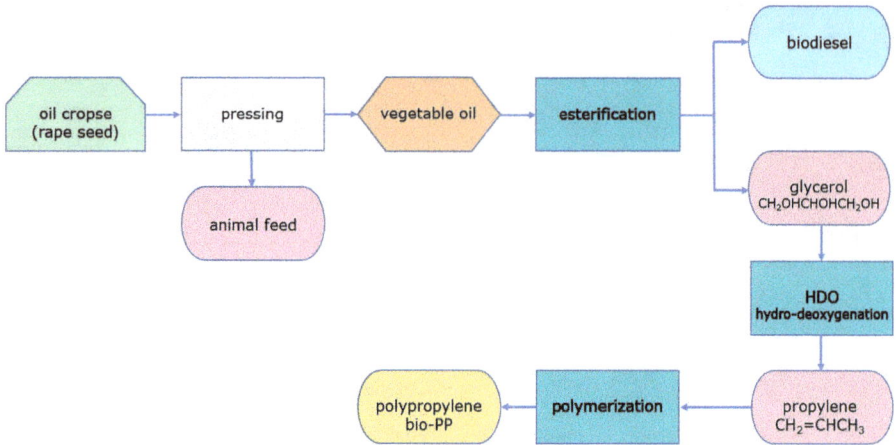

Figure 3.26: IEA process flow diagram for the production of bio-PP (route 2).

(hydrogenolysis) of vegetable oils, which are triglycerides, into the corresponding alkanes and propane. The glycerol part of the triglyceride is hydrogenated to propane. There is no glycerol side-stream. The NExBTL™ process removes oxygen from the oil, so the diesel is not an oxygenate-like traditional fatty acid methyl ester (FAME), which is obtained by transesterification of fats or oils with methanol. Catalytic isomerization into branched alkanes is then carried out to adjust the cloud point to meet winter-operability requirements. As it is chemically identical to ideal conventional diesel, and it requires no modification or special precautions for the engine.

Borealis' propane hydro-dehydrogenation (PDH) process and its PP production plant in Kallo (Finland) will enable the company to start offering bio-based propylene and consequently bio-PP in which the bio-based content can be physically verified and measured. See Figure 3.27 for the detailed process flow diagram.

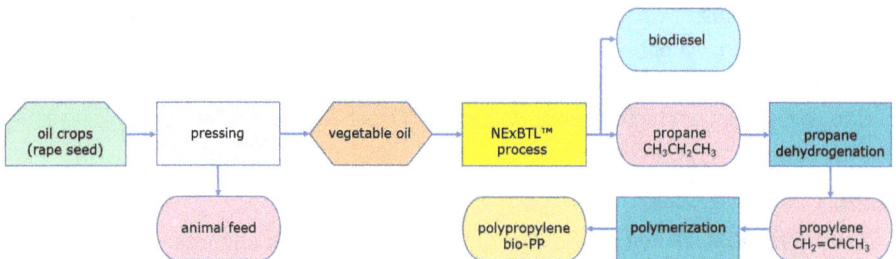

Figure 3.27: IEA process flow diagram for the production of bio-PP (route 3).

Lyondell Basell and Neste announced in June 2019 the commercial-scale production of bio-PP and bio-PE. These bio-based plastics have been approved for the production of food packaging and being marketed under the brand names Circulen™ and Circulen Plus™, the new family of Lyondell Basell circular economy product brands.

Route 4

Acetone-butanol-ethanol (ABE) fermentation is a process that uses bacterial fermentation to produce a mixture of acetone, 1-butanol, and ethanol from carbohydrates such as glucose. The anaerobic ABE fermentation produces the solvents in a ratio of three parts of acetone, six parts of 1-butanol, and one part of ethanol. The resulting acetone can be used to produce, after a reduction step, 2-propanol (isopropanol) and, subsequently, 2-propanol gives in a dehydration step bio-based propylene (Figure 3.28). The ABE fermentation usually uses a strain of bacteria from the class of Clostria (family Clostridiaceae). *Clostridium acetobutylicum* is the most well-studied and widely used species, although *Clostridium beijerinckii* has also been used with good results.

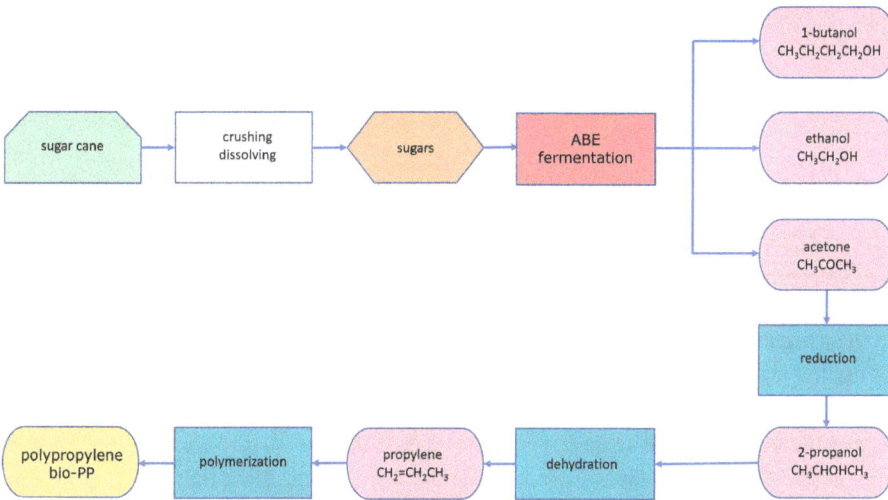

Figure 3.28: IEA process flow diagram for the production of bio-PP (route 4).

In the ABE *fermentation* process, 1-butanol is produced in significant amounts. 1-Butanol has been proposed as a substitute for diesel fuel and gasoline. Compared with ethanol, 1-butanol has a couple of advantages: lower CO_2 emissions and higher energetic value. 1-Butanol is a chemical building block for other products and materials, including 1-butylene ($CH_2 = CHCH_2CH_3$).

Cathay Industrial Biotech produces 1-butanol ($CH_3CH_2CH_2CH_2OH$) for chemical applications using corn starch in an ABE fermentation process with a capacity of 100,000 metric tons of 1-butanol a year. Cathay started production in 2009 with an output of about 65–70% of 1-butanol, 20–25% of acetone, and 5–10% of ethanol. In near future,

Cathay will use corn cobs and corn stover as feedstock, and production capacity will be increased up to 130,000 tons of 1-butanol a year.

Other bio-based 1-butanol producers include Cobalt Technologies, BranBio, Gevo, Metabolic Explorer, and Eastman Renewable Materials (acquired TetraVitae Bioscience in 2011).

Route 5

Olefin metathesis can also be applied to produce bio-based propylene, starting from both bio-based ethylene ($CH_2 = CH_2$) and bio-based 2-butylene ($CH_3CH = CHCH_3$). Olefin metathesis is an organic reaction that entails the redistribution of fragments of alkenes (olefins) by the scission and regeneration of carbon–carbon double bonds. The reaction requires special types of catalysts. The reaction scheme of olefin metathesis is shown in Figure 3.29.

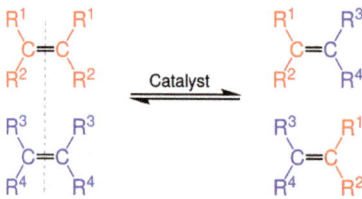

Figure 3.29: Olefin metathesis: metathesis of ethylene and 2-butylene (R^1, R^2, and R^3 are -H, and R^4 is -CH_3) results in 2 molecules of propylene.

In the process, bio-based ethylene is obtained from sugar fermentation. And, bio-based 2-butylene can be produced from bio-based ethylene, via dimerization into 1-butylene and subsequent isomerization into 2-butylene. Bio-based 2-butylene can also be produced from 1-butanol, obtained from the ABE fermentation process. In this way, 1-butanol is dehydrated into 1-butylene and, subsequently, isomerized into 2-butylene. Figure 3.30 shows the process flow diagram for the production of bio-PP (route 5).

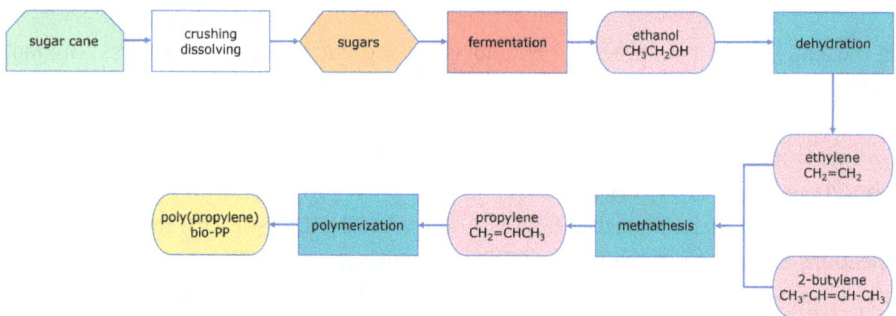

Figure 3.30: IEA process flow diagram for the production of bio-PP (route 5).

Note that bio-based 1-butylene can also be used as a comonomer to produce linear low-density polyethylene (LLDPE). Figure 3.31 shows two routes for the synthesis of bio-based 2-butylene. One route starts with an ABE fermentation where 1-butanol is dehydrated to 1-butylene. In the next step, 1-butylene is converted via an isomerization step to 2-butylene. 1-Butylene can also be produced via a dimerization of bio-based ethylene.

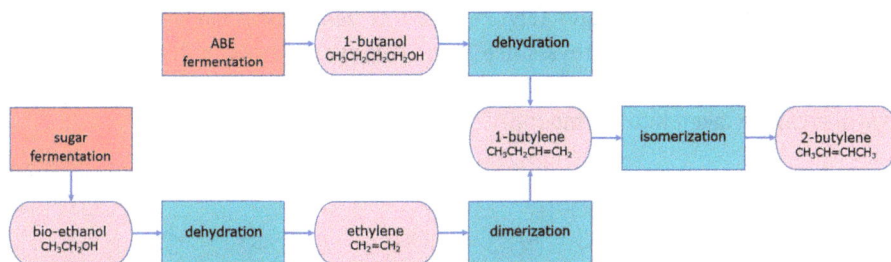

Figure 3.31: IEA process flow diagram for the production of 2-butylene.

Route 6
France-based Global Bioenergies (GBE) announced a process for the direct conversion of glucose to propylene using the same artificial metabolic pathway that the company uses to produce bio-based isobutylene (or 2-methylpropene). The Global Bioenergies pathway uses enzymatic activities and metabolic intermediates that are not found in nature but only occur in metabolic engineered strains. In a direct fermentation, bio-based propylene is obtained in one step and can be used to produce bio-PP, as shown in Figure 3.32.

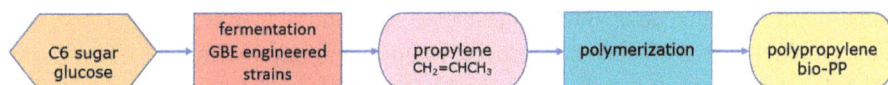

Figure 3.32: IEA process flow diagram for the production of bio-PP (route 6).

3.3.3 Polyvinyl chloride (PVC)

3.3.3.1 Fossil-based polyvinyl chloride (PVC)
Polyvinyl chloride (PVC) is made from the polymerization of vinyl chloride monomer. It is an odorless solid plastic which is naturally white and brittle in nature. It is a syn-

thetic polymer made of 57% of chlorine (derived from industrial salt), and the remaining 43% comprises carbon (derived from oil and gases). *Polyvinyl chloride* is produced in two different forms: first is the rigid or unplasticized PVC resin and second is the flexible or plasticized PVC.

Flexible PVC is soft due to the addition of plasticizers such as diisononyl phthalate (DINP). DINP is typically a mixture of chemical compounds consisting of various diisononyl esters of phthalic acid.

The use of some phthalates (Figure 3.33) has been restricted in the European Union for use in children's toys since 1999. For example, diethylhexyl phthalate (DEHP) is restricted for all toys, while diisononyl phthalate (DINP) is restricted only in toys that can be taken into the mouth.

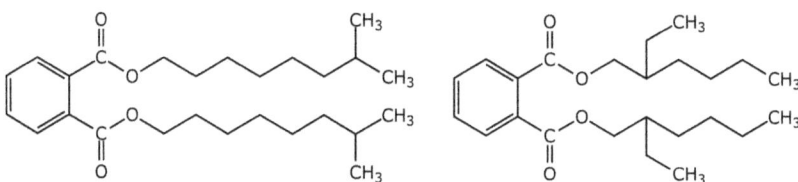

Figure 3.33: Chemical structures of DINP (left) and DEHP (right).

In the past few years, several new families of phthalate-free plasticizers were developed for the PVC industry. As an example, Dow introduced Dow Ecolibrium™ bio-based *plasticizer*, made from vegetable oil derivatives, on the market for the PVC wire and cable industry. Also, the France-based company Roquette offers a phthalate-free plasticizer for the flexible PVC market. Their product (Polysorb™ ID 37) is a mixture of isosorbide diesters produced from fatty acids of vegetable origin and bio-based isosorbide.

The largest single producer of PVC as of 2018 is Shin-Etsu Chemical of Japan, with a global share of around 30%. Vinyl chloride itself is produced in a single stage from ethylene combining two processes: direct chlorination and oxychlorination (Figures 3.34 and 3.35). This process is referred to as the balanced process. In the direct chlorination process, both pure chlorine (Cl_2) and ethylene ($CH_2 = CH_2$) are used to produce 1,2-dichloroethane ($ClCH_2CH_2Cl$), and in the oxychlorination process ethylene reacts with hydrogen chloride (HCl) and oxygen to form 1,2-dichloroethane. 1,2-Dichloroethane is then converted to vinyl chloride by thermal cracking (EDC *pyrolysis*), and the hydrogen chloride (HCl) by-product can be recycled to the oxychlorination plant.

3.3.3.2 Bio-based polyvinyl chloride (bio-PVC)

Recently, Inovyn, a business unit of Ineos, launched so-called "bio-attributed" PVC with the trade name Biovyn™. Biovyn™ is produced from 100% renewable ethylene derived from renewable feedstock that does not compete with the food chain (Figure 3.36).

Direct chlorination	$CH_2=CH_2 + Cl_2 \rightarrow Cl-CH_2CH_2-Cl$
Oxychlorination	$CH_2=CH_2 + 2\ HCl + \frac{1}{2}\ O_2 \rightarrow Cl-CH_2CH_2-Cl + H_2O$
EDC pyrolysis	$2\ Cl-CH_2CH_2-Cl \rightarrow 2\ CH_2=CH-Cl + 2\ HCl$
Overall reaction	$2\ CH_2=CH_2 + Cl_2 + \frac{1}{2}\ O_2 \rightarrow 2\ CH_2=CH-Cl + H_2O$

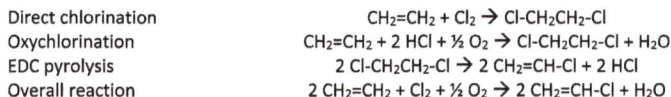

Figure 3.34: Industrial production of vinyl chloride.

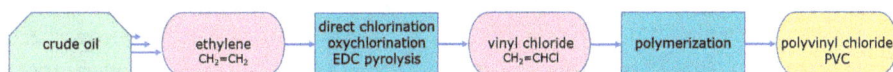

Figure 3.35: IEA process flow diagram for the production of PVC.

Biovyn™ will be the first PVC made without fossil fuels. In the near future, Biovyn™ will be used by the Tarkett company for a new PVC flooring collection.

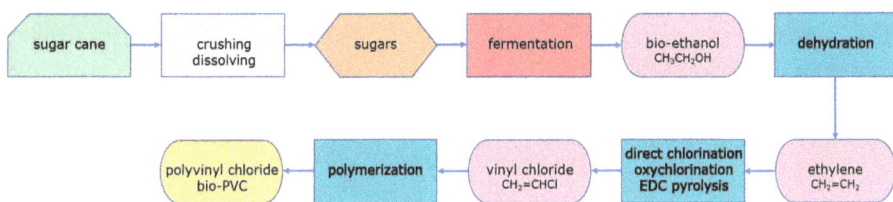

Figure 3.36: IEA process flow diagram for the production of bio-PVC.

3.3.4 Polystyrene (PS)

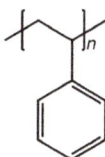

3.3.4.1 Fossil-based polystyrene (PS)

Polystyrene (PS) is made by the polymerization of styrene ($C_6H_5CH = CH_2$). *Polystyrene is a natural transparent thermoplastic that is available in two forms, solid as well as a rigid foam material (named *expanded polystyrene,* or EPS). Generally used PS material is clear, hard, and brittle in properties.

Polystyrene is produced primarily in two forms: atactic and syndiotactic PS. To produce atactic polystyrene (PS-at), the phenyl groups are randomly distributed on both sides of the polymer chain. This plastic polymer has a glass transition temperature of 100 °C, and the polymerization process is strongly initiated by free radicals.

In syndiotactic polystyrene (PS-st), the phenyl groups are positioned on the alternate sides of the hydrocarbon backbone, as explained in Chapter 2. PS-st is a highly crystalline form of polystyrene with a glass-transition temperature of 100 °C and a crystalline melting point of 270 °C. These types of polystyrene are currently produced under the brand name of XAREC™ by the Japan-based company Idemitsu Corporation by using metallocene catalysts in the polymerization reaction.

In the production of polystyrene (PS), three types of polymerization processes are brought into use. These are mass suspension polymerization, *emulsion polymerization*, and *solution polymerization*. The top producers of polystyrene (PS) in the global market are BASF (Germany), Styrosolution (Germany), Videolar (Brazil), Sabic (Saudi Arabia), and Formosa Plastic Corporation (Taiwan), among others.

General-purpose polystyrene (GPPS) is also termed Crystal-clear PS. This form of polystyrene is fully transparent, rigid, and usually a low-cost polystyrene made from styrene monomer. GPPS is a solid form of PS manufactured in form of 2–5 mm pellets. High-impact polystyrene (HIPS) contains usually 5–10% rubber (polybutadiene) and is used to produce products which require higher impact resistance. HIPS is usually a kind of graft copolymer. The grafting process occurs when some radicals react with the double bond of polybutadiene.

Route 1

Styrene ($C_6H_5CH = CH_2$) is the precursor to polystyrene (PS) and several copolymers (e. g. ABS). A common way to produce the styrene monomer is through the dehydrogenation of ethylbenzene. Ethylbenzene ($C_6H_5CH_2CH_3$) is produced on a large scale by combining benzene (C_6H_6) and ethylene ($CH_2 = CH_2$) in an acid-catalyzed chemical reaction. The IEA process flow diagram is shown in Figure 3.37.

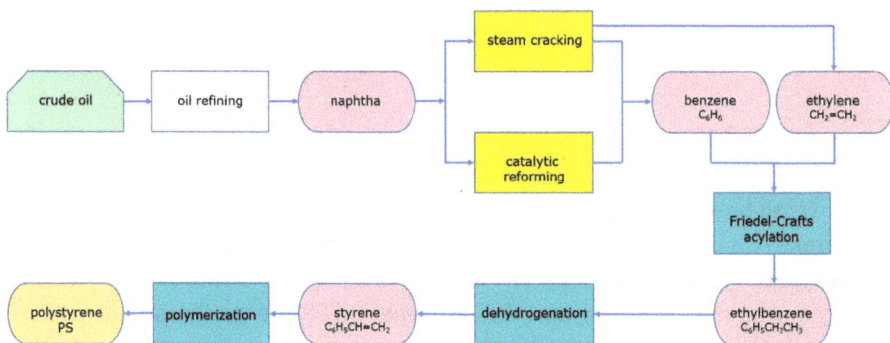

Figure 3.37: IEA process flow diagram for the production of PS (route 1).

Route 2

Styrene is also co-produced commercially in a process known as propylene oxide – styrene monomer (POSM, Lyondell Chemical Company) or styrene monomer/propylene oxide (SM/PO) (Shell). As shown in Figure 3.38, ethylbenzene ($C_6H_5CH_2CH_3$) is treated with oxygen (O_2) to form the ethylbenzene hydroperoxide. This hydroperoxide is then used to oxidize propylene ($CH_2 = CHCH_3$) to propylene oxide. The resulting 1-phenylethanol is dehydrated to give styrene (Figure 3.39).

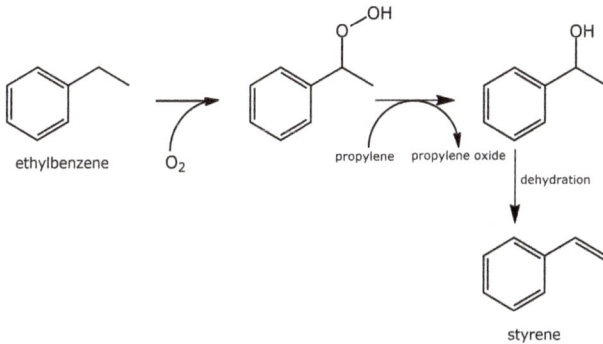

ethylbenzene O_2 propylene   propylene oxide   dehydration   styrene

Figure 3.38: POSM and SM/PO processes.

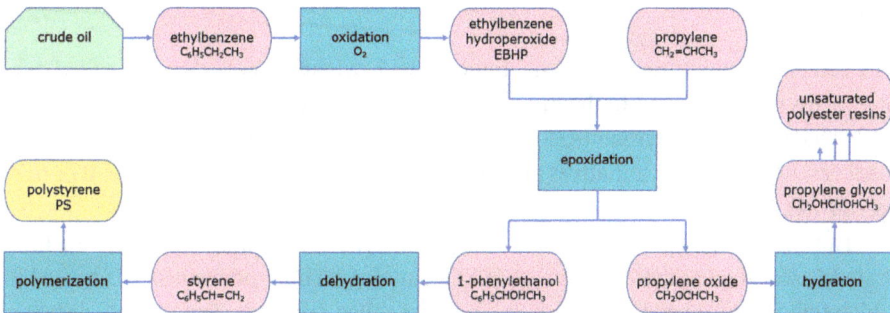

Figure 3.39: IEA process flow diagram for the production of PS (route 2).

3.3.4.2 Bio-based polystyrene (bio-PS)

Bio-based polystyrene (PS) is not yet commercially available on the market. Several process routes are described in the literature however.

Route 1

Bio-based 1,3-butadiene ($CH_2 = CHCH = CH_2$) is used to synthesize the bio-based styrene monomer. Note that bio-based 1,3-butadiene can be easily produced from sugars. In a Diels–Alder cyclic dimerization reaction 1,3-butadiene can be converted, using a Cu-zeolite catalyst, into 4-vinylcyclohexene in high yield and purity. In the next step, 4-vinylcyclohexene is oxidative dehydrogenated in the gas phase to styrene (Figure 3.40). Dow Chemical studied this pathway (Figure 3.41) to styrene in the past.

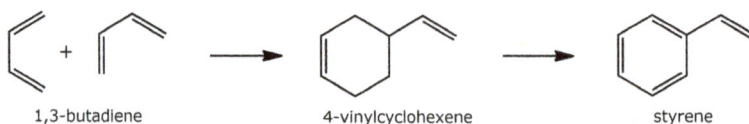

| 1,3-butadiene | 4-vinylcyclohexene | styrene |

Figure 3.40: Diels–Alder cyclic dimerization of 1,3-butadiene.

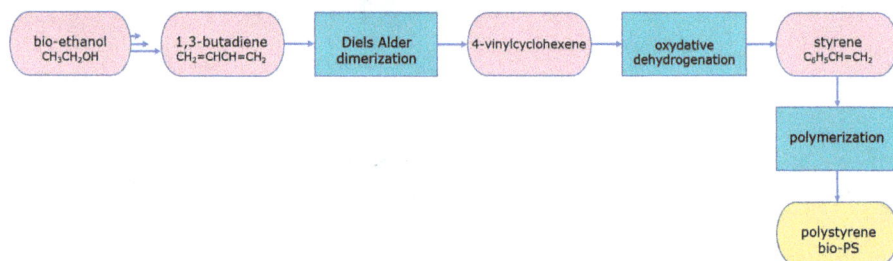

Figure 3.41: IEA process flow diagram for the production of bio-PS (route 1).

Route 2

BTX refers to a mixture of benzene, toluene, and xylenes. These three so-called aromatics are mostly applied in the production of plastics like PET, polystyrene (PS), or acrylonitrile-butadiene–styrene (ABS). Traditionally, BTX is derived from non-renewable fossil resources. Several initiatives are going on to develop non-fossil routes to BTX (*bio-based BTX*).

The BioBTX company, based in Groningen (The Netherlands) is using an Integrated Cascading Catalytic Pyrolysis (ICCP) technology. First, the feedstock is heated to a high temperature inside a reactor in a process called *pyrolysis*. The vapors are transferred into a second reactor where they are directly transformed into aromatics by a catalytic conversion. Finally, the liquid BTX product is collected, thus completing the process (Figure 3.42). Currently, BioBTX uses two groups of feedstocks to produce BTX: bio-based feedstock and plastic waste.

The resulting bio-based components, benzene, toluene, and three isomers of xylene are building blocks for several polymers.

Bio-based benzene can be used, together with bio-based ethylene (from bioethanol), to produce in a Friedel-Crafts acylation bio-based ethylbenzene. According to Figure 3.37, ethylbenzene can be dehydrogenated to bio-based styrene. The resulting bio-based styrene could be the basis for bio-based polystyrene (bio-PS).

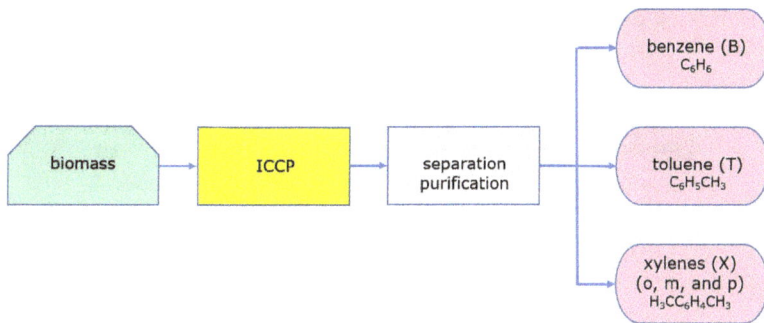

Figure 3.42: IEA process flow diagram for the production of bio-BTX.

Bio-based toluene can be used to produce stilbene ($C_6H_5CH = CHC_6H_5$) in a catalyzed oxidative coupling. The resulting bio-based stilbene can be used in a metathesis-type reaction with bio-based ethylene ($CH_2 = CH_2$). Two molecules of bio-based styrene ($C_6H_5CH = CH_2$) are obtained in this way. The resulting bio-based styrene (Figure 3.43) could be the basis for bio-based polystyrene (bio-PS). In the past, this process got serious industrial interest.

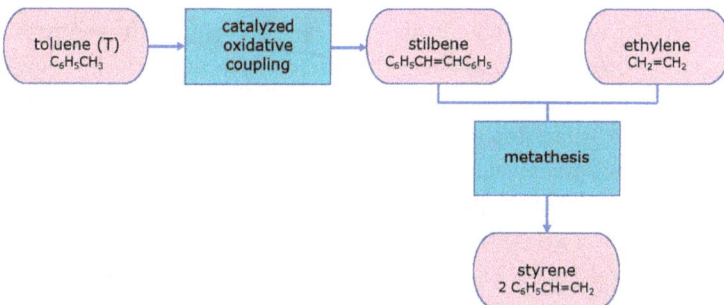

Figure 3.43: IEA process flow diagram for the production of bio-based styrene.

Xylene exists in three isomeric forms. The isomers can be distinguished by the designations *ortho-* (*o-*), *meta-* (*m-*) and *para-* (*p-*), which specify to which carbon atoms (of the benzene ring) the two methyl groups are attached. On industrial scale, xylenes are produced by methylation of toluene or benzene. *p*-Xylene is the principal precursor to TA and dimethyl terephthalate (DMT), both monomers used in the production of PET. *o*-Xylene is an important precursor to phthalic anhydride.

So, in particular, bio-based *para*-xylene received serious interest in the past few years because, after oxidation of *para*-xylene to TA, the way is open to produce 100% bio-based polyethylene terephthalate (bio-PET). Bio-based ethylene glycol can be easily produced from bio-based ethanol. The process flow diagram for the production of bio-PET is shown in Figure 3.44.

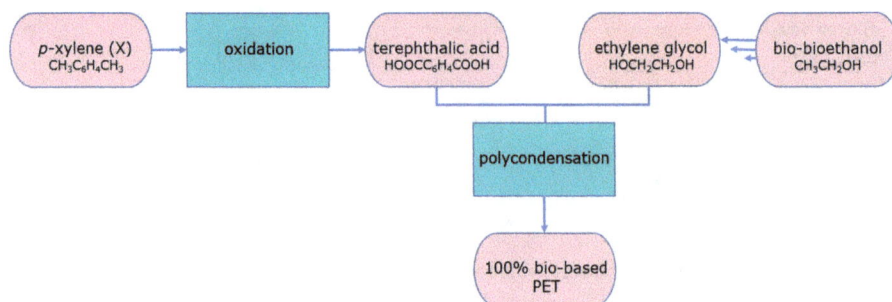

Figure 3.44: IEA process flow diagram for the production of bio-PET.

Route 3

Styrene can also be produced using a biochemical route where microorganisms are utilized as cellular factories for the production of styrene (Figure 3.45). First, phenylalanine, an essential α-amino acid, is produced by a fermentation process that use glucose or sucrose as the carbon source using specially engineered micro-organisms for the biotransformation. For example, L-phenylalanine can be produced from glucose via the glycolysis pathway and the subsequent shikimate pathway. In the *glycolysis* pathway, glucose is converted into pyruvate (CH_3COCOO^-). The shikimate pathway (shikimic acid pathway) is a seven-step metabolic pathway for the biosynthesis of aromatic amino acids (such as L-phenylalanine). Subsequently, a co-culture system can be used consisting of L-phenylalanine ammonia lyase (PAL, wherein NH_3 is released) and phenylacrylic acid decarboxylase (FDC1, wherein CO_2 is released). A key factor in styrene production using microbes (Figure 3.46) is the recovery of volatile styrene, however.

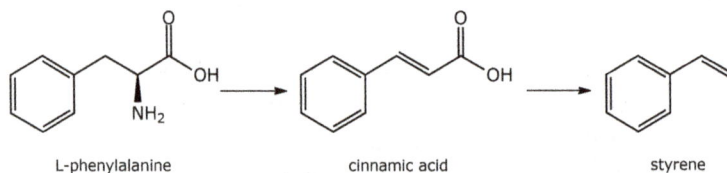

Figure 3.45: Synthesis of styrene from L-phenylalanine.

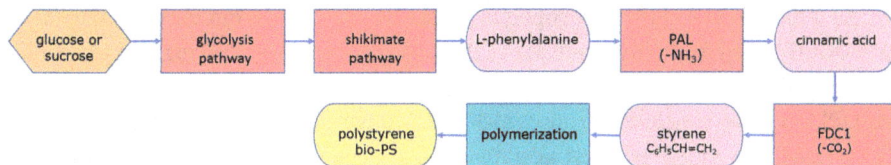

Figure 3.46: IEA process flow diagram for the production of bio-PS (route 3).

3.3.5 Polyvinyl acetate (PVAc)

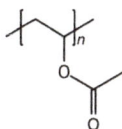

3.3.5.1 Fossil-based polyvinyl acetate (PVAc)

Polyvinyl acetate (PVAc) is a synthetic resin prepared by a *free radical polymerization* of vinyl acetate monomer (VAc). Vinyl acetate ($CH_3COOCH = CH_2$) is prepared from ethylene ($CH_2 = CH_2$) by reaction with oxygen and acetic acid (CH_3COOH) over a palladium catalyst. In this reaction, an acetoxylation reaction, an acetoxyl group is added to ethylene. Ethylene is obtained from the petrochemical industry, and acetic acid is obtained via a vapor-phase methanol carbonylation over a palladium-supported catalyst. Carbonylation refers to reactions that introduce carbon monoxide into organic substances. The process towards PVAc is presented in Figure 3.47.

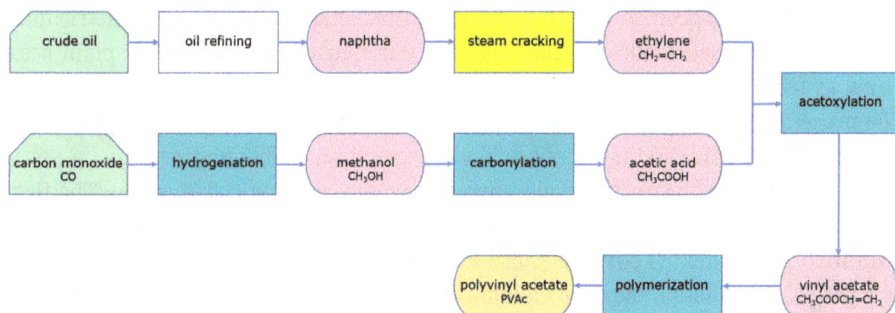

Figure 3.47: IEA process flow diagram for production of PVAc.

Polyvinylacetate can be hydrolyzed to polyvinyl alcohol (PVA), as discussed in Section 2.3.5. Sometimes, a base-catalyzed *transesterification* is applied with ethanol (CH_3CH_2OH) so that PVA and ethyl acetate ($CH_3COOCH_2CH_3$) are formed.

As indicated in Figure 3.48, copolymers can be produced from vinyl acetate (VAc, $CH_3COOCH = CH_2$) and ethylene ($CH_2 = CH_2$) in a high-pressure reactor. These copolymers, so-called ethylene-vinyl acetate (EVAc) or sometimes poly(ethylene-*co*-vinyl acetate) (PEVAc), are elastic materials which are "rubber-like" in softness and flexibility. The weight percentage of vinyl acetate usually varies from 10 to 40%, with the remainder ethylene.

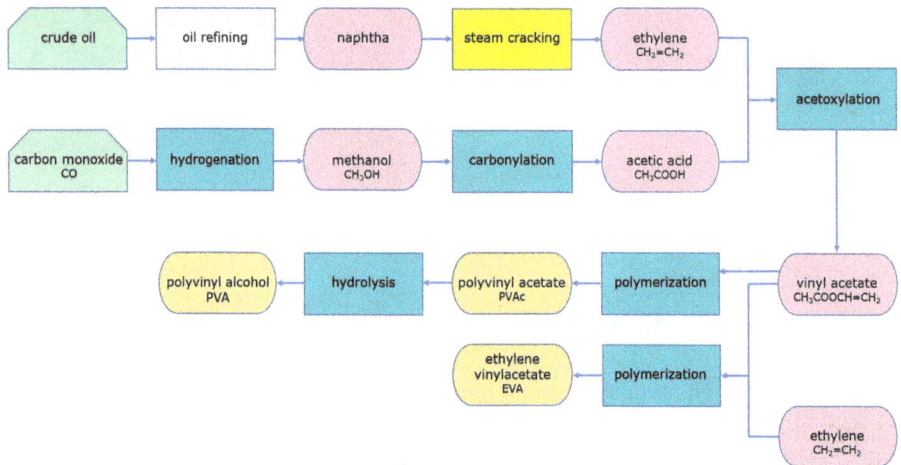

Figure 3.48: IEA process flow diagram for production of EVA and PVA.

3.3.5.2 Bio-based polyvinyl acetate (bio-PVAc)

German chemical company Wacker Chemie has developed a fermentation-based acetic acid (CH_3COOH) process, depicted in Figure 3.49. Wacker started its 500 metric tons a year of a bio-based acetic acid pilot plant in Burghausen (Germany). The plant uses straw as feedstock, but it can use other sugar-based biomass raw materials as well. The process is called ACEO and uses an enzymatic fermentation route, first producing ethanol (CH_3CH_2OH) and then acetic acid via a gas-phase oxidation process.

Wacker is also looking to develop two more bio-based acetic acid routes. The first is by converting bio-based 2,3-butanediol, obtained via a sugar fermentation, in a dehydration step to methyl ethyl ketone. Subsequently, methyl ethyl ketone is converted in a gas-phase oxidation into two equivalents of acetic acid. First, methyl ethyl ketone is oxidized to diacetyl. Diacetyl is hydrolyzed to both acetic acid (CH_3COOH) and acetaldehyde (CH_3CHO). Subsequently, acetaldehyde is oxidized to acetic acid. In the other process, acetic acid is obtained in an acetogen microbial fermentation step from sugar (Figure 3.49).

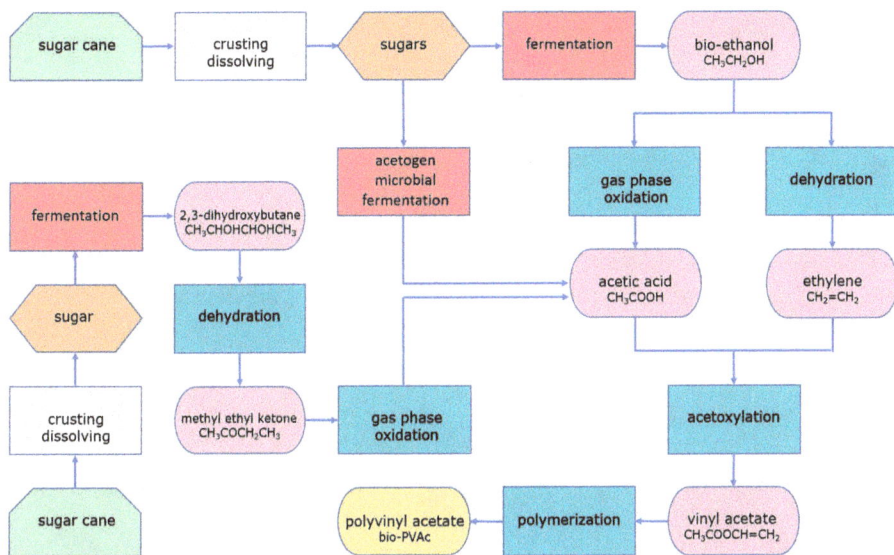

Figure 3.49: IEA process flow diagram for production of bio-PVAc.

3.3.6 Polyacrylic acid (PAA)

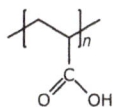

3.3.6.1 Fossil-based polyacrylic acid (PAA)

Polyacrylic acid (PAA) is a synthetic polymer produced via a free radical polymerization from acrylic acid monomer ($CH_2 = CHCOOH$). Acrylic acid is the simplest unsaturated carboxylic acid, consisting of a vinyl group connected directly to a carboxylic acid group, systematically named prop-2-enoic acid. Top producers of acrylic acid are BASF, Dow Chemical, Arkema, Evonik, Nippon Shokubai, LG Chemical, Mitsubishi Chemical Corp., Jiansu Jurong Chemical, and Formosa Plastics.

In practice, acrylic acid is produced by a two-stage oxidation of propylene. In the first stage, propylene ($CH_2 = CHCH_3$) is oxidized to acrolein ($CH_2 = CHCHO$). In the next catalyzed oxidation step, acrolein is converted to acrylic acid ($CH_2 = CHCOOH$). The oxidation of propylene to acrylic acid is presented in Figure 3.50.

The majority of acrylic acid is used to make acrylate esters ($CH_2 = CHCOOR$), followed by the use of acrylic acid for the production of *polyacrylic acid* (PAA), as shown in Figure 3.51. Polyacrylic acid (PAA) and its derivatives are used in disposable diapers, ion exchange resins, adhesives, and detergents. They are also popular as thick-

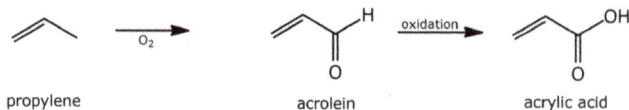

Figure 3.50: Oxidation of propylene to acrylic acid.

ening, dispersing, suspending, and emulsifying agents in pharmaceuticals, cosmetics, and paints. Acrylic acid and its esters readily combine with themselves (to form, e. g., polyacrylic acid) or with other monomers (e. g., acrylamide, acrylonitrile, styrene, and butadiene) forming copolymers.

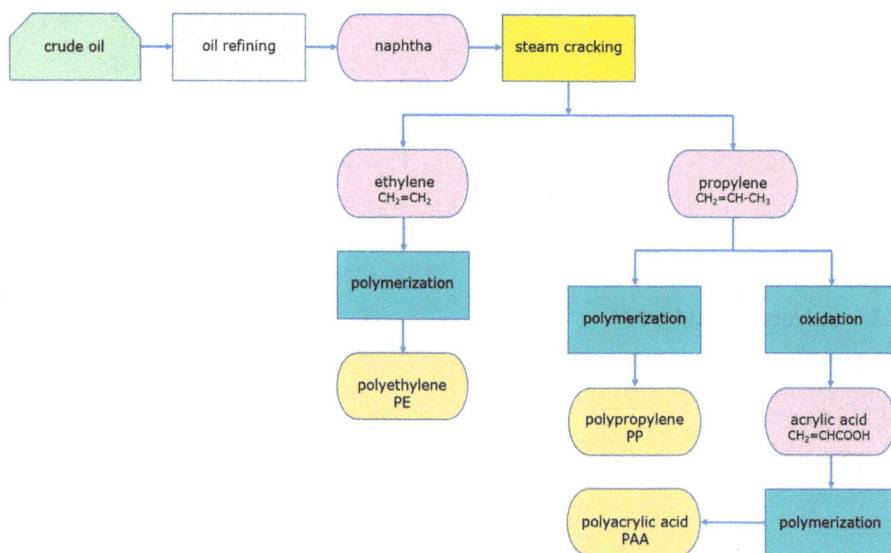

Figure 3.51: IEA process flow diagram for production of PAA (and PE).

3.3.6.2 Bio-based polyacrylic acid (bio-PAA)

Bio-based polyacrylic acid (bio-PAA) is not commercially available yet. Several routes to bio-based acrylic acid (bio-AA) and subsequently bio-PAA are described in the literature, however.

Route 1

US-based company Genomatica developed a process based on a cross metathesis reaction of bio-based fumaric acid ($HOOCCH_2 = CH_2COOH$) and bio-based ethylene ($CH_2 = CH_2$). Both materials can be obtained from a glucose fermentation. The resulting bio-based acrylic acid can be polymerized to bio-PAA (Figure 3.52).

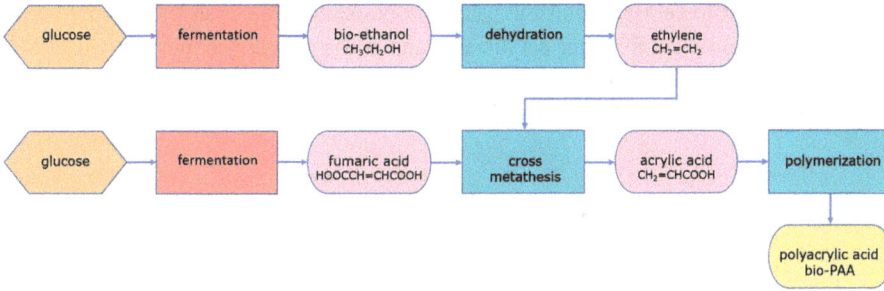

Figure 3.52: IEA process flow diagram for production of bio-PAA (route 1).

Route 2

A collaboration between BASF, Cargill, and Novozymes investigated the development and commercialization of bio-based acrylic acid, which is derived through the dehydration of 3-hydroxypropionic acid (3-hydroxypropanoic acid). 3-Hydroxypropionic acid (3-HPA, see Table 3.2) is derived via fermentation of sugar using bio-engineered organisms (Figure 3.53). BASF is the world's largest producer of acrylic acid. BASF's primary interest is the use of bio-based acrylic acid to manufacture superabsorbent polymer (SAP). SAP is mainly used in diapers and other hygienic products.

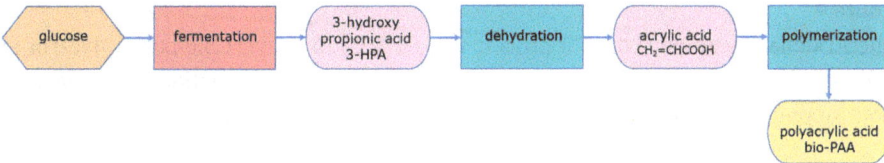

Figure 3.53: IEA process flow diagram for the production of bio-PAA (route 2).

In early 2015, BASF exited the collaboration and Cargill acquired OPX Biotechnologies' proprietary fermentation-based processes and systems. Another partnership of Dow Chemical and OPX Biotechnologies (OPXBio) is also working on the development of bio-based acrylic acid via 3-hydroxypropionic acid. In the collaboration, OPX Biotechnologies is contributing their "Efficiency Directed Genome Engineering" (EDGE) platform as well as the 3-HPA bioprocess.

Route 3

Novomer is working on a process (Figure 3.54) using carbon monoxide (CO) with ethylene oxide (EO) to produce ß-propiolacton, using a proprietary catalyst. Bio-based ethylene oxide can be obtained from bio-ethanol and carbon monoxide can be obtained from *gasification* of biomass leading to syngas. Alternatively, CO_2 (from industrial gas production) can be converted to carbon monoxide using a solid oxide electrolysis process. ß-Propiolacton is then converted to either acrylic acid and subsequently polymerized to bio-PAA or ß-propiolacton can be polymerized via a *ring opening polymerization* (ROP) into polypropiolacton (bio-PPL). PPL is a *biodegradable polymer*.

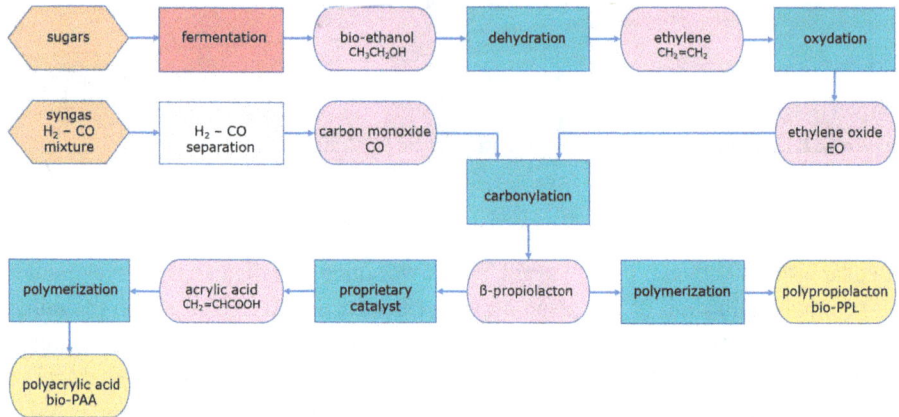

Figure 3.54: IEA process flow diagram for production of bio-PAA (route 3).

Route 4

The Arkema Group developed a process (Figure 3.55) to convert bio-based glycerol to acrylic acid via catalytic dehydration to acrolein followed by oxidation to acrylic acid. Both steps can also be covered in a single oxydehydration reaction.

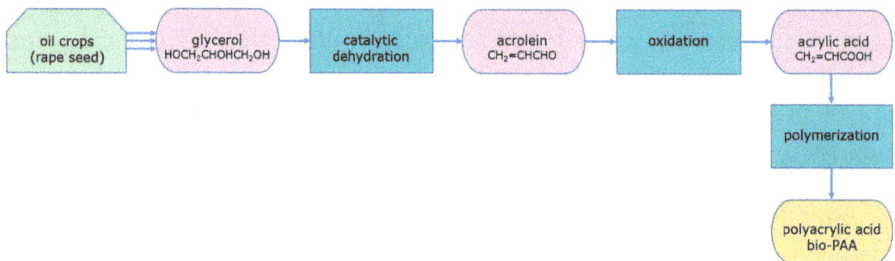

Figure 3.55: IEA process flow diagram for the production of bio-PAA (route 4).

Route 5

Metabolix is producing bio-based polypropiolacton (PPL, or so-called poly(3-hydroxypropionate), P3HP) using special engineered microbes. PPL can be heated (thermolysis, the FAST technology to chemical intermediates), and PPL vaporizes into acrylic acid. Metabolix's bio-based chemicals platform utilizes its novel "FAST" recovery process (abbreviation for "fast-acting, selective thermolysis") to enable the production of cost-effective, "drop-in" replacements for petroleum-based industrial chemicals (Figure 3.56).

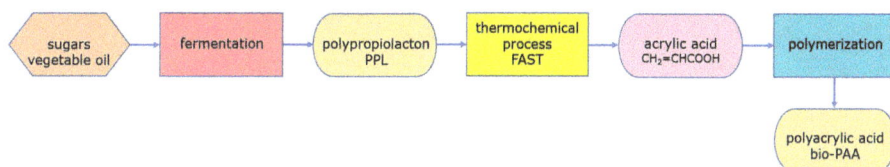

Figure 3.56: IEA process flow diagram for the production of bio-PAA (route 5).

Route 6

Both US-companies Myriant and SGA Polymers developed processes to produce bio-based acrylic acid from sugar-derived lactic acid (IUPAC name 2-hydroxypropanoic acid) via a dehydration step, as demonstrated in Figure 3.57.

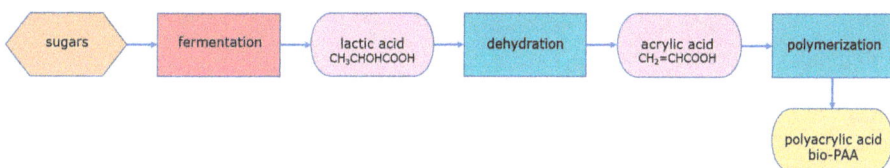

Figure 3.57: IEA process flow diagram for the production of bio-PAA (route 6).

Myriant officially changed its name to GC Innovation America in August 2018. GC Innovation America is a subsidiary of PTT (Thailand).

3.3.7 Polymethyl methacrylate (PMMA)

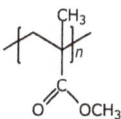

3.3.7.1 Fossil-based polymethyl methacrylate (PMMA)

Polymethyl methacrylate (PMMA), also known as *plexiglass* as well as by the trade names Plexiglass or Perspex, is a transparent thermoplastic polymer often used in sheet form as a lightweight or shatter-resistant alternative to glass. PMMA is routinely produced by emulsion polymerization, solution polymerization, or bulk polymerization from methyl methacrylate (MMA, $CH_2 = C(CH_3)COOCH_3$). PMMA, produced by *radical polymerization*, is atactic and completely amorphous.

Two principal routes appear to be commonly practiced to produce fossil-based MMA: the acetone cyanohydrin (ACH) route (Route 1) and the methyl propionate route (Route 2).

Route 1

ACH is produced by the condensation of acetone ($CH_3(CO)CH_3$) and hydrogen cyanide (HCN). In the next step, the cyanohydrin is hydrolyzed in the presence of sulfuric acid to a sulfate ester of methacrylamide. *Methanolysis* of this ester gives ammonium bisulfate (NH_4HSO_4) and, after an esterification with methanol, MMA. Although widely used, the ACH route (Figure 3.58) coproduces substantial amounts of ammonium sulfate and toxic and dangerous HCN is necessary. The ACH route accounts for roughly 60% of world's total MMA production capacity.

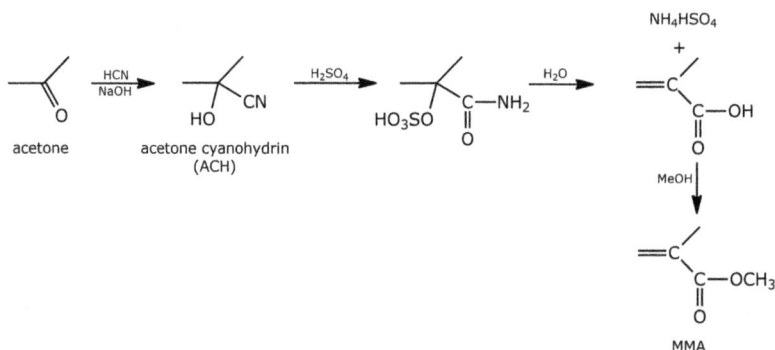

Figure 3.58: Cyanohydrin route to methyl methacrylate (MMA).

Mitsubishi Gas Chemical (MGC) has developed a recycle version of the ACH route (Figure 3.59) in which ACH is made as usual from acetone and HCN and is then hydrolyzed to α-hydroxyisobutyramide, which is reacted with carbon monoxide (CO) and methanol under pressure to yield formamide ($HCONH_2$) and methyl-α-hydroxyisobutyrate. The latter compound is dehydrated to MMA, while the co-product formamide is dehydrated to HCN for *recycling*. One commercial plant is operating in Japan.

Route 2

In the methyl propionate route, displayed in Figure 3.60, ethylene ($CH_2 = CH_2$) is used in a carboalkoxylation reaction with carbon monoxide (CO) and methanol (CH_3OH) to produce methyl propionate ($CH_3CH_2COOCH_3$). In the second stage, methyl propionate is condensed with formaldehyde (CH_2O) in a single step heterogeneous reaction to give MMA.

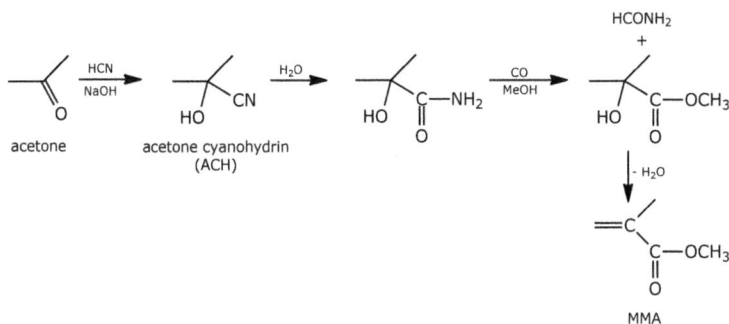

Figure 3.59: Mitsubishi Gas Chemical (MGC) route to methyl methacrylate (MMA).

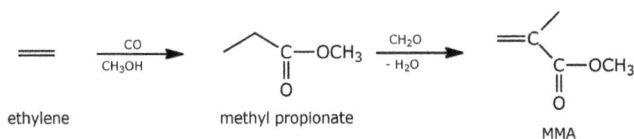

Figure 3.60: The methyl propionate route to methyl methacrylate (MMA).

The reaction of methyl propionate and formaldehyde takes place over a fixed bed of a special cesium oxide catalyst (Cs_2O).

Other routes are described in the literature to produce MMA from propionaldehyde (CH_3CH_2CHO), isobutyric acid (($CH_3)_2CHCOOH$), propyne, or isobutylene.

3.3.7.2 Bio-based polymethyl methacrylate (bio-PMMA)

Major players in the global synthetic and bio-based MMA (bio-MMA) market include BASF, Dow Chemicals, Arkema Group, Asahi-Kasei, Mitsubishi Rayon, and Evonik. Both bio-MMA and bio-based polymethyl methacrylate (bio-PMMA) are not yet commercially available. However, several routes are described in the literature. Two *fermentation* routes that might be of interest for industrial scale up are described next.

Route 1

In this process, sugars such as glucose, sucrose, or fructose, are fermented by using micro-organisms of the genus Aspergillus, and more especially those of *Aspergillus terreus* and *Aspergillus itaconicus*, to produce bio-based itaconic acid (bio-IA, see Table 3.2). In the next step, itaconic acid is decarboxylated to bio-based methacrylic acid (bio-MA). Esterification with methanol (CH_3OH) yields bio-MMA. Bio-PMMA is obtained in a radical polymerization from bio-MMA (Figure 3.61).

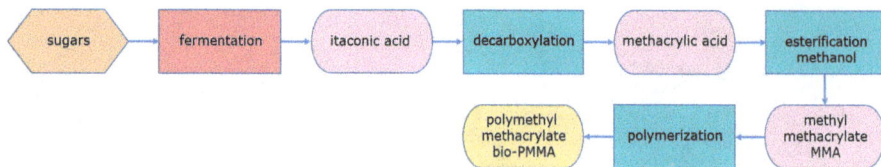

Figure 3.61: IEA process flow diagram for the production of bio-PMMA (route 1).

Route 2

In cooperation with Lucite International, Mitsubishi Rayon is developing a biotechnical process (Figure 3.62) for producing bio-based methacrylic acid (bio-MA) via fermentation of biomass with bacteria. Esterification yields bio-MMA. A patented process of Lucite International describes the biologically conversion of isobutyryl-CoA (isobutyryl-coenzyme A) into methacrylyl-CoA by the action of an oxidase and the conversion of methacrylyl-CoA into bio-based methacrylic acid (bio-MA).

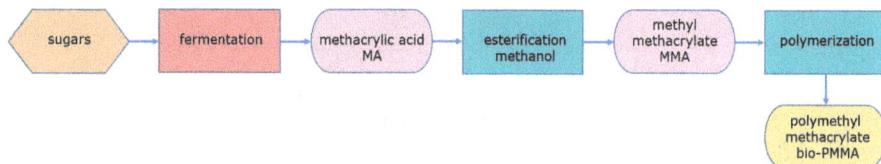

Figure 3.62: IEA process flow diagram for the production of bio-PMMA (route 2).

3.3.8 Polyacrylonitrile (PAN)

3.3.8.1 Fossil-based polyacrylonitrile (PAN)

Polyacrylonitrile (PAN), also known as polyvinyl cyanide, is a semi-crystalline organic polymer resin, produced from acrylonitrile ($CH_2 = CHCN$) monomer. Though it is thermoplastic, it does not melt under normal conditions. It degrades before melting. Almost all PAN resins are copolymers made from mixtures of monomers with acrylonitrile as the main monomer. It is a component in several important copolymers, such as styrene-acrylonitrile (SAN) or acrylonitrile-butadiene-styrene (ABS).

Acrylonitrile ($CH_2 = CHCN$) is produced by a catalytic ammoxidation of propylene ($CH_2 = CHCH_3$), also known as the SOHIO process (Figure 3.63). In the SOHIO process (developed by Standard Oil of Ohio), propylene, ammonia, and air (oxidant) are passed through a fluidized-bed reactor containing the catalyst (bismuth phosphomo-lybdate supported on silica as a heterogeneous catalyst).

$$2\ CH_2{=}CHCH_3 + 2\ NH_3 + 3\ O_2 \rightarrow 2\ CH_2{=}CHCN + 6\ H_2O$$

Figure 3.63: Ammoxidation of propylene.

In chemistry, ammoxidation is an industrial process for the production of nitriles using ammonia and oxygen using special catalysts.

3.3.8.2 Bio-based polyacrylonitrile (bio-PAN)

Various green chemistry routes are being developed for the synthesis of bio-based ac-rylonitrile (bio-AN) from renewable feedstocks.

Route 1

Glucose, for example obtained from *lignocellulosic biomass*, can be used in a fermenta-tion process to produce 3-hydroxypropionic acid (3-HPA). In the next step, 3-hydroxy-propionic acid (3-HPA) can be esterified with ethanol to ethyl 3-hydroxypropanoate (ethyl 3-HPA). Acrylonitril is obtained in high yield via dehydration and nitrilation with ammonia over an inexpensive titanium dioxide (TiO$_2$) solid acid catalyst (Figures 3.64 and 3.65).

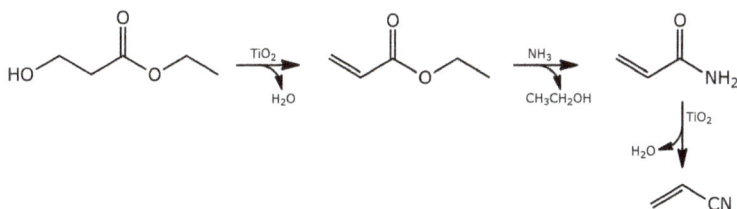

Figure 3.64: Synthesis of bio-based acrylonitrile from ethyl 3-HPA.

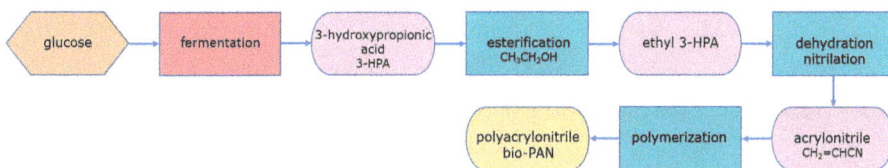

Figure 3.65: IEA process flow diagram for the production of bio-PAN (route 1).

Route 2

Glycerol can also be used as a renewable feedstock for the synthesis of bio-based acrylonitrile (bio-AN). Glycerol is a by-product generated in significant amounts during biodiesel production. Dehydration of glycerol yields 1-propen-1,3-diol. In the next step, tautomerization gives 3-hydroxypropionaldehyde (3-hydroxypropanal) and, subsequently, dehydration yields acrolein. In the ammoxidation reaction of acrolein with NH_3 and O_2 bio-based acrylonitrile (bio-AN) is produced. In Figure 3.66, the oxidative transformation of acrolein to bio-based acrylic acid (bio-AA) and bio-based methyl acrylate (bio-MA) are also shown.

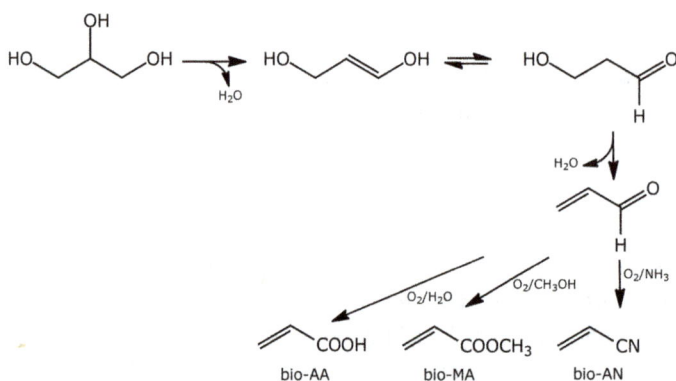

Figure 3.66: Synthesis of bio-based acrylonitrile from glycerol.

Polymerization of bio-based acrylonitrile (bio-AN) yields bio-based polyacrylonitrile (bio-PAN). The complete process is described in Figure 3.67.

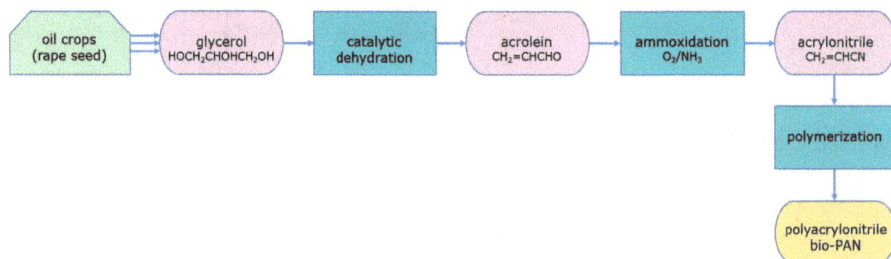

Figure 3.67: IEA process flow diagram for the production of bio-PAN (route 2).

Route 3

Glutamic acid can also be used in the synthesis of acrylonitrile. Glutamic acid is an α-amino acid that is used by almost all living beings in the biosynthesis of proteins. Glutamic acid is synthesized by the aerobic fermentation of sugars and ammonia, with the organism *Corynebacterium glutamicum* (also known as *Brevibacterium flavum*) being the most widely used for production. The carbon atom adjacent to the amino group is chiral (connected to four distinct groups), so glutamic acid can exist in two optical isomers, D(–) and L(+). L-glutamic acid is the most widely occurring in nature.

Oxidative decarboxylation of glutamic acid yields 3-cyanopropanoic acid. In a subsequent decarbonylation-elimination reaction of 3-cyanopropanoic acid ($HOOCCH_2CH_2CN$) bio-based acrylonitrile (bio-AN) is obtained, as shown in Figure 3.68.

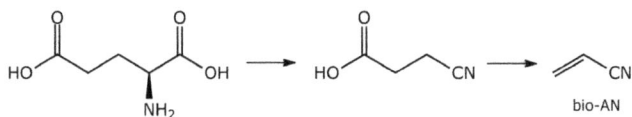

Figure 3.68: Synthesis of bio-based acrylonitrile from L-glutamic acid.

Of these three routes, the glycerol route (route 2) is broadly considered to be the most viable, although current methods are still unable to compete with the SOHIO process in terms of cost.

3.3.9 Polyethylene terephthalate (PET)

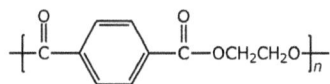

3.3.9.1 Fossil-based polyethylene terephthalate (PET)

Polyethylene terephthalate is commonly known as PET. It is a naturally transparent and semi-crystalline plastic used widely for products used in our day to day life. PET polymer is better termed as *"polyester"* in the textile industry. It is widely used as a fiber for clothing because it has an excellent moisture barrier, and it is also used for bottling and packaging on large scale. The major share of PET plastic produced comprises synthetic fiber (up to 60%) and remaining 30% is for *packaging* applications (e.g., PET bottles) of the total global demand. As a raw material, PET is regarded as a safe, strong, lightweight, flexible, and non-toxic material which can be easily recycled.

PET is produced by polymerization of ethylene glycol (EG) ($HOCH_2CH_2OH$) and TA ($HOOCC_6H_4COOH$). In general, an ester bond can be produced via a polycondensation reaction of a diol (HO-R-OH) and a dicarboxylic acid (HOOC-R-COOH), a hydroycarboxylic acid (HO-R-COOH) or via a ring opening polymerization of a lactone (a cyclic ester monomer). Several polymerization routes towards PET are discussed in Chapter 2 in more detail.

Fossil-based ethylene glycol (EG)

Ethylene glycol (EG) is also abbreviated as monoethylene glycol (MEG). The next diols in the series are diethylene glycol (DEG) and triethylene glycol (TEG), displayed in Figure 3.69.

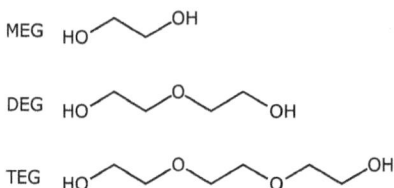

MEG HO⌒OH

DEG HO⌒O⌒OH

TEG HO⌒O⌒O⌒OH

Figure 3.69: Chemical structures of MEG, DEG, and TEG.

On a large scale, ethylene glycol (EG) is produced from ethylene ($CH_2 = CH_2$), as shown in Figure 3.70. In the first step, ethylene is oxidized to ethylene oxide (EO), and in the second step EO is hydrolyzed with water to produce EG. The highest yields of ethylene glycol are obtained at acidic or neutral pH with a large excess of water. In this way, ethylene glycol is obtained with yields of about 90%. Major by-products are DEG and TEG.

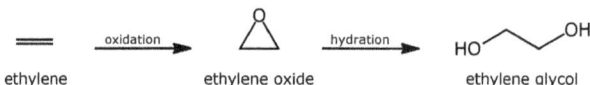

ethylene   →oxidation→   ethylene oxide   →hydration→   ethylene glycol

Figure 3.70: Industrial synthesis of ethylene glycol (EG) from ethylene.

A higher selectivity of ethylene glycol (EG) is obtained by using Shell's Omega process (Figure 3.71). In the process, ethylene oxide (EO) reacts with carbon dioxide (CO_2) to ethylene carbonate. In the next step, ethylene carbonate is hydrolyzed to produce ethylene glycol in 98% selectivity. The CO_2 released in this step can be fed back in the process.

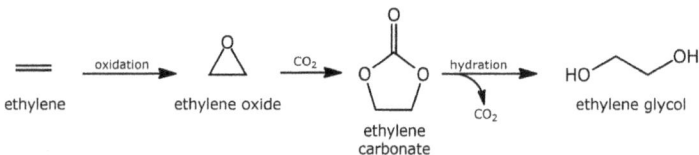

ethylene   →oxidation→   ethylene oxide   →CO_2→   ethylene carbonate   →hydration→   ethylene glycol

Figure 3.71: Industrial synthesis of ethylene glycol (EG) from ethylene via Shell's Omega process.

Today, more than 99% of ethylene glycol (EG) is produced from fossil resources, and the market demand for this product is expected to grow from 28 million to 50 million tons in the next 20 years.

Fossil-based terephthalic acid

In the past, TA ($HOOCC_6H_4COOH$) was produced by oxidation of p-xylene ($CH_3C_6H_4CH_3$) with dilute nitric acid (HNO_3). In another process, air oxidation of p-xylene gives p-toluic acid ($CH_3C_6H_4COOH$), which resists further air-oxidation. Conversion of p-toluic acid to methyl p-toluate ($CH_3C_6H_4COOCH_3$) opens the way for further oxidation to mono-methyl terephthalate ($HOOCC_6H_4COOCH_3$), which is further esterified with methanol to DMT ($CH_3OOCC_6H_4COOCH_3$). For this reason, DMT was in the early beginning the pre-ferred starting material for the production of PET (Figure 3.72).

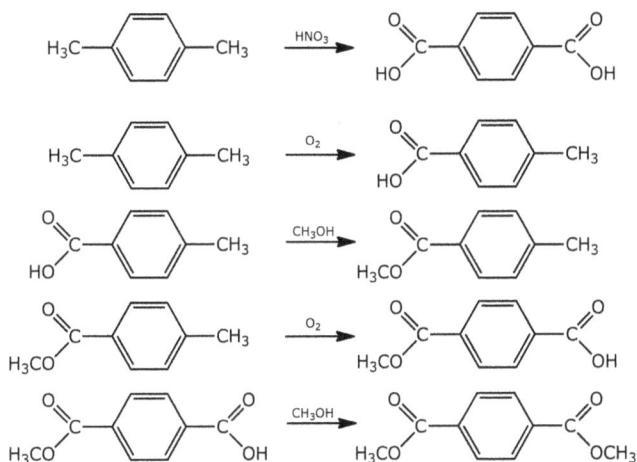

Figure 3.72: Industrial synthesis of terephthalic acid (TA) and dimethyl terephthalate (DMT).

In 1955, Mid-Century Corporation and ICI announced the bromide-promoted oxidation of p-toluic acid to TA. This innovation enabled the conversion of p-xylene to TA with-out the need to isolate intermediates (such as DMT). Amoco (as Standard Oil of Indi-ana) purchased the Mid-Century/ICI technology.

The Amoco process, which is widely adopted worldwide now, produces purified TA (PTA) by oxidation of p-xylene using oxygen in air. Therefore, nowadays, PTA is the preferred starting material for PET, most often in continuous production installa-tions. DMT is still used on small scale. To conclude, PET is produced from ethylene glycol together with DMT or PTA. The former is a transesterification reaction, whereas the latter is a direct esterification reaction (Figure 3.73).

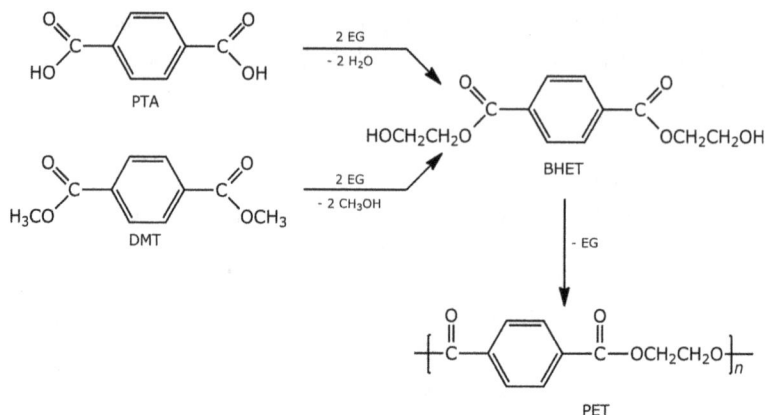

Figure 3.73: Synthesis of polyethylene terephthalate (PET).

3.3.9.2 Bio-based polyethylene terephthalate (bio-PET)

The global bio-based PET market is one that is growing fast. To produce a 100% bio-based PET, one needs both bio-based ethylene glycol (bio-EG) and bio-based terephthalic acid (bio-TA) (or bio-based dimethyl terephthalate (bio-DMT)).

A first step towards bio-based PET is the replacement of fossil-based ethylene glycol with bio-based ethylene glycol (bio-EG), which is discussed in the next paragraphs. Bio-TA is not yet commercially available. For that reason, partially bio-based polyethylene terephthalate (bioPET30) is available on the market. However, several global companies are involved in the development of bio-TA.

Bio-based ethylene glycol (bio-EG): Route 1

As described in Section 3.4.1, bio-based ethylene ($CH_2 = CH_2$) can be easily derived from bio-ethanol (CH_3CH_2OH) via a fermentation process starting from sugarcane. Bio-based ethylene can be oxidized to bio-based ethylene oxide (bio-EO) in the presence of a silver catalyst. In the next step, bio-based ethylene oxide (bio-EO) reacts with water to produce bio-based ethylene glycol (bio-EG). India Glycols has commercialized the production (Figure 3.74) of bio-EG from renewable agricultural resources. Note that India Glycols has been manufacturing bio-EG, derived from bio-ethanol, since 1989.

Greencol Taiwan Corporation (GTC) operates also a plant in Taiwan to manufacture bio-based ethylene glycol (bio-EG). GTC is a 50:50 joint venture between Nagoya-based Toyota Tsusho Corporation (TTC) and Taipei-based China Man-Made Fiber Corporation. GTC will use bio-ethanol made from sugarcane as feedstock. This ethanol will be secured by Toyota Tsusho from Petrobras (Brazil).

Using GTC's bio-based ethylene glycol (bio-EG), Toyota Tsusho Corporation (TTC) brings a partly bio-based PET to the market, trade-named Globio®. This PET contains 70% fossil-based TA and 30% bio-based ethylene glycol (bio-EG). So, in Globio®, the

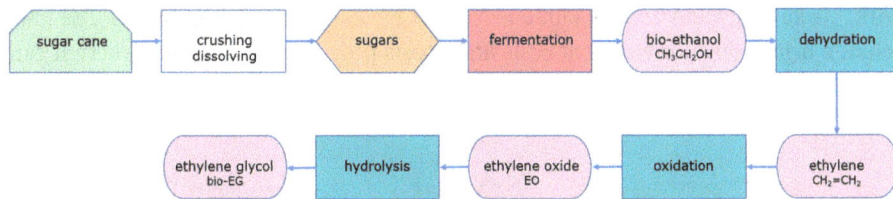

Figure 3.74: IEA process flow diagram for the production of bio-EG (route 1).

30% of ethylene glycol is replaced by bio-EG, made from sugarcane. Sometimes, this partially bio-PET is abbreviated as bioPET30. Bio-PET is similar in quality to fossil-based PET, but is more environmentally friendly (lower CO_2 footprint). Globio® is used in bottled mineral water packaging. This bottled mineral water is marketed by Japan-based Suntory Beverage & Food Ltd.

Also the Coca-Cola Company is using bioPET30 for their *PlantBottle®*. In 2009, the PlantBottle® was introduced on the market. Coca-Cola is also exploring the possibilities of using 2G feedstocks for the production of bio-EG (instead of sugarcane). Some years ago, Coca-Cola licensed the PlantBottle technology to H. J. Heinz.

Coca-Cola entered agreements with three biotechnology companies, including Virent, an energy technology company based in the USA, Gevo, a biofuel company based in the USA, and Avantium, a technology company based in the Netherlands, in December 2011 to commercially develop completely bio-based material for the next-generation PlantBottle packaging.

Coca-Cola distributed more than 35 billion of its 1G, 30% plant-based PlantBottle bottles in approximately 40 countries in June 2015, prior to the launch of its 100% plant-based PlantBottle packaging.

Teijin developed bio-PET (bioPET30) fibers under the brand ECO Circle™ Plant-Fiber. These fibers were selected for use in the seats and interior trim surface of the 100% electric Nissan LEAF automobile. The bio-PET fibers can also be recycled using Teijin's ECO Circle closed-loop polyester recycling system, where the polyester is chemically decomposed at the molecular level and then recycled as new DMT that offers purity and quality comparable to material derived directly from crude oil.

Bio-based ethylene glycol (bio-EG): Route 2
Both Braskem and Haldor Topsoe started up a demo-plant to develop bio-based ethylene glycol (bio-EG) from sugar (Figure 3.75). Haldor Topsoe is a world leader in catalysts and technology for the chemical and refining industries. In their newly developed process, feedstock glucose is introduced in the MOnoSAccharide Industrial Cracker (MOSAIK™ technology) and is converted in high yield to glycolaldehyde ($HOCH_2CHO$). Other C_1–C_3 oxygenates are formed in minor quantities (such as glyoxal). Hydrogenation of glycolaldehyde yields ethylene glycol. Current processes to produce ethylene glycol (EG) from biomass involve several steps. This can be reduced to two simple steps with MO-

SAIK™ technology and Topsoe's unique catalyst for the production of bio-EG. The new solution brings down investment costs and boosts productivity to a level where it can compete on commercial terms with traditional production from fossil feedstock (mostly naphtha).

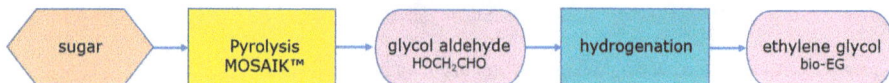

Figure 3.75: IEA process flow diagram for the production of bio-EG (route 2).

Bio-based ethylene glycol (bio-EG): Route 3

The Dutch-based company Avantium recently started construction of a new demonstration plant for the production of EG made directly via a 1-step direct *hydrogenolysis* of renewable *sugars*. Note that hydrogenolysis is a chemical reaction whereby a carbon–carbon or carbon–heteroatom single bond is cleaved or undergoes a breakdown by hydrogen. The heteroatom may vary, but it usually is oxygen, nitrogen, or sulfur. A related reaction is hydrogenation, where hydrogen is added to the molecule, without cleaving bonds. Usually hydrogenolysis is conducted catalytically using hydrogen gas.

The demonstration plant in Delfzijl, in the Netherlands, will use Avantium's Ray Technology™ to convert renewable sugars into bio-based ethylene glycol (bio-EG) (Figure 3.76). Avantium will use sugars from various 1G feedstocks such as sugar beet, sugarcane, wheat, and corn, as well as 2G non-food feedstock such as forestry or agricultural residues which is converted into sugars by Avantium's Dawn Technology™. This technology produces high-purity glucose and lignin from non-food biomass. This biorefinery is also located in Delfzijl.

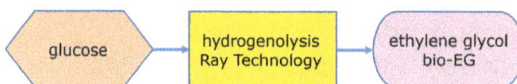

Figure 3.76: IEA process flow diagram for the production of bio-EG (route 3).

Instead of glycose, xylose can also be used. In the first step, depicted in Figure 3.77, xylose is hydrogenated (with H_2) into xylitol and, in the second step, hydrogenolysis yields ethylene glycol (EG) with small amounts of propylene glycol (PG). In the hydrogenolysis step, a special Cu-based catalyst is used for a high conversion.

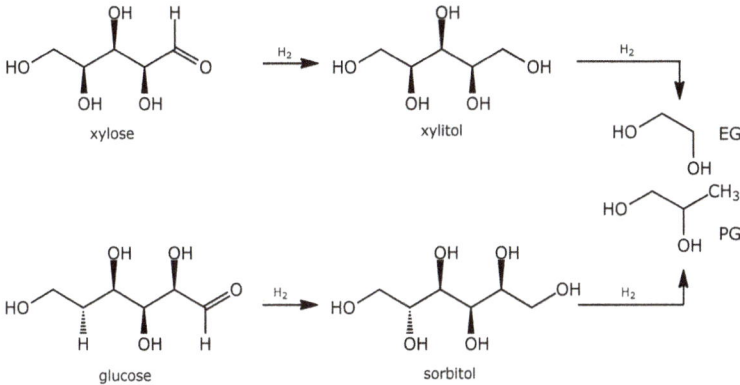

Figure 3.77: Hydrogenolysis of sugars (glucose or xylose).

Bio-based ethylene glycol (bio-EG): Route 4

Glycerol, obtained as a by-product in the biodiesel production process, can also be used in a selective hydrogenolysis process to produce ethylene glycol (EG) and propylene glycol (PG). The process is schematically presented in Figure 3.78.

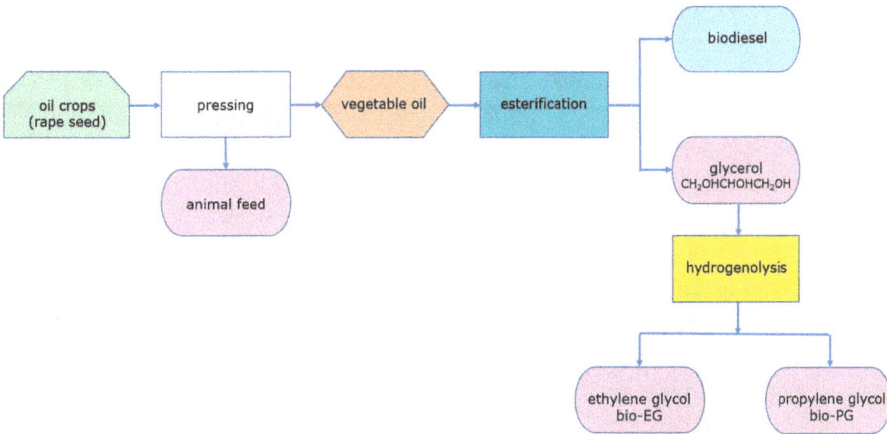

Figure 3.78: IEA process flow diagram for the production of bio-EG (route 4).

Bio-based terephthalic acid (bio-TA): Route 1

The former Draths Corp. disclosed a process based on genetically modified organisms (GMOs) that are produced by altering several genes in Escherichia coli. The GMOs produce *cis,cis*-muconic acid and *cis,trans*-muconic acid from non-aromatic, renewable carbon sources such as glucose. *Trans-trans* muconic acid can be obtained by using an isomerase enzyme. Via a Diels–Alder addition reaction of bio-based *trans,trans*-muconic acid and ethylene, 2-cyclohexene-1,4-dicarboxylic acid is obtained. This intermediate product is dehydrogenated in a consecutive aromatization reaction in the presence of a Pd/C catalyst to form TA (Figure 3.79). Draths Corp. was acquired by Amyris Biotechnologies in late 2011.

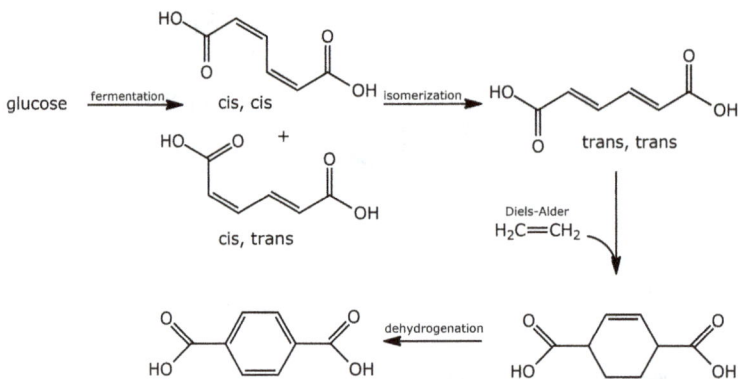

Figure 3.79: Synthesis of terephthalic acid (TA) via *trans,trans*-muconic acid (route 1).

Bio-based terephthalic acid (bio-TA): Route 2

In the past years, Virent has developed a new technology, called the BioForming™ process, that uses soluble C5/C6 carbohydrates (e. g., pentose and hexose sugars) and converts them into a BTX mixture (mixture of benzene, toluene, and xylenes), rich in *para*-xylene (pX). This mixture serves as an aromatic feedstock for separation and purification of *para*-xylene.

Virent's process requires the following steps: solubilizing sugar and introducing hydrogen for hydrodeoxygenation (HDO) of the sugars in the presence of a deoxygenation catalyst to produce a variety of oxygenated intermediates (oxygenates: such as alcohols, aldehydes, and ketones (the addition of hydrogen removes oxygen from the carbohydrates in the form of water)). Then, the oxygenated intermediates react over an acid condensation catalyst to produce an aromatics stream comprising benzene, toluene, xylenes, and ethyl benzene; and oxidation of *para*-xylene to TA. In fact, the existing infrastructure for fossil-derived TA can be used. In the Virent process (Figure 3.80), the hydrodeoxygenation reaction occurs between about 100 and 600 °C.

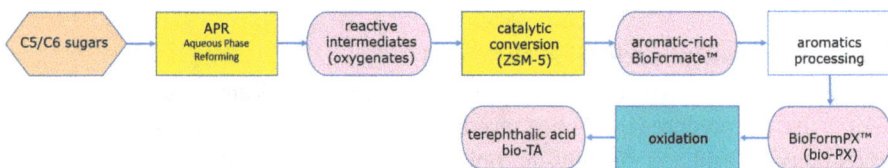

Figure 3.80: IEA process flow diagram for the production of bio-PX (route 2).

Virent is partnering with world class companies including Marathon Petroleum Corporation (MPC), Johnson Matthey, BP, and Toray Industries, on scaling-up and commercializing their technology. In early 2020, Toray announced to start the mass production of world's first 100% bio-based polyester (bioPET100). In addition, Far Eastern New Century (FENC) worked with Virent to convert the BioFormPX™ to bioPET100, and to produce 100% bio-based PET textile fibers, a fabric, and shirts. Virent is today a wholly owned subsidiary of MPC.

Bio-based terephthalic acid (bio-TA): Route 3

US-based Annelotech has developed an approach (Figure 3.81) to produce a BTX aromatics stream directly from lignocellulosic biomass in a single-stage reactor. Anellotech's bio-based aromatics production technology is called Bio-TCat. This technology uses a thermochemical catalytic fast pyrolysis (CFP) approach for rapid heating of the biomass under anaerobic conditions and immediate conversion of the resulting gases to aromatics using zeolite-based catalysts in a fluidized-bed reactor. The Bio-TCat™ technology produces a liquid product containing over 98% C6 + aromatic chemicals directly from the pretreated feedstock. After mild hydro-treating, separation, and purification, AnelloMate™ products – the family of liquid products made through Bio-TCat™ – meet all specifications. Reaction products are separated, and the *p*-xylene is oxidized to TA (bio-based TA) using the existing infrastructure for fossil-based TA.

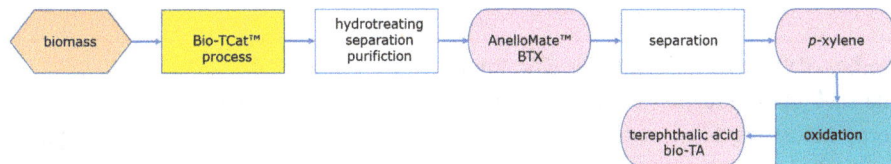

Figure 3.81: IEA process flow diagram for the production of bio-TA (route 3).

Recently, Anellotech introduced their Plas-TCat™ technology which transforms mixed plastic waste, including multilayer food packaging and other non-PET waste plastics, directly into functional chemicals.

Bio-based terephthalic acid (bio-TA): Route 4

Gevo has developed a new technology to produce *p*-xylene directly from isobutanol without going through the BTX route and associated separation processes to remove *p*-xylene from the BTX mixture. Isobutanol can be produced from C5/C6 sugars through a fermentation process using proprietary microorganisms. From a carbon efficiency perspective, one or two carbons from the sugar are lost during fermentation to CO_2. The Gevo process uses the following steps to go from isobutanol to *p*-xylene (Figure 3.82): dehydration of isobutanol to isobutylene; dimerization to C8 diisobutylenes; dehydrocyclization producing *p*-xylene; and oxidation of *p*-xylene to TA, using the existing infrastructure for fossil-derived TA.

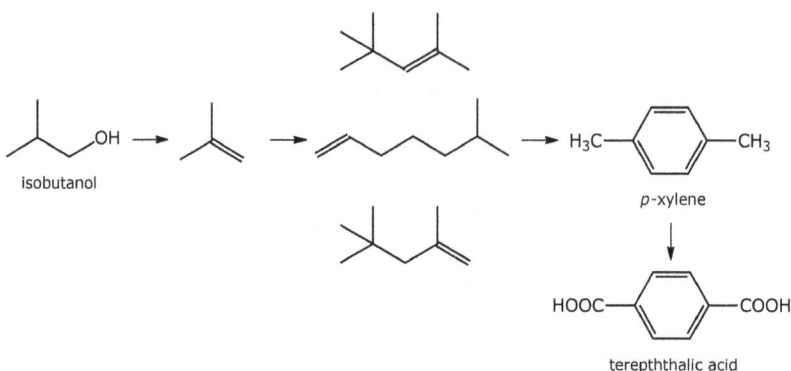

Figure 3.82: Synthesis of terephthalic acid from isobutanol (route 4).

In a joint cooperation, Gevo supplied bio-based p-xylene to Toray. Toray is one of the world's leading producers of fibers, plastics, films, and chemicals. Using bio-TA, (synthesized from Gevo's bio-based p-xylene), and commercially available bio-EG, Toray succeeded in lab-level PET polymerization to produce 100% bio-based PET (bioPET100) fibers and films.

Bio-based terephthalic acid (bio-TA): Route 5

The *Diels–Alder reaction* of 2,5-methylfuran (DMF) with ethylene has been proposed by Toray Industries, Micromidas, and UOP (Universal Oil Products, a Honeywell company) as part of the pathway to make bio-based p-xylene (Figure 3.83). The overall DMF technology from fructose to bio-TA includes the following five steps: dehydration of fructose to 5-hydroxymethylfurfural (HMF), hydrodeoxygenation of HMF to DMF, Diels–Alder cycloaddition of DMF and ethylene (as the dienophile) to produce 1,4-dimethyl-7-oxa-

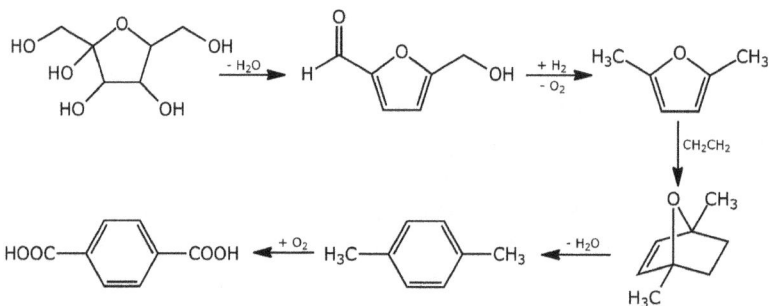

Figure 3.83: Synthesis of bio-based terephthalic acid from fructose (route 5).

bicyclo[2,2,1]hept-2-ene (a bicyclic ether intermediate), dehydration of the bicyclic ether to *p*-xylene, and oxidation of *p*-xylene to TA, using the existing infrastructure for fossil-based TA. Note that fructose can be derived from glucose via isomerization using enzymatic or chemical catalysis.

Bio-based terephthalic acid (bio-TA): Route 6
The Dutch-based company BioBTX transforms non-food biomass and plastic waste into BTX using the Integrated Cascading Catalytic Pyrolysis (ICCP) technology. First, the feedstock is heated to a high temperature inside a reactor in a process called *pyrolysis*. The vapors released in this process are transferred to a second reactor where they are directly transformed into aromatics by catalytic conversion. Finally, the liquid BTX product is collected. BTX is a mixture of benzene, toluene, and xylenes. These are important building blocks for several new products, including bio-TA. Using this bio-TA, BioBTX demonstrated the production of 100% bio-based PET (bioPET100) in a polycondensation with bio-based ethylene glycol (bio-EG). The 100% bio-based PET was used to produce a cosmetic container.

3.3.10 Polyethylene furanoate (PEF)

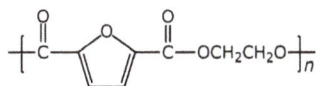

Polyethylene furanoate (PEF) is typically produced by polymerization of ethylene glycol (EG) ($HOCH_2CH_2OH$) and 2,5-furandicarboxylic acid (FDCA). PEF exists since 1951 and has gained renewed attention since the US department of Energy (US DOE) proclaimed its chemical *building block*, FDCA, as a potential bio-based replacement for fossil-based TA. Together with bio-based ethylene glycol (bio-EG), the production of a 100% bio-based polyester is among the possibilities.

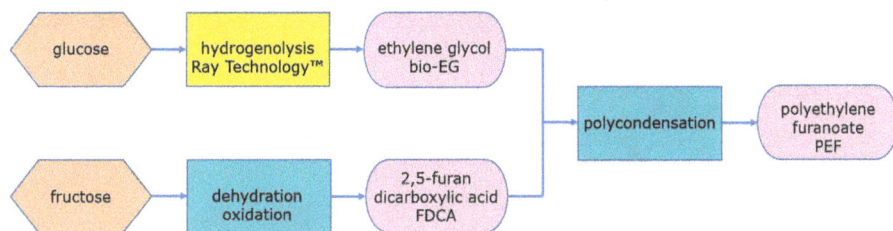

Figure 3.84: IEA process flow diagram of bio-based polyethylene furanoate (PEF).

Dutch-based Avantium developed the YXY technology. In this process (Figure 3.84), plant-based sugar (fructose) is converted into a wide range of plant-based chemicals (including 2,5-furandicarboxylic acid (FDCA)) and plastics (PEF). PEF is 100% bio-based and 100% recyclable with superior performance properties compared to well-known fossil-based packaging materials. Avantium's Ray Technology™ can be used in the supply of bio-based ethylene glycol (bio-EG). In bottles, PEF shows improved barrier properties for carbon dioxide and oxygen, leading to a longer shelf life of packaged products. It also offers higher mechanical strength, which means that thinner PEF packaging can be produced and fewer resources are required. In combination with the plant-based feedstock, the added functionality gives PEF all the attributes required to become the next-generation polyester, superior to the conventional PET.

3.3.11 Polytrimethylene terephthalate (PTT)

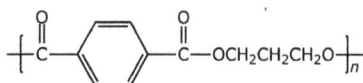

3.3.11.1 Fossil-based polytrimethylene terephthalate (PTT)

Similar to PET, polytrimethylene terephthalate (PTT) is prepared by the *direct esterification* of 1,3-propanediol ($HOCH_2CH_2CH_2OH$) with TA ($HOOCC6H4COOH$), or by *transesterification* with DMT ($CH_3OOC_6H_4COOCH_3$). See Section 3.3.9.1 for more information about those monomers.

Fossil-based 1,3-propanediol (1,3-PDO)

Degussa used to produce 1,3-propanediol via acrolein (Figure 3.85). This process, a cross-aldol condensation, involves the condensation of formaldehyde and acetaldehyde into acrolein. In the next step, acrolein is hydrated to give 3-hydroxypropionaldehyde which is further hydrogenated to 1,3-propanediol (1,3-PDO). After being taken over by RAG in 2006, Degussa was incorporated into Evonik Industries as Evonik Degussa.

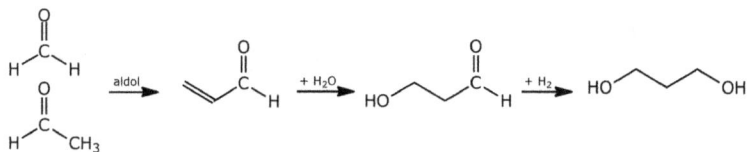

Figure 3.85: Degussa process to produce 1,3-PDO.

In the Shell process, shown in Figure 3.86, ethylene oxide (EO) is converted in a two-step process, involving a hydroformylation, into 3-hydroxypropionaldehyde which is further hydrogenated to 1,3-PDO.

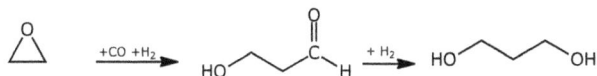

Figure 3.86: Shell process to produce 1,3-PDO.

Fossil-based terephthalic acid (TA)
As described in detail in Section 3.3.9.1, TA is produced on a large scale from crude oil.

3.3.11.2 Bio-based polytrimethylene terephthalate (bio-PTT)
To produce a 100% bio-based polytriethylene terephthalate (PTT), one needs both bio-based 1,3-propanediol (bio-1,3-PDO) and bio-TA (or even bio-based dimethyl terephthalate (bio-DMT)).

Bio-based terephthalic acid (bio-TA)
As indicated in Section 3.3.9.1, bio-TA is not yet commercially available, although serious investments have been made in research and development (R&D) to develop an economic process, starting from biomass feedstocks.

Bio-based 1,3-propanediol (bio-1,3-PDO)
DuPont Tate & Lyle BioProducts developed a fermentation process (Figure 3.87) with selected micro-organisms to produce bio-based 1,3-propanediol (bio-1,3-PDO). Conversion from corn syrup effected by a genetically modified strain of *E. coli* by DuPont Tate & Lyle BioProducts yields two grades of 1,3-propanediol: Susterra® and Zemea®. Susterra® propanediol is the building block that delivers high performance in a variety of applications, from polyurethanes and unsaturated polyester resins to heat-transfer fluids. Zemea® propanediol is the multifunctional, preservative-boosting humectant and the ingredient that delivers high performance in a variety of consumer applications, from cosmetics and personal care to food, flavor, and pharmaceuticals, and laundry and household cleaning.

DuPont is using Susterra® propanediol in the production of Sorona®. Sorona® is the polyester of bio-based 1,3-propanediol (obtained by fermentation) and fossil-based TA or DMT. Sorona® has been used in the manufacture of clothing, residential carpets, and automotive fabrics and plastic parts. Mohawk Industries is currently the exclusive North American carpet manufacturer making carpets using DuPont Sorona fiber. Sometimes, Sorona® is abbreviated as bioPTT37 indicating that Sorona® is partially (37%) bio-based.

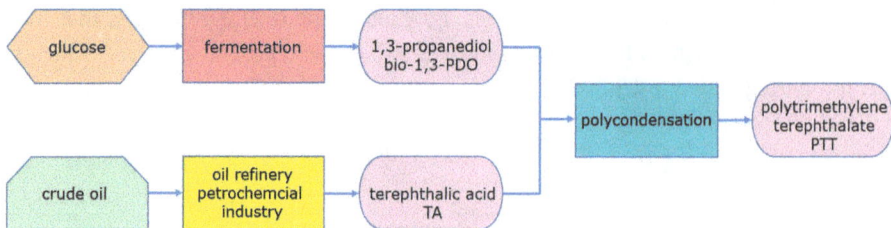

Figure 3.87: IEA process flow diagram for the production of (bio-)PTT.

3.3.12 Polybutylene terephthalate (PBT)

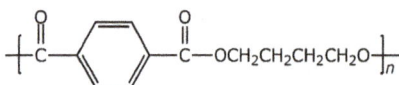

3.3.12.1 Fossil-based Polybutylene terephthalate (PBT)

Similar to PET, *polybutylene terephthalate* (PBT) is prepared by the *direct esterification* of 1,4-BDO ($HOCH_2CH_2CH_2CH_2OH$) with TA ($HOOCC_6H_4COOH$), or by *transesterification* with DMT ($CH_3OOC_6H_4COOCH_3$). See also Section 3.3.9.1. Currently, regular PBT resin is produced using fossil-based ingredients.

Fossil-based 1,4-butanediol (1,4-BDO): Route 1

1,4-BDO is a diol with four carbon atoms. 1,4-BDO is used in the manufacture of several types of plastics, such as polyurethanes, polybutylene terephthalate (PBT), and thermoplastic polyetherester (TPEE) elastomers. In the petrochemical industry, 1,4-BDO can be produced from acetylene (Reppe process, Figure 3.88). Acetylene ($HC \equiv CH$) reacts with two equivalents of formaldehyde (H_2CO) to form 1,4-butynediol ($HOCH_2C \equiv CCH_2OH$). *Hydrogenation* of 1,4-butynediol gives 1,4-BDO.

$$HC{\equiv}CH \ + 2\ CH_2O \longrightarrow HOCH_2{-}C{\equiv}C{-}CH_2OH \longrightarrow HOCH_2CH_2CH_2CH_2OH$$

Figure 3.88: Synthesis of 1,4-butanediol from acetylene (route 1).

Fossil-based 1,4-butanediol (1,4-BDO): Route 2

1,4-BDO can also be manufactured from maleic anhydride (Davy process, Figure 3.89). Maleic anhydride (MAN), obtained via an oxidative reaction from butane, is esterified to form a dimethyl maleate (DMM) which is then hydrogenated in the vapor phase to produce dimethyl succinate (DMS). Further hydrogenation yields γ-butyrolactone (GBL) and then 1,4-BDO.

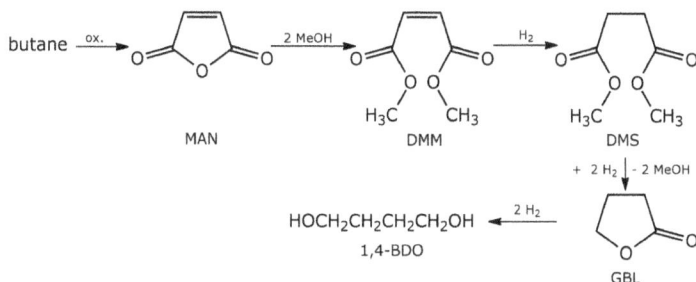

Figure 3.89: Synthesis of 1,4-butanediol from butane (route 2).

The Reppe process covers 42% of the global production capacity of 1,4-BDO, the maleic anhydride process 28%, the propylene oxide process 20%, and the butadiene process 7%.

Fossil-based terephthalic acid (TA)

As described in detail in Section 3.3.9.1, TA is produced on a large scale from crude oil.

3.3.12.2 Bio-based polybutylene terephthalate (bio-PBT)

To produce a 100% bio-based polybutylene terephthalate (bio-PBT), one needs both bio-based 1,4-BDO (bio-1,4-BDO) and bio-TA or even bio-DMT.

Bio-1,4-BDO can be a direct drop-in replacement for fossil-based 1,4-BDO. Bio-based 1,4-BDO production can either take place via direct fermentation of sugars or via the hydrogenation of bio-based succinic acid (bio-SA, $HOOCCH_2CH_2COOH$).

Bio-based 1,4-butanediol (bio-1,4-BDO): Route 1

Genomatica is a US-based biotech company that has developed the GENO BDO™ process, which uses a dedicated engineered micro-organism, for bio-based 1,4-BDO production directly via fermentation of sugars. The 1,4-BDO biosynthetic pathway, used in *E. coli* production strains, is presented in Figure 3.90. Genomatica's process technology for 1,4-BDO is now commercial. The GENO BDO process has been licensed by BASF and Novamont.

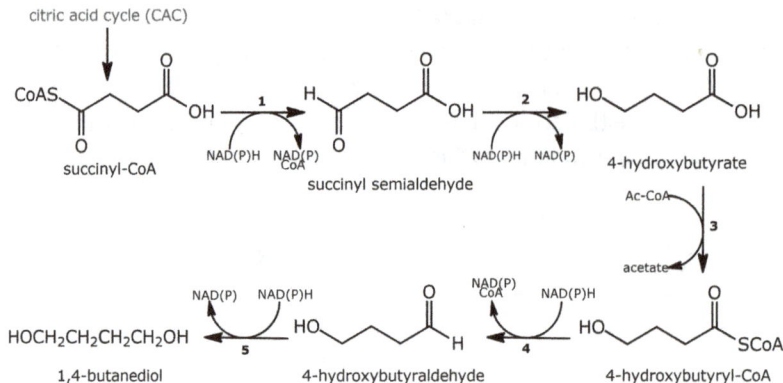

Figure 3.90: Synthesis of bio-based 1,4-butanediol from sugars (route 1).

Several companies (BASF and Novamont) are active in fermenting 1,4-BDO directly from *sugars*, based on Genomatica's GENO BDO™ process (Figure 3.91).

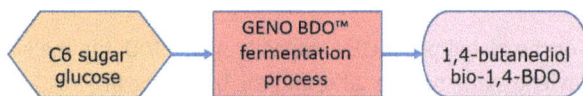

Figure 3.91: IEA process flow diagram for the production of bio-1,4-BDO (route 1).

In 2016, Novamont started production of bio-based 1,4-BDO (capacity 30 ktons). The 1,4-BDO is produced from sugars derived from the *hydrolysis* of starch, the so-called glucose syrup. Bio-1,4-BDO is produced through a single-step fermentation by a metabolically engineered strain of an *E. coli* type bacteria developed by Genomatica. Novamont will use the bio-based 1,4-BDO internally to produce bioplastics (including polybutylene-*co*-adipate-*co*-terephthalate (PBAT)).

Both DSM and Toray have approved Genomatica's bio-based 1,4-BDO for their PBT polyester product lines. In this way, partially bio-based PBT could be produced because TA is still not yet commercially available. Both companies stated that the resulting partially bio-based PBT has equivalent properties to PBT made from fossil-based 1,4-BDO.

Bio-based 1,4-butanediol (bio-1,4-BDO): Route 2
GC Innovation America, the former Myriant, developed a yeast fermentation process where sugars are fermented into succinic acid ($HOOCCH_2CH_2COOH$). The JM Davy Technologies (JM Davy) can be applied to produce bio-based 1,4-BDO (see Figure 3.92). GC Innovation America is a subsidiary of PTT Global Chemical Public Company Ltd., the leading Thai chemicals company.

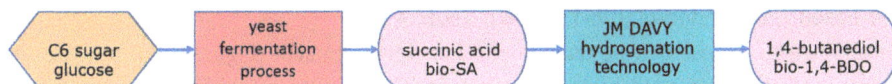

Figure 3.92: IEA process flow diagram for the production of bio-1,4-BDO (route 2).

Also, the former BioAmber developed a process to produce bio-based 1,4-BDO from their bio-based succinic acid using a hydrogenation technology license with JM Davy Technologies (JM Davy). JM Davy is the global leader in 1,4-BDO and THF technology. BioAmber officially ceased operations in August 2018.

Bio-based 1,4-butanediol (bio-1,4-BDO): Route 3

Bio-ethanol can be used to produce in a few steps bio-based 1,3-butadiene ($CH_2 = CH_2CH_2 = CH_2$). In the first step, bio-ethanol is dehydrogenated to acetaldehyde. In an aldol-condensation, acetaldol is obtained from acetaldehyde. Dehydration of acetaldol, followed by a Meerwein–Ponndorf–Verley reduction yields crotyl alcohol. In a final dehydration step of crotyl alcohol, 1,3-butadiene is obtained. Figure 3.93 gives the reaction scheme to 1,3-butadiene from bio-ethanol.

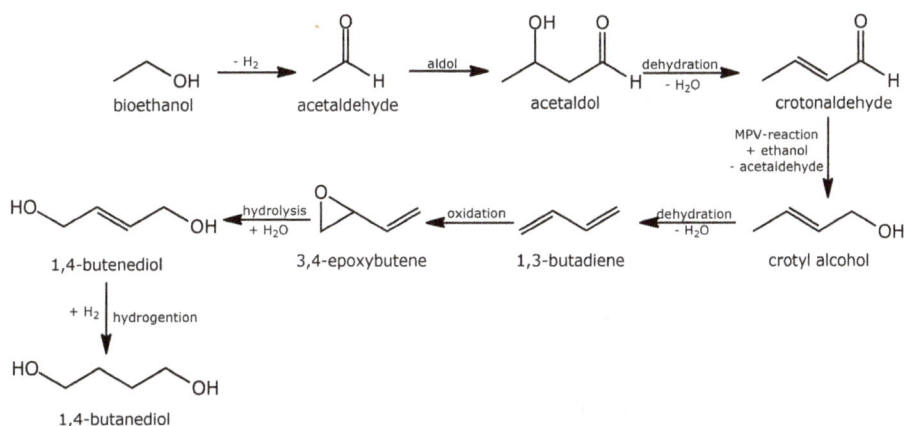

Figure 3.93: Synthesis of bio-based 1,3-butadiene from bio-ethanol (route 3).

In the next step, 1,3-butadiene is oxidized to 3,4-epoxybutene (EPB), based on Eastman Chemical patents. Then EPB is hydrolyzed to 1,4-butenediol. Subsequently, 1,4-butenediol is hydrogenated to 1,4-butanediol. A by-product of the process is 1,2-butanediol.

Bio-based 1,4-butanediol (bio-1,4-BDO): Route 4

In this route, glucose can be transferred into levulinic acid (Figure 3.94). In a subsequent oxidation step with hydrogenperoxide (H_2O_2), succinic acid is obtained which can be reduced to bio-1,4-BDO.

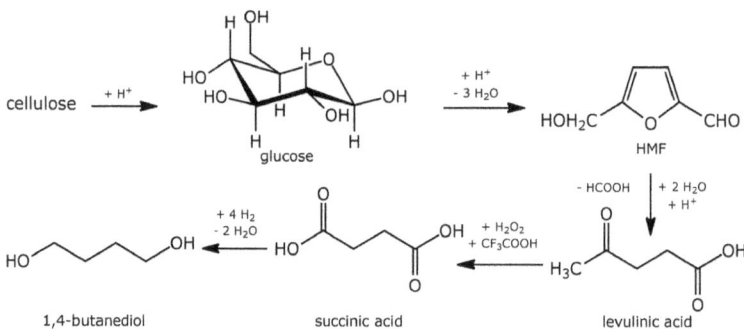

Figure 3.94: Synthesis of bio-based 1,4-butanediol from cellulose (route 4).

Bio-based terephthalic acid (bio-TA)

As indicated in Section 3.3.9.1, bio-TA is not commercially available yet, although serious investments have been done in research and development (R&D) to develop an economic process, starting from biomass feedstocks.

3.3.13 Thermoplastic polyetherester (TPEE)

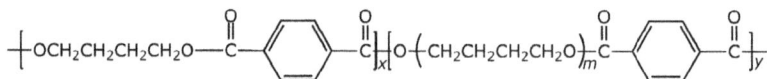

3.3.13.1 Fossil-based thermoplastic polyetherester (TPEE)

Thermoplastic polyetheresters (TPEEs) are a class of copolyesters that consist of materials with both thermoplastic and elastomeric properties. These materials are also abbreviated as TPC-ET, a thermoplastic polyester elastomer. The benefit of using thermoplastic polyester elastomers is the ability to be stretch moderately and return to nearly the original shape, creating a longer life and better physical range than other materials. In general, thermoplastic polyetheresters (TPEEs) consist of a hard crystalline segment of polybutylene terephthalate (PBT) and a soft amorphous segment based on a polyether glycol, such a poly(tetramethylene ether)glycol (PTMG, $HO-(CH_2CH_2CH_2CH_2O)_n-H$). PTMG is a linear polyether glycol, based on tetrahydrofuran (THF), with hydroxyl groups on both ends. For that reason, PTMG is sometimes abbreviated as PTHF (polytetrahydrofuran). Therefore, PTHF can be viewed as a polymer of tetrahydrofuran or as the polyether derived from 1,4-BDO.

As an alternative, polyethylene glycol (PEG, $HO-(CH_2CH_2O)_n-H$) or polypropylene glycol (PPG, $HO-(CH_2CH(CH_3)O)_n-H$) can be used instead of the soft segment PTMG ($HO-(CH_2CH_2CH_2CH_2O)_n-H$). PTMG is commercially available in molecular weights between 250 (PTMG250) up to 4,000 (PTMG4000). PTMG is sold under various trade names including Terathane® from Invista and PolyTHF® from BASF.

Production of thermoplastic polyetheresters (TPEEs) occurs in two steps: in a first reactor the transesterification reaction is carried out between DMT and 1,4-BDO (in excess). The transesterification product is transferred to a second reactor and, in the second stage, the polycondensation reaction is carried out by adding a polyether glycol, such as PTMG, in the presence of the polycondensation catalyst TBT (tetrabutyl titanate). The reaction scheme is given in Figure 3.95.

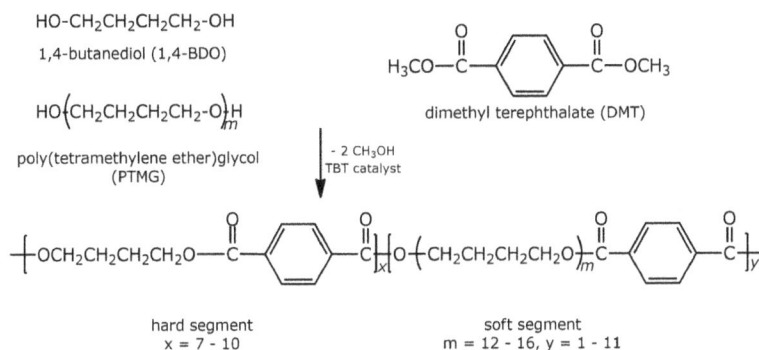

Figure 3.95: Synthesis of thermoplastic polyetheresters (TPEEs).

Commercial thermoplastic polyetheresters (TPEEs), based on PBT/PTMG, are, for example, special grades Hytrel™ from DuPont, Pelprene™ from Toyobo, Arnitel™ from DSM, and Riteflex™ from Celanese.

Sympatex™, produced by Sympatex Technologies, is a PBT/PEG thermoplastic polyetherester. Due to the hydrophilic properties of polyethylene glycol (PEG), a Sympatex™ membrane has unique properties in outdoor clothing, shoes, and many other applications. Due to the closed Sympatex™ membrane, water cannot penetrate, but water vapor molecules are transported through the membrane from the inside to the outside by way of an absorption and evaporation process. This moisture transfer through the membrane is what is referred to as the breathability of the fabric.

3.3.13.2 Bio-based thermoplastic polyetherester (bio-TPEE)

BASF produces bio-based poly(tetramethylene ether)glycol (bio-PTMG) with a molecular weight of 1,000 (PolyTHF® 1000), which is derived from bio-based 1,4-BDO (bio-1,4-

BDO). The bio-based 1,4-BDO has been produced under license from Genomatica. Recently, Mitsubishi Chemical Corporation (MCC) announced the development of bio-PTMG that is manufactured from bio-based feedstocks.

DSM, however, evaluated Arnitel PBT/PTMG grades (Figure 3.96) where the existing fossil-based 1,4-BDO in the PBT hard segment has been replaced by the bio-based 1,4-BDO (bio-1,4-BDO). Bio-based 1,4-BDO was produced with Genomatica's process technology.

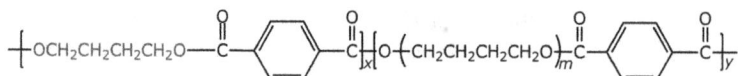

Figure 3.96: Chemical structure of Arnitel PBT/PTMG grade with bio-1,4-BDO.

3.3.14 Isosorbide-based polymers

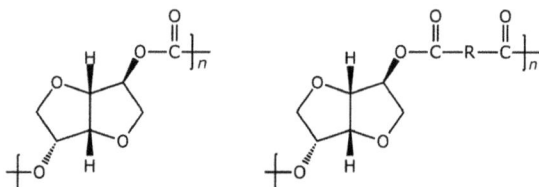

3.3.14.1 Bio-based isosorbide

Isosorbide (1,4:3,6-dianhydro-D-glucitol) is a bicyclic diol containing two fused furan rings. The starting material for isosorbide is D-sorbitol, which is obtained by catalytic hydrogenation of D-glucose, which is in turn produced by hydrolysis of *starch*. Isosorbide is currently of great scientific and technical interest as a monomer building block for new bio-based polymers, including polycarbonates (PCs, left) and polyesters (right).

In fact, isosorbide can be produced from D-glucose, obtained from cellulose or starch by using an acid or enzymatic hydrolysis. In Figure 3.97, both the open-chain representation (Natta projection) of D-glucose and the cyclic α-D-glucopyranose form (Haworth projection) are shown. D-glucose can be hydrogenated to D-sorbitol. Via an intramolecular dehydration, both 1,4-D-sorbitan (1,4-anhydrosorbitol) and 3,6-D-sorbitan (3,6-anhydrosorbitol) are obtained. In a further and subsequent dehydration, isosorbide is obtained.

The France-based company Roquette is the leading world producer of high-purity isosorbide, brand-named POLYSORB®.

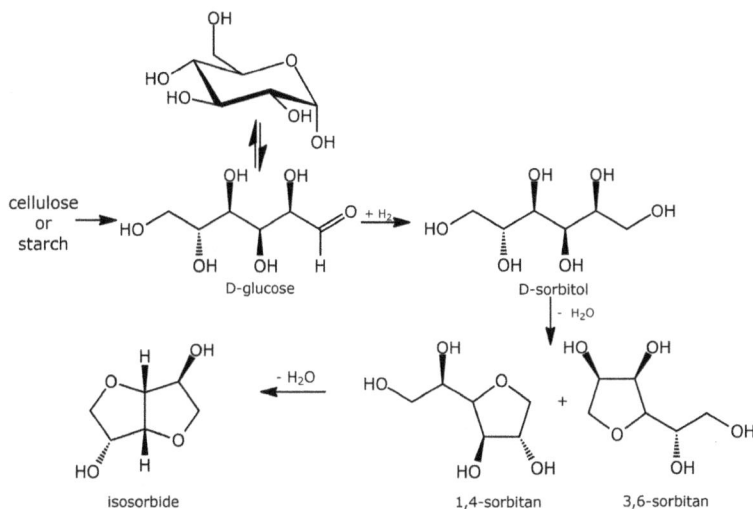

Figure 3.97: Synthetic pathway for the production of isosorbide from biomass.

3.3.14.2 Bio-based polyester, based on isosorbide

When ethylene glycol (EG) is replaced with isosorbide (I) in PET, polyisosorbide tere-phthalate (PIT) is obtained, which is characterized by an extreme thermal stability. Clearly, due to the presence of bio-based isosorbide, PIT is a partially bio-based poly-ester (Figure 3.98).

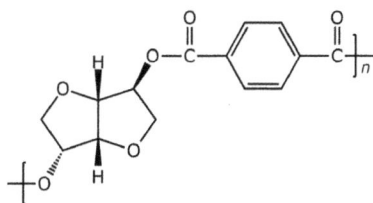

Figure 3.98: Chemical structure of polyisosorbide terephthalate (PIT).

However, the inherently lower reactivity of both secondary hydroxyl groups in isosor-bide cause lower molecular weights of the resulting polymers. Therefore, major appli-cations of isosorbide now are the incorporation as a comonomer in PET (abbreviated as PEIT, polyethylene-*co*-isosorbide-*co*-terephthalate, shown in Figure 3.99). Due to the rigid structure of isosorbide, the incorporation in the polyester main chain results in an increase of the glass-transition temperature (T_g), better heat resistance, and better optical properties (clarity). Due to the higher Tg's of the copolyesters, applications can be found in the area of hot-fill containers, optical discs, films, and fibers.

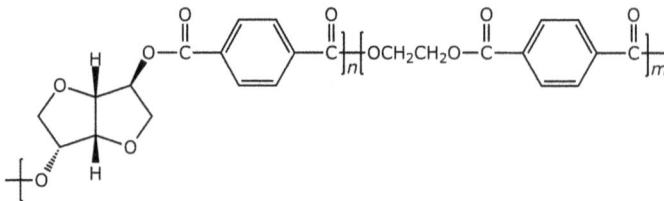

Figure 3.99: Chemical structure of polyethylene-*co*-isosorbide-*co*-terephthalate (PEIT).

3.3.14.3 Bio-based polycarbonate, based on isosorbide

Isosorbide is of particular interest as a monomer for bio-based aliphatic PCs (Figure 3.100). In general, PC is prepared from biophenol A, which was identified as xenoestrogen. Copolycarbonates based on isosorbide (Figure 3.101) have been recently developed by Mitsubishi Chemical under the trade name Durabio™ and by Teijin under the trade name Planext™.

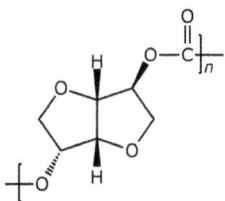

Figure 3.100: Chemical structure of a polycarbonate, based on isosorbide.

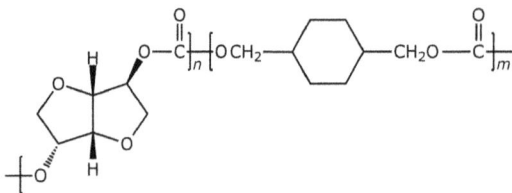

Figure 3.101: Copolycarbonate based on isosorbide and 1,4-cyclohexanedimethanol.

All PCs produced with isosorbide were reported to be amorphous and transparent. It has also been reported that the impact strength of the homopolymer with isosorbide is lower compared with copolymers, using, e. g., 1,4-cyclohexanedimethanol (CHDM) or bisphenol A (BPA). Clearly, Durabio™ combines most of the advantageous properties of PC and those of the polymethacrylate (PMMA).

3.3.15 Polyamide 6 (PA6)

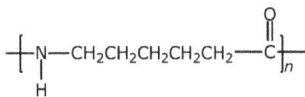

3.3.15.1 Fossil-based polyamide 6 (PA6)

Polyamide 6 (or nylon 6 or polycaprolactam) is a semi-crystalline *polyamide*, prepared by *ring opening polymerization* starting from caprolactam (a cyclic amide).

PA6 is prepared from benzene in a few steps. In the chemical industry, benzene is obtained from crude oil in a steam-cracking process. Subsequently, benzene is hydrogenated to cyclohexane in the presence of a Raney nickel catalyst. In the next step, a mixture of cyclohexanone and cyclohexanol ("KA"-oil, ketone-alcohol oil) is produced by the oxidation of cyclohexane in air, typically using cobalt catalysts. As an alternative, cyclohexanol can also be obtained from the hydrogenation of phenol. The mixture of cyclohexanone and cyclohexanol can be used in a nitric acid (HNO_3) oxidation to adipic acid (AA), feedstock for polyamide 6,6 (PA6,6) production. Otherwise, the mixture can also be used to produce cyclohexanone oxime (using NH_2OH). Using sulfuric acid (H_2SO_4), in a Beckmann rearrangement, cyclohexanone oxime is converted into caprolactam. Subsequently, in a (continuous) polymerization reaction caprolactam is processed into polyamide 6 (PA6). The complete synthesis scheme is depicted in Figure 3.102.

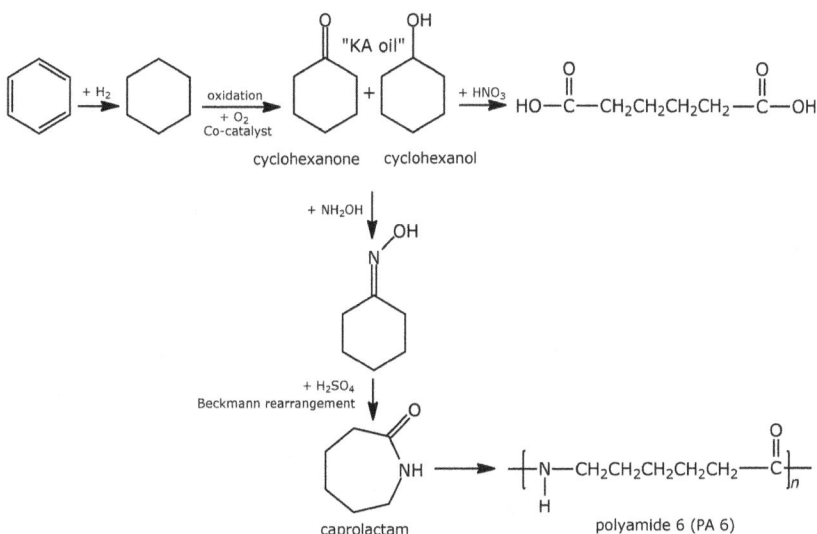

Figure 3.102: Synthesis of fossil-based PA6 from benzene.

PA6 can be modified using comonomers or stabilizers during polymerization to introduce new chain-end or functional groups, which changes the reactivity and chemical properties. At present, PA6 is the most significant construction material used in many industries, for instance, in the automotive industry, aircraft industry, electronic and electrotechnical industry, textile and carpet industry, and medicine.

3.3.15.2 Bio-based polyamide 6 (bio-PA6)

Route 1

Bio-based PA6 is not yet commercially available. However, several initiatives have been reported to produce bio-based caprolactam (bio-CPL). Genomatica, also involved in the development of bio-1,4-BDO, also reported a new fermentation technology (Figure 3.103), applying an integrated bioengineering platform (GENO-CPL™ technology), to produce bio-based caprolactam (bio-CPL) from using plant-based renewable ingredients, rather than the crude oil-derived materials traditionally used by the PA6 industry.

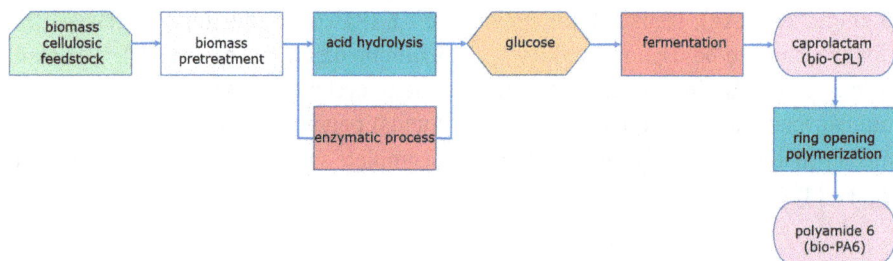

Figure 3.103: IEA process flow diagram for production of PA6 (route 1).

Recently, Genomatica and Aquafil announced a multi-year cooperation to create sustainable caprolactam. The Italian-based Aquafil is a leading producer of polyamide 6 (PA6) with expertise in converting caprolactam to the synthetic polymer PA6. Bio-PA6 synthetic fibers can be spun from the bio-PA6 polymer.

Bioamber was a Canadian sustainable chemicals company. Together with Japanese-based Mitsui, Bioamber developed a new industrial platform that uses biotechnology to produce bio-based chemical building blocks for the production of polyamide 6 (PA6) and polyamide 6,6 (PA6,6). In this platform, sugar is used to produce AA, hexamethylene diamine (HMD), and caprolactam (CPL), for use in everyday products such as engineering plastics and synthetic fibers. BioAmber officially ceased operations in August 2018.

Route 2

This alternative route describes a production process to bio-based caprolactone from an agricultural residue (*cellulose*) via glucose, fructose, 5-hydroxymethyl furfural

(HMF), THF dimethanol, 1,2,6-hexanetriol and 1,6-hexanediol. The reaction scheme is presented in Figure 3.104.

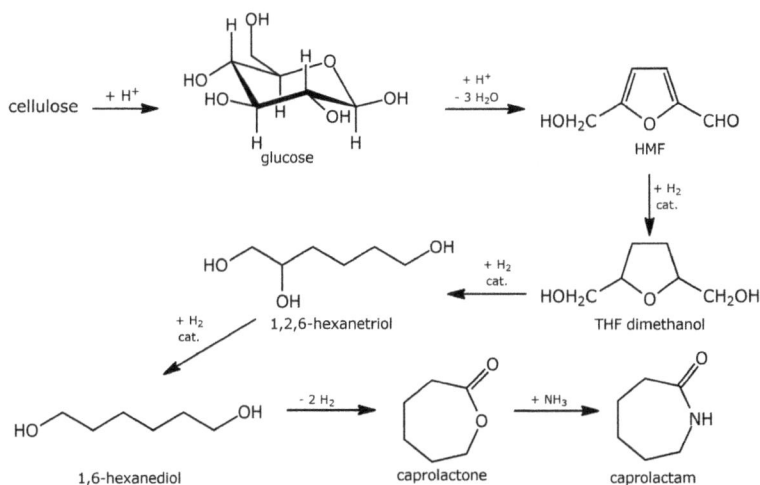

Figure 3.104: Synthesis of bio-based caprolactam from cellulose (route 2).

Fibrant, a global producer and supplier of high-quality caprolactam, recently announced the development of EcoLactam® Bio. EcoLactam® Bio is Fibrant's biomass-balanced caprolactam where the carbon is verifiably replaced by renewable resources, as certified by ISCC PLUS. EcoLactam® Bio is a high-quality caprolactam produced using an oil-like product derived from bio-based waste streams.

3.3.15.3 Bio-based polyamide 11 (bio-PA11)

Polyamide 11 (PA11) or nylon 11 is a polyamide, a bioplastic, and a member of the nylon family of polymers produced by the polymerization of an aminocarboxylic acid (H_2N-R-COOH). In the case of polyamide 11, the starting bio-based monomer is 11-aminoundecanoic acid. This bio-based monomer, vegetable castor oil, is produced from castor beans by pressing castor beans. Castor oil is a triglyceride in which approximately 90% of the fatty acid chains are ricinoleates. Oleate and linoleates are the other significant fatty acid components. Ricinoleic acid is a mono-unsaturated 18-carbon fatty acid with a hydroxyl functional group on the 12th carbon atom.

Transesterification of castor oil followed by *steam cracking* gives methyl 11-undecanoate, the precursor to 11-aminoundecanoic acid, and heptanal, a component in fragrances. The complete reaction scheme to 11-aminoundecanoic acid is given in Figure 3.105.

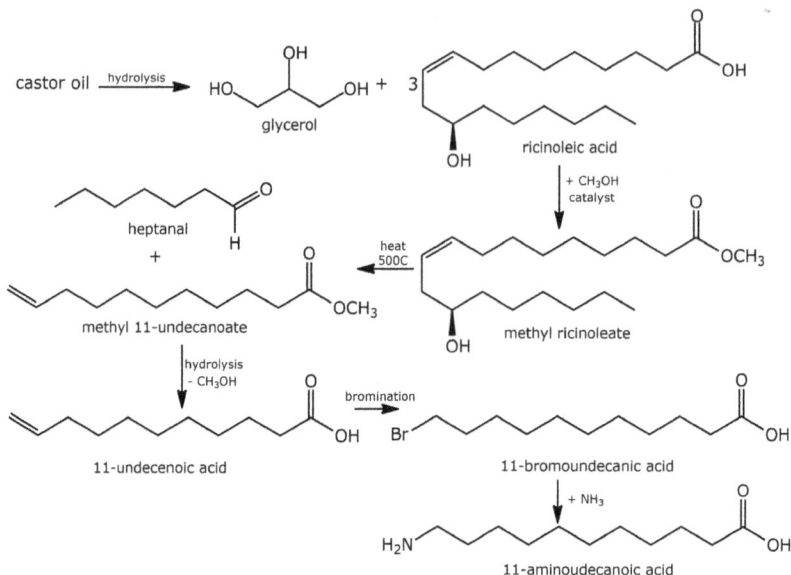

Figure 3.105: Reaction scheme to 11-aminoundecanoic.

The Arkema Group supplies bio-based polyamide 11 (bio-PA11) under the trade name Rilsan® 11 on the market. Polyamide 11 is applied in the fields of automotive, textiles, electronics, and sports equipment, frequently in tubing and wire sheathing.

3.3.15.4 Bio-based polyamide 12 (bio-PA12)

Like PA11, polyamide 12 (PA12) or nylon 12 is a polyamide and a member of the nylon family of polymers produced by the polymerization of an aminocarboxylic acid (H_2N-R-COOH). In the case of polyamide 12, the starting monomer is laurolactam. Laurolactam is synthesized from fossil-based 1,3-butadiene ($CH_2 = CHCH = CH_2$) in some steps via cyclododecanone. The complete process scheme is depicted in Figure 3.106. Ring opening polymerization (ROP) of laurolactam is used to polymerize this monomer to polyamide 12.

The Arkema Group supplies fossil-based polyamide 12 (PA12) under the trade name Rilsamid® 12 on the market. Rilsamid® 12 is mainly applied in the fields of fuel transfer solutions and braking systems. Evonik also brings polyamide 12 (PA12) to the market, which it branded Vestamid® L. Evonik also makes use of laurolactam (produced from 1,3-butadiene) to produce polyamide 12 (PA12).

When bio-based 1,3-butadiene ($CH_2 = CHCH = CH_2$) becomes available on the market in large quantities, this would provide the opportunity to produce fully bio-based polyamide 12 (bio-PA12).

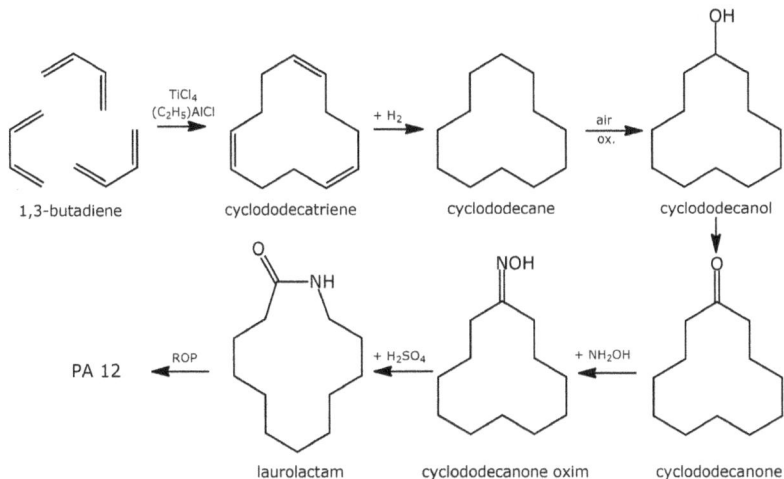

Figure 3.106: Synthesis of PA12 from 1,3-butadiene.

3.3.16 Polyamide 6,6 (PA6,6)

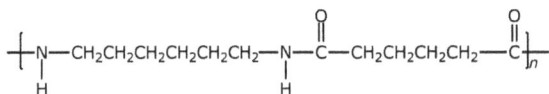

3.3.16.1 Fossil-based polyamide 6,6 (PA6,6)

Polyamide 6,6 (PA6,6) or nylon 6,6 is a type of polyamide prepared from two monomers: HMD (a 6 carbon aliphatic diamine) and AA (a six-carbon aliphatic dicarboxylic acid).

Polyamide 6,6 is synthesized by a polycondensation reaction of HMD and AA, as explained in Chapter 2. Equivalent amounts of HMD and AA are combined with water in a reactor. This is crystallized to make a *nylon salt* (so-called AH-salt, an ammonium/carboxylate mixture). The nylon salt goes into a reaction vessel where the polymerization process takes place either in batches or continuously. Polyamide 6,6 is frequently used when high mechanical strength, rigidity, good stability under heat, and/or chemical resistance are required. It is used in fibers for textiles and carpets and injection-molded parts.

HMD ($H_2N(CH_2)_6NH_2$) can be produced via a hydrogenation process of adiponitril, diluted with ammonia (NH_3) or via a Raney nickel *hydrogenation* of adiponitril. Adiponitril itself is produced by the nickel-catalyzed hydrocyanation of 1,3-butadiene, as discovered at DuPont or a hydrodimerization, starting from acrylonitrile, discovered at Monsanto.

AA ($HOOC(CH_2)_4COOH$), or hexanedioic acid, is produced from a mixture of cyclohexanone and cyclohexanol called "KA oil", the abbreviation of ketone-alcohol oil.

The KA oil is oxidized with nitric acid to give AA, via a multistep pathway. See also the reaction scheme in Figure 3.107.

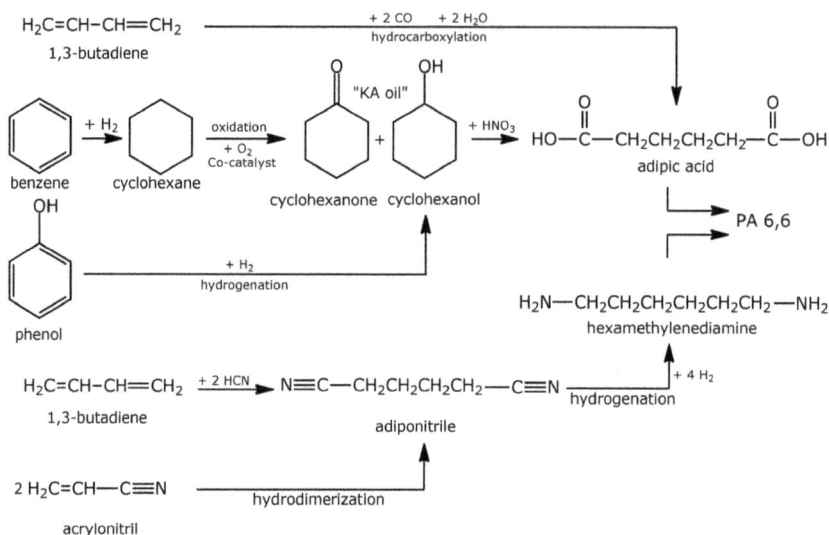

Figure 3.107: Synthesis of fossil-based polyamide 6,6.

3.3.16.2 Bio-based polyamide 6,6 (bio-PA6,6)

Route 1
Bio-based polyamide 6,6 (bio-PA6,6) is not yet commercially available on the market. However, several initiatives have been reported to produce bio-based hexamethylenediamine (bio-HMD) and bio-based AA (bio-AA). Genomatica, involved in the development of bio-1,4-BDO and bio-based caprolactam (bio-CPL), also reported a new fermentation technology, applying an integrated bioengineering platform (GENO technology), to produce both bio-based starting materials for the production of bio-based polyamide 6,6 – by using plant-based renewable ingredients, rather than the crude oil-derived materials traditionally used.

Route 2
Rennovia has been able to produce both AA (bio-AA) and hexamethylenediamine (bio-HMD) using glucose for feedstock by using their proprietary chemical catalytic process technology. Rennovia produces its bio-AA by a selective aerobic oxidation of glucose to produce glucaric acid, which then undergoes selective hydrogenation to produce AA. For Rennovia's bio-based HMD, it is produced by hydrodeoxygenation of glucose to produce 1,6-hexanediol (HDO), which then undergoes a chemocatalytic pro-

cess (a direct amination) to produce bio-HMD. Aerobic oxidation of HDO represents an alternative route to bio-AA. Rennovia announced that it has produced and shipped samples of their RENNLON™100% bio-based polyamide 6,6 polymer (Figure 3.108). The bio-based polyamide was made from the combination of Rennovia's RENNLON™ AA and RENNLON™ hexamethylenediamine (HMD). Despite those efforts, Rennovia ceased their operations in 2018.

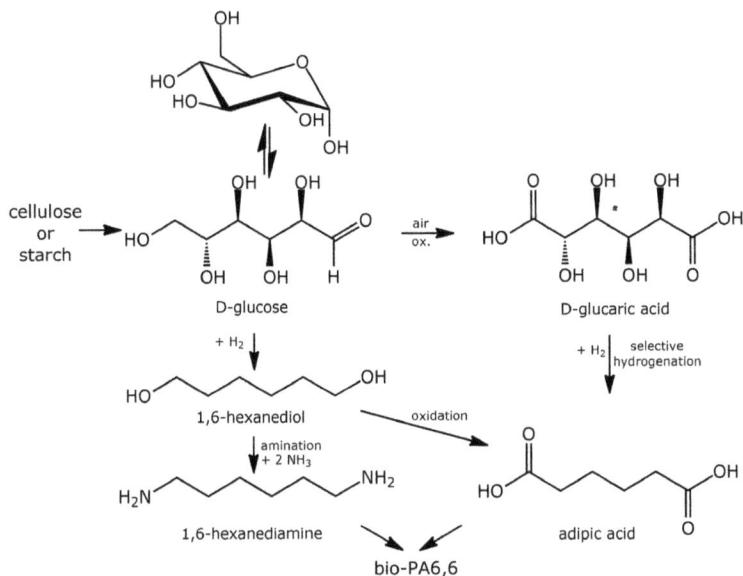

Figure 3.108: Synthesis of bio-based polyamide 6,6 (bio-PA6,6) from glucose (route 2).

3.3.16.3 Other Bio-based polyamides (bio-PAx,y)

Polyamide 6,6 (PA6,6) is a well-known aliphatic polyamide. In the past several years, a whole range of bio-based or partially bio-based aliphatic *polyamides* were introduced onto the market, based on newly developed (bio-based) diamines (H_2N-R_x-NH_2) or (bio-based) dicarboxylic acids (HOOC-R_y-COOH). Note that in polyamide PA(x,y), x indicates the number of carbon atoms in the diamine, and y indicates the total number of carbon atoms in the dicarboxylic acid. Some examples are presented in Table 3.6 and will be discussed in the remaining chapter sections.

1,4-Diaminobutane (DAB)

1,4-Diaminobutane (DAB, tetramethylenediamine, putrescine) is a 4-carbon-based aliphatic diamine. Bio-based 1,4-diaminobutane (bio-DAB) can be produced from bio-based succinic acid (bio-SA) via a reductive amination (Figures 3.109 and 3.110). Several companies are involved in the development and production of bio-based succinic acid (bio-SA).

Table 3.6: PAx,y raw materials.

Diamines	$H_2N\text{-}(CH_2)_x\text{-}NH_2$
Tetramethylenediamine 1,4-diaminobutane (DAB) Putrescine	$x = 4$
Pentamethylenediamine 1,5-diaminopentane (DAP) Cadaverine	$x = 5$
Decamethylenediamine 1,10-diaminodecane (DAD)	$x = 10$
Dicarboxylic acids	$HOOC\text{-}(CH_2)_y\text{-}COOH$
Azelaic acid 1,7-heptanedicarboxylic acid	$y = 7$
Sebacic acid 1,8-octanedicarboxylic acid	$y = 8$
1,10-dodecanedicarboxylic acid	$y = 10$

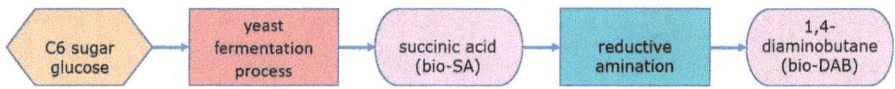

Figure 3.109: IEA process flow diagram for the production of bio-based 1,4-diaminobutane.

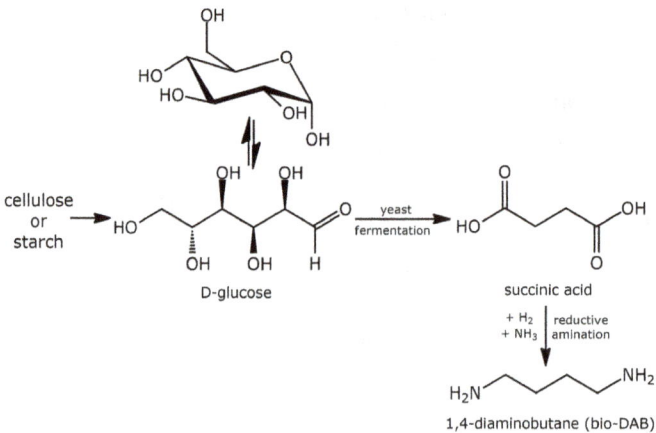

Figure 3.110: Synthesis of bio-based 1,4-diaminobutane from glucose.

Biotechnological production of 1,4-diaminobutane (bio-DAB) from renewable feedstock is also a promising alternative to the chemical synthesis (Figure 3.111). A metabolically engineered strain of *Escherichia coli* has been described that produces 1,4-diaminobutane by fermentation of glucose. An alternative route to bio-based 1,4-diaminobutane (bio-DAB) starts from the *α*-amino acid glutamine (Gln). Glutamine can be directly produced by microbial fermentation processes. Mutants of *Brevibacterium flavum* can be used with glucose as a carbon source.

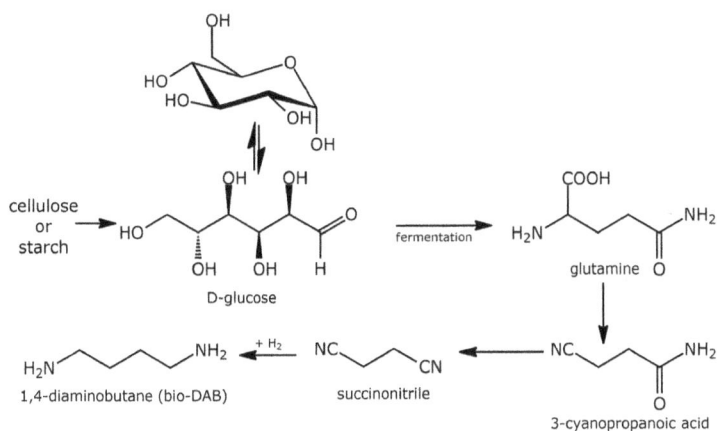

Figure 3.111: Synthesis of bio-based 1,4-diaminobutane from glucose.

Based on fossil-based 1,4-diaminobutane (DAB) and fossil-based AA, DSM developed polyamide 4,6 (PA4,6). The product, distributed by DSM as Stanyl®, has been demonstrated to possess mechanical and physical properties comparable, or even superior, to those of polyamide 6,6 (PA6,6). In addition, DSM also introduced polyamide 4,10 (PA4,10) onto the market. Polymerization of fossil-based 1,4-diaminobutane (DAB), together with bio-based sebacic acid, a 10-carbon dicarboxylic acid derived from castor plant oil, yields the 70% bio-based polyamide 4,10. The polymer is distributed on the market with the brand name EcoPaXX®. EcoPaXX® has a high melting point and high crystallization rate and is used, among other applications, as an engineering plastic in the automotive industry. In future, when bio-based 1,4-diaminobutane (bio-DAB) becomes commercially available, this gives the opportunity to introduce the diamine in both Stanyl® or EcoPaXX® (and make it 100% bio-based). Both sports specialist Salomon and DSM developed new light mountaineering shoes based on DSM's EcoPaXX®.

1,5-Diaminopentane (DAP)
1,5-Diaminopentane (DAP, pentamethylenediamine, cadaverine) is a 5-carbon-based aliphatic diamine. In contrast to 1,4-diaminobutane (DAB), there is no efficient petrochemi-

cal production route available yet. However, several process routes have been developed based on the enzymatic decarboxylation of lysine, using L-lysine decarboxylase.

Only recently, Cathay Industrial Biotech introduced a direct fermentation process to produce 1,5-diaminopentane (bio-DAP) from renewable resources. 1,5-diaminopentane (bio-DAP) is a key building block to produce a new series of 100% bio-based polyamides, including PA5,4 (using bio-based succinic acid) and polyamide PA5,10 (using bio-based sebacic acid). Clearly, a partly bio-based polyamide PA5,6 can be used by using AA. The 1,5-diaminopentane-based polyamides (bio-PA5,x) are the basis for newly developed textile and industrial yarn fibers, brand-named TERRYL®.

1,10-Diaminodecane (DAD)

1,10-Diaminodecane (DAD, decamethylenediamine) is a C10 carbon bio-based aliphatic diamine. Caustic fusion of ricinoleic acid produces the 10-carbon bio-based sebacic acid and 2-octanol as a by-product. In the next step, sebacic acid can be converted into bio-based 1,10-diaminodecane (bio-DAD). Sebacid is exposed to gaseous ammonia, so that the diammonium salt of sebacic acid (diammonium sebacate) is formed. In the next dehydration step, sebaconitrile ($NC(CH_2)_8CN$, decanedinitrile) is obtained which upon hydrogenation in the presence of potassium hydroxide (KOH) and a nickel catalyst yields 1,10-diaminodecane (bio-DAD), as shown in Figure 3.112.

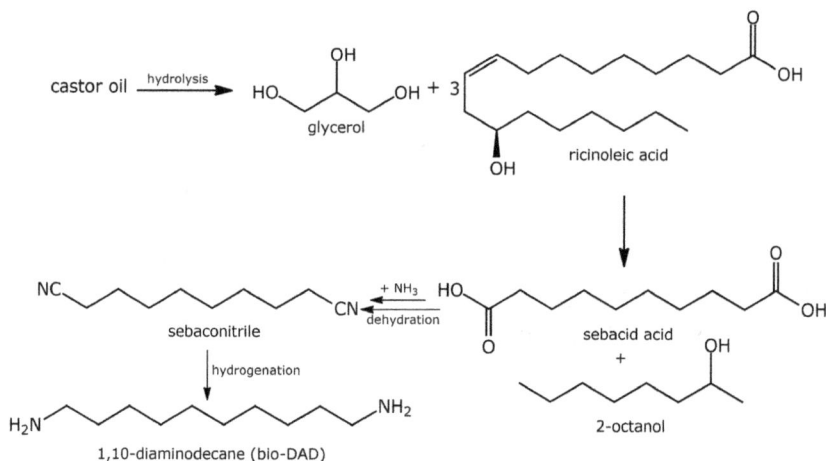

Figure 3.112: Synthesis of bio-based 1,10-diaminodecane (bio-DAD) from castor oil.

Together with the 10-carbon bio-based sebacic acid, polyamide 10,10 can be produced. Arkema introduced Rilsan T onto the market, being a PA10,10, produced from 100% renewable resources. In addition, Evonik makes use of bio-based 1,10-diaminodecane (bio-DAD) to produce both polyamide 10,10 (Vestamid Terra DS) and polyamide 10,12 (Vestamid Terra DD). Note that Vestamid Terra DS is 100% bio-based because of the use of bio-based sebacid acid. On the contrary, Vestamid Terra DD is partly bio-based (45% renewables) because the 1,10-dodecanedicarboxylic acid (C12) is still fossil-based. When applying bio-based 1,10-dodecanedicarboxylic acid (C12) it would be possible to produce the polyamide 10,12 from 100% renewable resources.

1,12-Diaminododecane (DADD)
1,10-Diaminododecane (DADD, dodecamethylenediamine) is a C12 carbon bio-based aliphatic diamine. This diamine can be synthesized from 1,10-dodecanedicarboxylic acid, in the same way as described for 1,10-diaminodecane (DAD).

1,7-Heptanedicarboxylic acid (HDA)
1,7-Heptanedicarboxylic acid (HDA, azelaic acid) is a saturated aliphatic dicarboxylic acid with the chemical formula $HOOC(CH_2)_7COOH$. Bio-based azelaic acid is industrially produced by the ozonolysis of oleic acid (Figure 3.113). Oleic acid is classified as a mono-unsaturated omega-9 fatty acid and is the most common fatty acid in nature. Oleic acid can be obtained from olive oil, pecan oil, and canola oil.

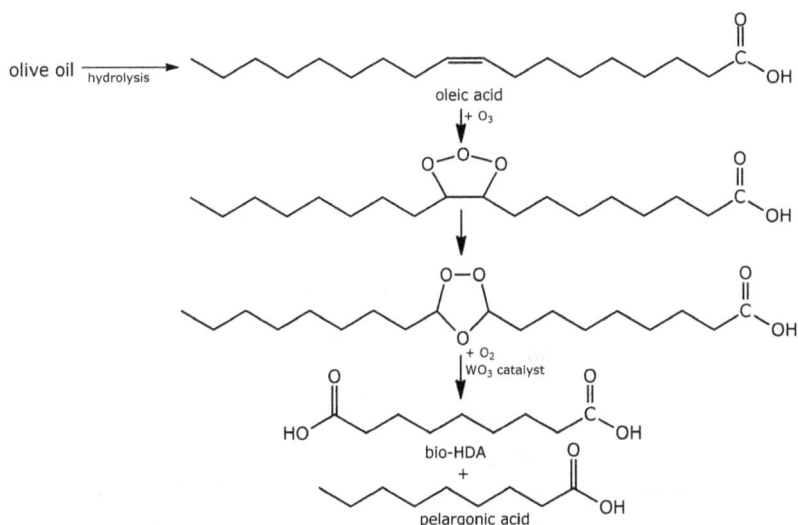

Figure 3.113: Synthesis of bio-based 1,7-heptanedicarboxylic acid (bio-HDA) from olive oil.

Polyamide 6,9 (PA6,9), produced from HMD and azelaic acid (bio-HDA), has been studied in detail, but it has never been actually commercially introduction.

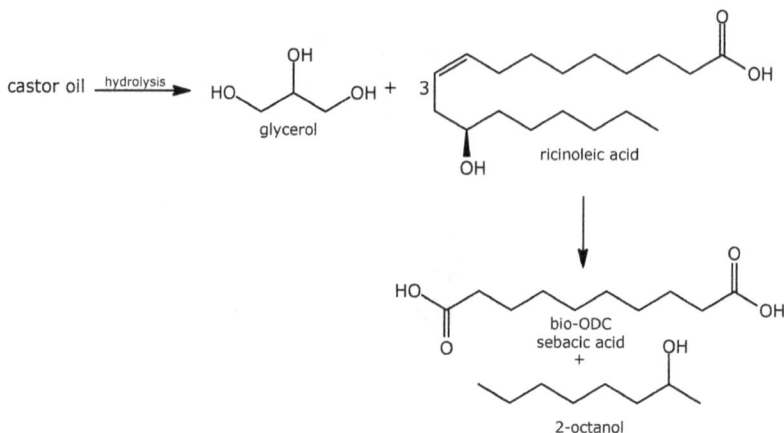

Figure 3.114: Synthesis of bio-based 1,8-octanedicarboxylic acid (bio-ODC) from castor oil.

1,8-Octanedicarboxylic acid (ODC)

1,8-Octanedicarboxylic acid (ODC, sebacic acid) is a bio-based aliphatic dicarboxylic acid. Sebacic acid is produced from castor oil by cleavage of ricinoleic acid, which is obtained from castor oil (Figure 3.114). 2-Octanol is a by-product.

1,10-Dodecanedicarboxylic acid (DDDA)

1,10-Dodecanedicarboxylic acid (DDDA) is an aliphatic dicarboxylic acid. DDDA is currently produced by a chemical process starting from 1,3-butadiene or via a biochemical route. The chemical route is shown in Figure 3.115.

A biochemical route was proposed by Verdezyne using palm kernel oil as the starting material. Palm kernel oil is an edible plant oil derived from the kernel of the oil palm *Elaeis guineensis*. The main component in palm kernel oil is the medium-chain length lauric acid. Lauric acid, or so-called dodecanoic acid, is a saturated fatty acid with a 12-carbon atom chain. Verdezyne developed a fermentation process by adapting yeast to produce DDDA from lauric acid in a three-step *enzymatic process*.

Based on the availability of bio-based 1,10-diaminodecane (bio-DAD), 1,8-octanedicarboxylic acid (bio-ODC), and 1,12-dodecanedicarboxylic acid (bio-DDDA), several new bio-based or partial bio-based polyamides have been developed by various companies such as Evonik, Arkema, EMS, DuPont and several others.

Evonik developed their Vestamid® *Terra* grades based on renewable raw materials. These series include the following grades:

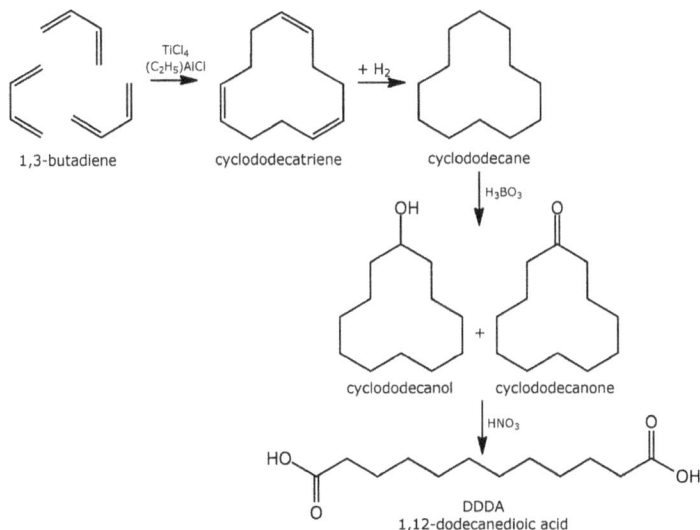

Figure 3.115: Synthesis of 1,10-dodecanedicarboxylic acid (DDDA) from 1,3-butadiene.

- Vestamid® *Terra* HS (PA6,10), 62% based on bio-renewables (being bio-based 1,8-octanedicarboxylic acid (bio-ODC)).
- Vestamid® *Terra* DS (PA10,10), 100% based on bio-renewables (being both bio-based 1,10-diaminodecane (bio-DAD) and bio-based 1,8-octanedicarboxylic acid (bio-ODC)). In this case, both monomers are derived from castor oil.
- Vestamid® *Terra DD* (PA10,12), 45% or 100% based on bio-renewables (being bio-based 1,10-diaminodecane (bio-DAD) and (bio-based) 1,10-dodecanedicarboxylic acid (DDDA or bio-DDDA)). The 100% bio-based PA10,12 is designated as Vestamid® *Terra DD-G*.

As discussed in the previous paragraph, Arkema supplies bio-based Rilsan® (PA11) and Rilsamid® (PA12) onto the market. Moreover, bio-based Rilsan T (PA10,10), based on castor oil, and Rilsan G (PA6,10) were recently introduced on the market.

Arkema has 60 years of expertise in the chemistry of castor oil. This position was expanded in 2012 by the acquisition of Chinese companies Casda, the world leader in sebacic acid (HOOC(CH$_2$)$_8$COOH) derived from castor oil, and Hipro Polymers, which produces polyamides also from castor oil (Hiprolon® grades PA6,10, PA6,12, PA10,10, PA10,12), as well as the recent purchase of a stake in Ihsedu Agrochem, a subsidiary of Jayant Agro in India which specializes in the production of castor oil.

Under the general term GreenLine, EMS-GRIVORY markets a wide range of bio-based polyamides which are manufactured partially or wholly from renewable raw materials. The GreenLine series is made up of products from the families Grilamid 1S (PA10,10, being 100% bio-based from castor oil) and Grilamid 2S (PA6,10, being 62% bio-based from castor oil).

DuPont developed Zytel RS™ renewably sourced polyamides (PA6,10, PA10,10, and PA10,12).

In addition, Radici recently developed Radilon® D, a partly bio-based PA6,10, produced from renewable sources.

3.3.17 Polyphthalamides (PPA)

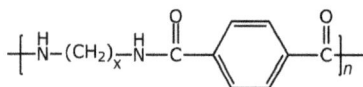

Polyphthalamide (PPA) is a subset of thermoplastic synthetic polyamides. Polyphthalamides are characterized by the replacement of at least 55 mol% of the dicarboxylic acid portion of the repeating unit in the polymer chain, and they are composed of a combination of TA and/or isophthalic acid (IPA). Substitution of aliphatic diacids by aromatic diacids in the polymer backbone of the polyamide increases the melting point, glass-transition temperature, chemical resistance, and stiffness.

In general, the diamines in PPAs are aliphatic. For example, the PA6T homopolymer, produced from HMD and TA, melts at 371 °C, which renders it intractable. To make usable polymers, it is necessary to lower the melting point, which can be achieved practically using either a longer diamine (with 9–12 carbon atoms) or by copolymerizing 6I (additional isophthalic acid) in the main chain.

In the past, several companies developed fossil-based grades of polyphthalamides. Evonik, however, developed a polyphthalamide (PPA) based on bio-based 1,10-diaminodecane (bio-DAD), obtained from castor oil. This partly bio-based polyamide PA10,T (Rilsan® HT*Plus*) has high-end properties, such as a very good dimensional stability.

3.3.18 Polyether block amides (PEBA)

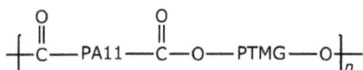

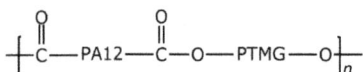

Polyether block amides (PEBA) are *thermoplastic elastomers* (TPEs). These materials are known under the tradenames of Pebax® (Arkema) or Vestamid® E (Evonik). A polyether block amide is a *block copolymer* obtained by the polycondensation of a carboxylic acid polyamide (PA6, PA11, PA12) with an alcohol-terminated polyether (such as PTMG or polyethylene glycol (PEG)). Polyether block amides are segmented-block copolymers with hard blocks consisting of the polyamide segments, while the soft blocks usually consist of flexible segments having a low glass-transition temperature. PEBAs are characterized by a high thermal stability, excellent mechanical performances, chemical resistance, and excellent processability.

Arkema introduced Pebax® Renew onto the market. Pebax® Renew grades are partially bio-based that use polyamide Rilsan® 11 as the hard segment. Castor oil, derived from the seeds of castor plants, is used to synthesize the basic bio-based building block for polyamide Rilsan® 11. Note that standard Pebax® grades have a polyamide 12 hard segment.

3.3.19 Aramides (aromatic polyamides)

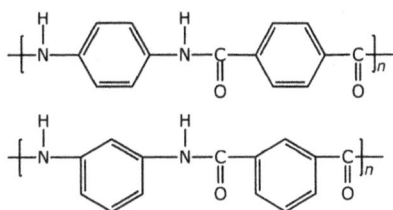

Aramids are synthetic full aromatic polyamides. In the main chain, aromatic phenyl rings are linked together by amide functional groups. Amide functionalities (-CO-NH-) form strong bonds that are resistant to solvents and heat. The phenyl rings prevent the polymer chains from rotating and twisting around their chemical bonds. As a result, aramids are rigid, straight, high-melting, and largely insoluble molecules that are ideal for fiber spinning into high-performance fibers.

The best-known *para*-aramid fibers are Kevlar® (DuPont), Twaron® (Teijin Aramid), Heracron® (Kolon Industries), and Alkex® (Hyosung). Commercial *meta*-aramid fibers are supplied by DuPont (Nomex®) and Teijin Aramid (Teijinconex®). In addition to the *para*-aramid and *meta*-aramid, Teijin Aramid developed an aromatic copolyamide based on terephthaloyl dichloride (TDC) and a mixture of *para*-phenylene diamine (PPD) and 3,4-oxydianiline (3,4'-ODA). This material's brand name is Technora®.

Both Teijin Aramid and the Dutch BioBTX worked together on the development of a bio-based para-aramid fiber. BioBTX has developed a new technology, the Integrated Cascading Catalytic Pyrolysis (ICCP), to produce bio-TA. Bio-TA is used to produce bio-based terephthaloyl dichloride (bio-TDC), one of the main components to produce *para*-aramid.

3.3.20 Elastane

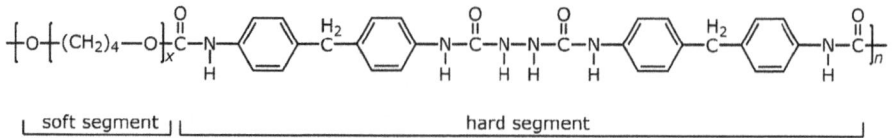

$$\left.\left[O\left[(CH_2)_4-O\right]_x\overset{\overset{O}{\|}}{C}-\overset{H}{N}-\bigcirc-\overset{H_2}{C}-\bigcirc-N-\overset{\overset{O}{\|}}{C}-N-N-\overset{\overset{O}{\|}}{C}-N-\bigcirc-\overset{H_2}{C}-\bigcirc-\overset{H}{N}-\overset{\overset{O}{\|}}{C}\right.\right]_n$$

|___ soft segment __||_____ hard segment _____|

3.3.20.1 Fossil-based elastane

Elastane (or spandex or Lycra®) is a lightweight, synthetic fiber that is used to make stretchable clothing such as sportswear. It is made up of a long-chain polymer called poly-urethane, which is produced by reacting a polyol with a diisocyanate. The polyol, for example PTMG (HO-(CH_2CH_2CH_2CH_2O)$_n$-H)) has hydroxyl groups (-OH) on both ends. The important feature of these molecules is that they are long and flexible. This part of the elastane fiber is responsible for its stretching characteristic. The other part of the polymer is a polymeric diisocyanate. This is a shorter-chain polymer, which has an isocyanate (-NCO) group on both ends. The principal characteristic of this molecule is its rigidity. In the fiber, this molecule provides strength in the fiber. The polymer is converted into a fiber using a dry spinning technique, which is explained in more detail in Section 2.5.3.

In the polymerization process, the high molecular weight diol (e. g. PTMG) is converted to a prepolymer that has an isocyanate group at each end by combining it with two molar equivalents of methylene diphenyl diisocyanate (MDI) as the capping agent. This prepolymer is then converted to high molecular weight polyurethane by combining it with a chain propagation agent such as a diamine or some other bifunctional active hydrogen compound such as a diol. The urea bonds (in the case of diamine) or urethane bonds (in the case of diol), which form in this reaction, produce hard segments. Hydrazine (H_2NNH_2) and ethylene diamine (H_2NCH_2CH_2NH_2) are the most commonly used diamines.

3.3.20.2 Bio-based elastane

Recently, Invista introduced Lycra® T162R onto the market. This material is 70% by weight based on renewable resources. In fact, the fossil-based PTMG has been replaced by a bio-based grade PTMG.

BASF is the world's leading provider of PTMG. BASF has made bio-based PTMG 1000 (PolyTHF® 1000) available for the first time. The company is now providing this intermediate to selected partners for testing various applications on a large scale. PTMG (bio-PTMG) is derived from 1,4-BDO (bio-1,4-BDO), which BASF has produced under license from Genomatica. Recently, MCC announced the development of bio-PTMG that is manufactured from bio-based feedstocks.

3.3.21 Polybutylene succinate (PBS)

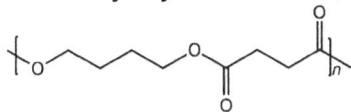

3.3.21.1 Fossil-based polybutylene succinate (PBS)

Polybutylene succinate, sometimes expressed as polytetramethylene succinate, is a thermoplastic aliphatic polyester. PBS is a biodegradable semi-crystalline polymer with properties that are comparable with PP.

Like other polyesters, two main routes exist for the synthesis of PBS: the *transesterification* process, starting from succinate diesters such as dimethyl succinate (DMS, $CH_3OOCCH_2CH_2COOCH_3$), and the direct esterification process starting from succinic acid (SA, $HOOCCH_2CH_2COOH$). The *direct esterification* of succinic acid (SA) with 1,4-BDO ($HOCH_2CH_2CH_2CH_2OH$) is the most common way to produce PBS. It consists of a two-step process. First, an excess of the diol is esterified with succinic acid to form PBS oligomers with the elimination of water. Then, these oligomers are trans-esterified under vacuum to form a high molecular weight PBS polymer. This step requires an appropriate catalyst such as a titanium-based catalyst.

At the moment, most of the commercial PBS grades are synthesized from fossil-based raw materials. Common industrial routes for succinic acid include hydrogenation of maleic acid, oxidation of 1,4-BDO, and a Reppe carbonylation of ethylene glycol (Figure 3.116).

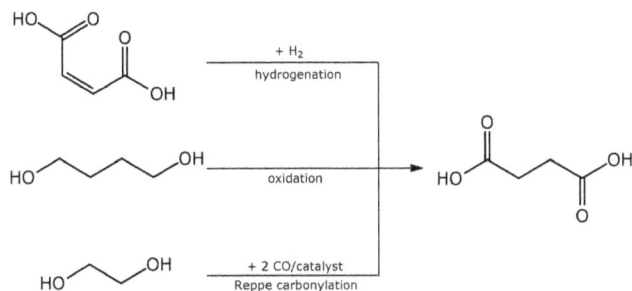

Figure 3.116: Synthesis of fossil-based succinic acid.

In the market, PBS is known by the brand names Bionolle (Showa Denko) and GSPla (Green and Sustainable Plastic, Mitsubishi Chemical). In 2016, Showa Denko announced termination of the production and sale of the Bionolle production. Mitsubishi Chemical entered the PBS market with fossil-based GSPla.

3.3.21.2 Bio-based polybutylene succinate (bio-PBS)

Mitsubishi Chemical established the PTT Biochem Company, a 50–50 joint venture with PTT Global Chemical Public Company. The Mitsubishi Chemical process know-how with respect to the production of PBS was licenced to this new joint-venture. Now, PTT MCC Biochem produces and sells bio-based PBS, brand named BioPBS™.

The current level of bio-content in BioPBS™ is 50%. When bio-1,4-BDO becomes commercially available in large quantities, PTT MCC Biochem will be targeting the 100% bio-based level for their BioPBS™.

Interestingly, BioPBS™ is both (partially) bio-based and *compostable* into biomass, carbon dioxide (CO_2) and water (H_2O), leaving no toxic by-products. BioPBS™ does not require a specialized composting facility with a film thickness of up to 200 µm. It is compostable at 30 °C and 90% relative humidity (RH). BioPBS™ has been certified by the Biodegradable Product Institute (BPI) for ASTM D6400 in North America, by Vinçotte for OK Compost (EN13432) and OK Home Compost marks in European Union, and by the Japan Bioplastics Association for GreenPla Mark in Japan.

The OK Compost Certificate (EN13432, composting conditions at 60 °C) means that a BioPBS™ film (using grade BioPBS™ FZ91) with a thickness of 67 µm composts in 12 weeks. Note that the period of dispose depends on the thickness and conditions. The OK Compost Home Certificate for a BioPBS™ film (using grade BioPBS™ FD92 (being PBSA)) with a thickness of 85 µm composts in 180 days at ambient temperature.

BioPBS™ products, such as BioPBS™-coated paper cups do not require composting facilities for decomposition. Some applications, however, contain other types of *biodegradable plastics* that may not possess the same compostable properties as BioPBS™. If other components require composting facilities, then the whole packages will have to be processed so. It is recommended to check instructions at the packages before selecting a suitable disposal method.

3.3.22 Poly(butylene-*co*-succinate-*co*-adipate) (PBSA)

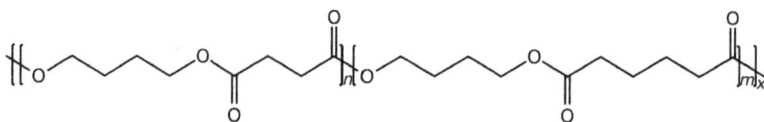

PTT MCC Biochem developed a special grade BioPBS™ (grade FD92, being poly(butylene-*co*-succinate-*co*-adipate), PBSA) with properties suitable for both blown- and cast-film extrusion. PBSA is a copolymer with AA. As a consequence, the material is softer than the standard BioPBS™. In addition, the melting point (84°) of PBSA is also lower than the melting point of BioPBS™ (115 °C). Because PBSA is a copolymer, it has a higher percentage of amorphous structure. Also, in addition, the heat deflection temperature (HDT) of

PBSA is lower than BioPBS™. Additionally, the glass-transition temperature (T_g) of PBSA is − 40 °C, compared to − 22 °C for standard BioPBS™.

Note that the chemical structure of PBSA is shown as a block copolymer due to the common synthetic method of first synthesizing two copolymer blocks and then combining them. However, it is important to note that the actual structure of the polymer is a random copolymer of the blocks shown.

PTT MCC Biochem's poly(butylene-*co*-succinate-*co*-adipate) (PBSA), BioPBS™ FD92, is partially bio-based because bio-based succinic acid has been used in the production of the material. For this reason, the bio-based content is lower than for standard BioPBS™.

3.3.23 Polyhydroxyalkanoates (PHAs)

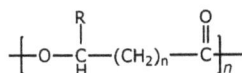

PHAs are a diverse group of biodegradable polymers produced from renewable resources. *Polyhydroxyalkanoate* polymers are naturally produced by bacteria and generally cultivated on agricultural raw materials, such as carbohydrates or *fatty acids*. They can be processed to make a variety of useful products. In fact, PHAs is a family of aliphatic polyhydroxy carboxylic acids (PHCAs). A well-known PHA is the homopolymer PHB. PHB is the most widespread and best characterized member of the family of PHAs. PHB is a linear polyester of 3-hydroxybutyric acid. Another linear PHA homopolymer is poly(3-hydroxyvalerate) (PHV). General structures of PHAs are represented in Table 3.7.

Table 3.7: Well-known polyhydroxyalkanoates.

n	R	Chemical name
1	Hydrogen	poly(3-hydroxypropionate)
	Methyl	poly(3-hydroxybutyrate) (PHB)
	Ethyl	poly(3-hydroxyvalerate) (PHV)
	Propyl	poly(3-hydroxyhexanoate)
	Pentyl	poly(3-hydroxyoctanoate)
	Nonyl	poly(3-hydroxydodecanoate)
2	Hydrogen	poly(4-hydroxybutyrate)
3	Hydrogen	poly(5-hydroxyvalerate)

In Figure 3.117, the difference between the chemical structures of poly(3-hydroxybuty-rate) and poly(4-hydroxybutyrate) is presented, and in Figure 3.118 between poly(3-hydroxyvalerate) and poly(5-hydroxyvalaterate).

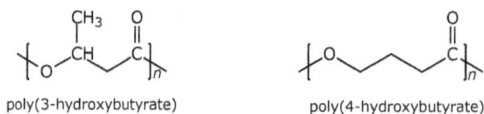

poly(3-hydroxybutyrate)   poly(4-hydroxybutyrate)

Figure 3.117: Chemical structures of poly(3-hydroxybutyrate) and poly(4-hydroxybutyrate).

poly(3-hydroxyvalerate)   poly(5-hydroxyvalerate)

Figure 3.118: Chemical structures of poly(3-hydroxyvalerate) and poly(5-hydroxyvalaterate).

One of the approaches to improve the properties of PHAs is the production of PHB co-polymers via the biosynthesis route. Two copolyesters of PHB include poly(3-hydroxybu-tyrate-*co*-3-hydroxyvalerate (PHBV) or poly(3-hydroxybutyrate-*co*-3-hydroxyhexanoate) (PHBH), with different molar ratios of hydroxycarboxylic acids (HCAs). The PHA copoly-mer approach has been studied in detail. Chemical structures of PHBV and PHBH are presented in Figure 3.119.

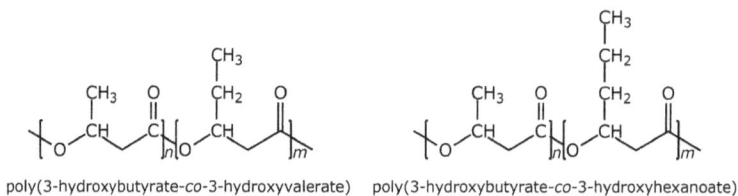

poly(3-hydroxybutyrate-*co*-3-hydroxyvalerate)   poly(3-hydroxybutyrate-*co*-3-hydroxyhexanoate)

Figure 3.119: Chemical structures of poly(3-hydroxybutyrate-*co*-3-hydroxyvalerate (PHBV) or poly(3-hydroxybutyrate-*co*-3-hydroxyhexanoate) (PHBH).

Based on the number of carbon atoms in the repeating units, PHAs can be divided into three main types: the short-chain length polyhydroxyalkanoates (scl-PHAs), which con-sist of 3–5 carbon atoms; the medium-chain length polyhydroxyalkanoates (mcl-PHAs) consisting of 6–14 carbon atoms; and the long-chain length polyhydroxyalkanoates (lcl-PHAs) with 15 or more carbon atoms. Also, it has been observed that some microorgan-isms produce copolymers including both scl and mcl monomers; these are referred to as scl-mcl PHAs. These types of PHAs are a consequence of the PHA synthase substrate specificity, which accepts precursors of a certain range of carbon length. So far, more than 150 different PHA building blocks have been identified.

PHAs are accumulated in intracellular granules by a wide variety of microorganisms under conditions of a nutrient limitation other than the carbon source. The molecular weight of PHA differs depending on the organism, conditions of growth, and method of extraction, and can vary from about 50,000 to well over a million.

As an example, PHB can be produced by microorganisms, such as *Ralstonia eutrophus* or *Bacillus megaterium* in response to conditions of physiological stress and can be produced either by pure culture or a mixed culture of bacteria. So, the manufacturing involves providing a microorganism a carbon feed source such as dextrose or glucose along with suitable nutrients, such as nitrogen, phosphorus, or oxygen which encourage growth and multiplication of the *microorganisms*. Once the number of microorganisms reaches the required point, the nutrients are reduced to create an imbalance, which puts the microorganisms under stress. The microorganism then begins to convert the extracellular carbon source through a series of enzymatic pathways to a reserve energy source in the form of polymeric inclusions within their cells (Figure 3.121). Under ideal conditions, typically, from 80% to 90% of the cell can comprise the polymeric form of the hydroxy esters, conventionally referred to as PHB. When the mass of the polymer within the cell reaches the maximum level, the process is terminated, and the polymeric material is extracted from the cells.

Microbial biosynthesis of PHB starts with the condensation of two molecules of acetyl-CoA to give acetoacetyl-CoA, which is subsequently reduced to hydroxybutyryl-CoA. This latter compound is then used as a monomer to polymerize PHB (Figure 3.120).

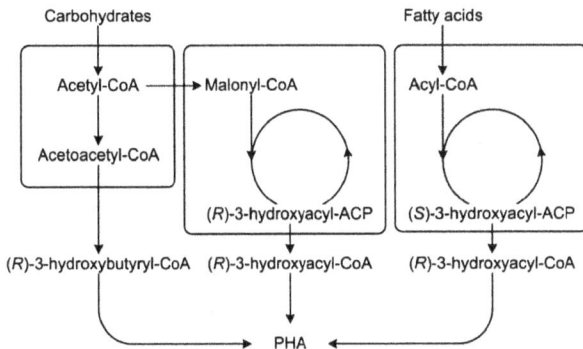

Figure 3.120: PHA synthesis in bacteria. Different pathways provide different monomers to be joined by the polymerase to PHA [44].

Biologically produced, PHB is a semicrystalline isotactic stereo regular polymer with 100% R-configuration that allows a high level of degradability.

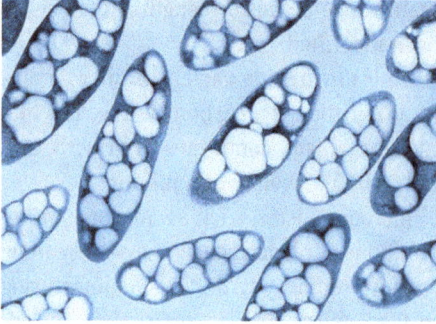

Figure 3.121: Granules of PHA within bacteria.

So, PHAs are an intracellular (energy storage) product of the bacteria. Approximately 250 different types of bacteria were found to produce PHAs. The PHAs are then harvested through the destruction of the bacteria and are separated from the microbial cell matter (centrifugation/filtration and PHA extraction using solvents such as chloroform). PHAs are degradable in soil, compost, and in seawater. PHAs degrade fastest in anaerobic sewage and slowest in seawater. The degrading microbes colonize the polymer surface and secrete PHA depolymerases. PHAs meet the standard specification ASTM D7081 for marine degradability.

Most commercial PHAs are injection-molding grades. Limitations to their commercial success are the brittleness, narrow processing window, a slow crystallization rate, and their sensitivity to thermal degradation. At this very moment, several companies are involved in the development and production of PHAs as will be discussed in the remainder of this section.

Danimer Scientific

In 2007, Danimer Scientific purchased the PHA intellectual property rights from Procter & Gamble, adding to their bioplastic technology platform. This included a specific type of PHA known as a medium-chain length polyhydroxyalkanoate (mcl-PHA), or Nodax™. As a consequence, Danimer Scientific now owns 125 patents in nearly 20 countries. Recently, Danimer Scientific received Vinçotte certifications and statements of aerobic and anaerobic compostability and biodegradability of Nodax™ PHA in soil, freshwater, saltwater, and industrial and home compost, besides a US FDA approval of Nodax™ PHA for food contact and classification of the substance, after disposal, as nonhazardous waste. Nodax™ PHA biodegrades as fast as cellulose powder or wood pulp in a proper waste-management setting and degrades over six months in ocean water. The PHA is processed into a powder and mixed with other biopolymers resulting in a biodegradable plastic resin. This material can be used to manufacture drinking straws, food packaging cups, bottles, shopping bags, plates, trash bags, labels, and more.

Nodax™ PHA copolymers (Figure 3.122) consist of (R)-3-hydroxyalkanoate comonomer units with medium-chain length groups (3 HA), containing at least three carbon atoms, and (R)-3-hydroxybutyrate (3HB). The simplest form of Nodax™ PHA is a copolymer comprising 3-hydroxyhexanoate ($n = 2$) and 3-hydroxybutyric acid ($n = 0$). Other Nodax™ PHA copolymers are described with 3-hydroxyoctanoate ($n = 4$) or 3-hydroxydecanoate ($n = 6$) comonomers (instead of 3-hydroxyhexanoate ($n = 2$)).

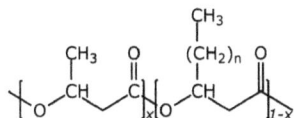

Figure 3.122: Molecular structure of Nodax™ PHA copolymers ($x = 0.01–0.50$, $n = 2–14$).

Kaneka

Kaneka developed and introduced bio-based and biodegradable Kaneka Biodegradable Polymer PHBH™ onto the market. Its raw materials are biomasses such as plant oils, which are renewable resources. Polymers are accumulated in the bodies of microorganisms through strain development and fermentation technology. The chemical structure of Kaneka Biodegradable Polymer PHBH™ is shown in Figure 3.123.

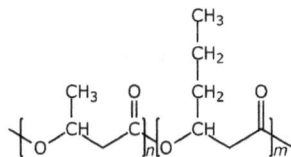

poly(3-hydroxybutyrate-*co*-3-hydroxyhexanoate)

Figure 3.123: Chemical structure of Kaneka Biodegradable Polymer PHBH™.

Kaneka's PHBH shows excellent biodegradability under aerobic, anaerobic, aquatic, and composting conditions and has proven to be an environmental friendly plastic. In Europe, Kaneka has received Vinçotte certifications for OK bio-based, OK compost, HOME OK compost, and MARINE OK biodegradable.

TianAn Biopolymer

TianAn Biopolymer is world's largest producer of poly(3-hyroxybutyrate-*co*-3-hydroxyvalerate) (PHBV). TianAn Biopolymer has an annual production capacity of 2,000 metric tons. The chemical structure of Enmat PHBV is shown in Figure 3.124.

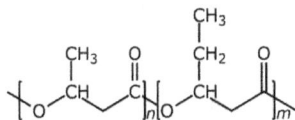

poly(3-hydroxybutyrate-*co*-3-hydroxyvalerate)   **Figure 3.124:** Chemical structure of Enmat PHBV.

The manufacturing of PHBV involves providing a microorganism a carbon feed source, such as dextrose or glucose, and a small amount of propionic acid along with suitable nutrients, such as nitrogen, phosphorus, or oxygen that encourage growth and multiplication of the microorganisms. TianAn makes use exclusively of the non-genetically modified natural strain *Ralstonia Eutropha*. In addition, TianAn uses a patented water-based extraction technology to recover the PHBV from the fermentation broth. In this way, hazardous organic solvents like chloroform can be avoided. PHBV can be consumed by microbes in the soil or water at ambient temperature, breaking down into carbon dioxide and water, hence ideally suited for home composting or in regions where industrial composting infrastructure is lacking. PHBV can also be digested anaerobically, producing methane which can be recovered and used as an energy source.

Besides the standard grade Enmat Y1000P (PHBV) TianAn supplies ENMAT Y3000P (PHB) on the market.

Biomer

Biomer is a producer of PHB. This PHB is an isotactic thermoplastic homopolyester built of 3-hydroxy butyric acid. PHB is a highly crystalline and absolutely linear (60–70% crystallinity) polymer. Biomer polyesters liquify when heated and freeze when refrigerated. Crystallization speed is fast, between 80 °C and 100 °C. Below 60 °C or above 130 °C, the speed of crystallization is rather slow. The material then remains amorphous and sticky for hours.

CJ Bio

Recently, CJ Bio, based in Seoul (South Korea), introduced poly(3-hydroxybutyrate-*co*-4-hydroxybutyrate) (P(3HB-*co*-4HB)) copolymers in the market. The product's brand name is PHACT. PHACT is produced by bacterial fermentation of sugar. The chemical structure is given in Figure 3.125.

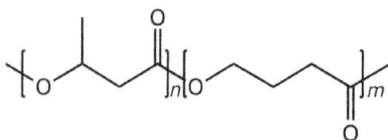

Figure 3.125: Chemical structure of poly(3-hydroxybutyrate-*co*-4-hydroxybutyrate) (P(3HB-*co*-4HB)) copolymers.

P(3HB-*co*-4HB) copolymers have decreased melting temperatures and crystallinity compared with PHB. P(3HB-*co*-4HB) copolymers are considered as the most popular PHAs that have a significant application potential, especially in the biomedical and pharmaceutical fields. So, PHACT is a 100% bio-based, marine biodegradable plastic.

Several other companies are working to develop methods for producing PHA from wastewater, including Bluepha and Paques Biomaterials.

3.3.24 Polylactic acid (PLA)

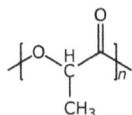

Polylactic acid (PLA), or polylactide, is a thermoplastic aliphatic polyester. PLA's basic monomer lactic acid (LA, or also named 2-hydroxy propionic acid (HOCH(CH₃)COOH)) is chiral, consisting of two enantiomers. *Enantiomers* are mirror images of each other, they are not superposable (Figure 3.126). Enantiomers have identical chemical and physical properties except for their ability to rotate plane-polarized light by equal amounts, but in opposite directions. Such compounds are called optically active. Dextrorotatory lactic acid (D-(–)-lactic acid or (R)-lactic acid) rotates light in a clockwise direction while levorotatory lactic acid (L-(+)-lactic acid or (S)-lactic acid) rotates light in a counterclockwise direction. The R/S system is an important nomenclature system used to denote distinct enantiomers, as discussed in Chapter 2.

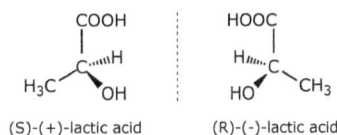

(S)-(+)-lactic acid   (R)-(-)-lactic acid

Figure 3.126: (S)-(+)-Lactic acid (left) and (R)-(–)-lactic acid (right) are non-superposable mirror images of each other.

A mixture of equal amounts of both enantiomers is called a racemic mixture or racemate. The racemic mixture is also called DL-lactic acid.

3.3.24.1 Fossil-based polylactic acid (PLA)
Racemic lactic acid (DL-lactic acid) can be produced industrially from acetaldehyde by addition of hydrogen cyanide (HCN), forming cyanohydrin (also called lactonitril), and subsequent hydrolysis to lactic acid and ammonium chloride (Figure 3.127).

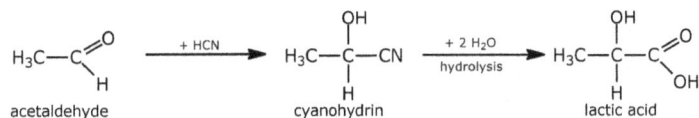

acetaldehyde   cyanohydrin   lactic acid

Figure 3.127: Synthesis of lactic acid from acetaldehyde.

Synthesis of both racemic lactic acid and enantiopure lactic acids is also possible from other starting materials, such as vinyl acetate (CH₃COOCH = CH₂) and glycerol, using catalytic procedures. Vinyl acetate is reacted with carbon monoxide (CO) and hydrogen in the presence of a hydroformylation catalyst to an α-substituted propionaldehyde. This material is further oxidized and hydrolyzed to lactic acid, as shown in Figure 3.128.

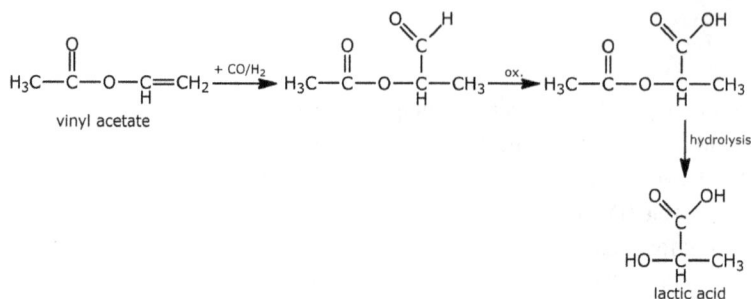

Figure 3.128: Synthesis of lactic acid from vinyl acetate.

Clearly, chemo-catalytic routes to lactic acid lead to racemic mixtures of lactic acid. New procedures have been developed to separate both lactic acid enantiomers by using membrane techniques in combination with a chiral selector molecule or via enzymatic kinetic resolution.

3.3.24.2 Bio-based polylactic acid (bio-PLA)

As a starting material for industrial *fermentation* of lactic acid, almost all carbohydrate sources containing C5 or C6 sugars can be used. Pure sucrose, glucose from starch, cassava, or beet juice can be applied. Lactic acid producing bacteria can be divided in two classes: home fermentative bacteria producing two molecules of lactic acid from one molecule glucose, and heterofermentative bacteria producing one molecule of lactic acid and carbon dioxide (CO_2) and acetic acid/ethanol. Through fermentation, mainly L-lactic acid is produced. The PLA production process can make highly efficient use of the renewable feedstocks: to make one kg of PLA, one need just 1.6 kg of sugar.

Nowadays, lactic acid is predominantly produced by bacterial fermentation of carbohydrates. By using bacterial fermentation, L-lactic acid with an optical purity of 99.9% can be obtained. Industrial fermentation to optical pure D-lactic acid is much more difficult, however.

In general, PLA resins are made from renewable resources and are 100% bio-based according to European standard NEN-EN 16785-1:2016 or ASTM D6866. PLA is obtained by the polycondensation of lactic acid or by the *ring opening polymerization* of lactide, the cyclic dimer of lactic acid. Both routes are depicted in Figure 3.129.

Since the polycondensation is an equilibrium reaction, there are difficulties in removing sufficient water during the later stages of polymerization, which limits the molecular weight of PLA obtained by this method. So, water removal by application

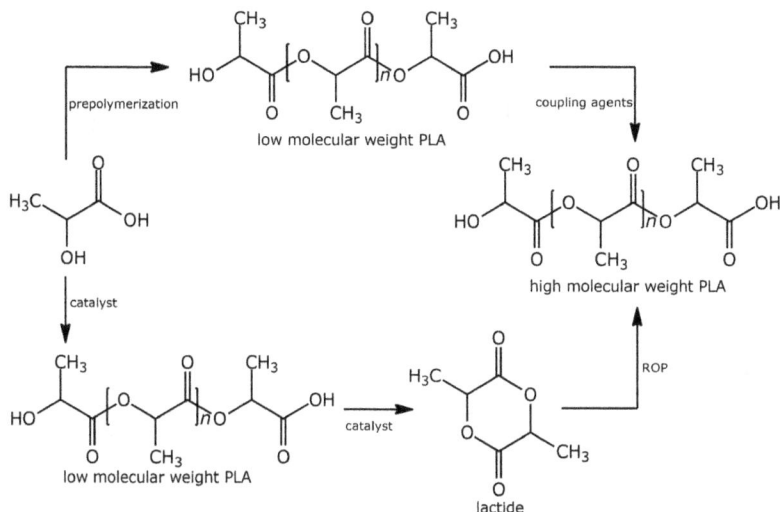

Figure 3.129: Synthesis of polylactic acid (PLA).

of vacuum or by azeotropic distillation is required to drive the reaction toward poly-condensation. As a consequence, most of the research has focused on the method of ring-opening of lactide. In this method, a low molecular weight PLA is formed via a polycondensation from lactic acid. In the next step, a purified cyclic lactide is obtained by a depolymerization reaction. Finally, the catalyzed (tin(II) 2-ethylhexanoate) ring-opening polymerization (ROP) gives a high molecular weight PLA.

While PLA is one of the most common *bioplastics* used today, the process for this material to be degraded is very specific and has to occur in the appropriate composting facilities, i. e., if it ends up in a landfill, it will remain there for a long period like a normal plastic. Compostability indicates the biodegradation of a plastic in a certain amount of time, maintaining specific conditions (moisture, elevated temperatures, and the presence of certain microorganisms, such as bacteria, fungi, and algae) to achieve it. PLA is *compostable* according to European standard NEN-EN 13432:2000 or ASTM D6400. Compostable applications such as bin liners help to divert valuable organic waste from landfill. At the end of its useful life, PLA applications can be mechanically or chemically recycled as well. More information on those recycling routes can be found in Chapter 4.

Using composting conditions, degradation occurs mainly through scission of ester bonds in the polymer main chain. This is how the long polymeric chains are broken down into shorter oligomers, dimers or monomers. Specifically, the ester bonds of PLA fragment into carboxylic acid and alcohol by chemical hydrolysis. These shorter units are small enough to pass through the cell walls of microorganisms and be used as substrates for their biochemical processes and thus can be degraded by microbial enzymes. Additional factors that influence the biodegradability of PLA are, for example, molecular weight and *crystallinity*. Clearly, amorphous chains are far more susceptible to hydrolysis, and crystalline regions are more resistant to degradation.

Three major suppliers of PLA are NatureWorks LLC (capacity of 150,000 metric tons in the USA), Total Corbion (75,000 metric tons produced in Thailand), and Zhejiang Hisun Biomaterials (45,000 metric tons produced in China).

PLA polymers range from *amorphous* glassy polymer to semi-crystalline and highly *crystalline polymer* with a glass transition (T_g) of 60–65 °C and a melting temperature (T_m) of 130–180 °C. Total Corbion developed several grades of PLA, such as standard PLA for general-purpose applications, high-heat PLA for demanding applications, low-heat PLA typically used as seal layer, and PDLA used either as a nucleating agent or to create full stereocomplex PLA compounds.

Total Corbion's standard-grade PLA (LX175)
The standard grade of PLA has an optical purity of 96% of the L-enantiomer of the lactic acid. Its melting point is 155 °C; its glass-transition temperature 60 °C. This grade can be used for fiber spinning, extrusion/thermoforming processes, and monofilament production as ink for 3D printing.

Total Corbion's high-heat-grade PLA
Total Corbion developed a few high heat grades of PLA with an optical purity of > 99% of the L-enantiomer. Melting temperatures are 175 °C, glass-transition temperatures 60 °C. These grades are available in a range of melt viscosities and deliver improved heat resistance over the standard PLA grade. To obtain improved heat resistance over standard PLA, these resins need to be crystallized during processing. In this way, a heat-distortion temperature (HDT B) of 105 °C can be realized. As a result, coffee cups made with this material, can be filled with hot drinks without damage, as shown in Figure 3.130.

Standard PLA   Luminy® high heat PLA
Coffee at 100°C   Coffee at 100°C   **Figure 3.130:** High-heat PLA [45].

Total Corbion's low-heat-grade PLA

Two low-heat development grades of PLA have been designed for use as heat-seal layers in film applications or for use as a low melting-point component in sheat-core configurations. The optical purities of both materials are quite low (88 and 90%). As a consequence, the melting points of both materials decrease to 130 °C, while glass-transition temperatures are still around 60 °C.

PLA bioplastics offer a significantly reduced carbon footprint versus traditional oil-based plastics (Figure 3.131). This is important for the health of our planet and is a growing concern among consumers, who examine the sustainability aspects of their purchases ever more critically. As media attention increases and regulatory activity gains momentum, biocontent in plastic will become a more and more relevant issue for producers to address.

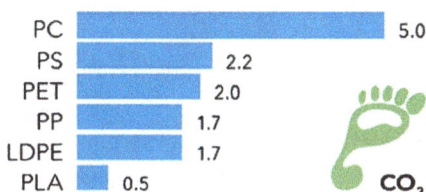

Figure 3.131: Emissions from production of common polymers (in kg CO_2 eq per kg of polymer – cradle to gate [45].

3.3.25 Polyglycolic acid (PGA)

Polyglycolic acid (PGA), or polyglycolide, is a biodegradable thermoplastic polyester. PGA is the simplest linear aliphatic polyester, prepared from glycolic acid (GA, $HOCH_2COOH$), also called hydroxyacetic acid.

3.3.25.1 Fossil-based glycolic acid (PGA)

Several industrial routes are known to produce glycolic acid. A well-known industrial process, shown in Figure 3.132, is the carbonylation of formaldehyde with carbon monoxide (syngas) and water, using strong acid catalysts. Another commercial process makes use of monochloroacetic acid (MCA). The predominant route to MCA involves the chlorination of acetic acid, with acetic acid anhydride as a catalyst.

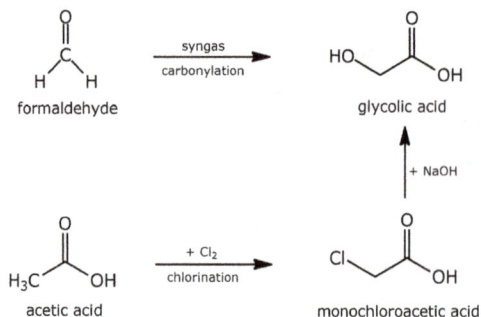

Figure 3.132: Industrial synthesis of glycolic acid (GA).

3.3.25.2 Bio-based glycolic acid (bio-PGA)

Glycolic acid can also be prepared using an enzymatic biochemical process that requires less energy and produces fewer impurities compared to traditional fossil-based chemical processes.

METabolic EXplorer (METEX) developed a new fermentation pathway for the biological production of glycolic acid from lignocellulosic C6 sugars, such as glucose (Figure 3.133). A recombinant *E. coli* is used to produce glycolic acid in high yield. This new fermentation pathway is also called the glycoptimus pathway and produces glycolic acid without carbon loss.

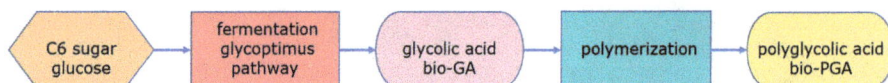

Figure 3.133: IEA process flow diagram for the production of bio-PGA.

A thermo-chemical route has been developed by Danish company Haldor Topsoe. In their new technology, sugar is converted into glycolaldehyde ($HOCH_2CHO$) in a single industrial unit. Primary interest is to use glycolaldehyde to produce in a reduction-step bio-based ethylene glycol (bio-EG) for PET production. The goal is for the plant to convert various raw materials, such as sucrose, dextrose, and 2G sugars, into ethylene glycol (EG). Both Braskem and Haldor Topsoe work together in the development of the EG production process. In the same way, glycolaldehyde (bio-GA) can be used to produce in an oxidation step bio-based glycolic acid for PGA production.

Haldor Topsoe's process is named the MOSAIK, MOnoSAccharide Industrial Cracker. MOSAIK is a solution for cracking sugars to an intermediary product, which can be further converted to functional chemicals, such as ethylene glycol (EG), methyl vinyl glycolate (MVG) or glycolic acid (GA), using Haldor Topsoe's patented processes and catalysts. Methyl vinyl glycolate (MVG) is an interesting platform molecule for the production of a variety of functional chemicals. Methyl vinyl glycolate (MVG) can be produced from gly-colaldehyde or erythrose (Figure 3.134).

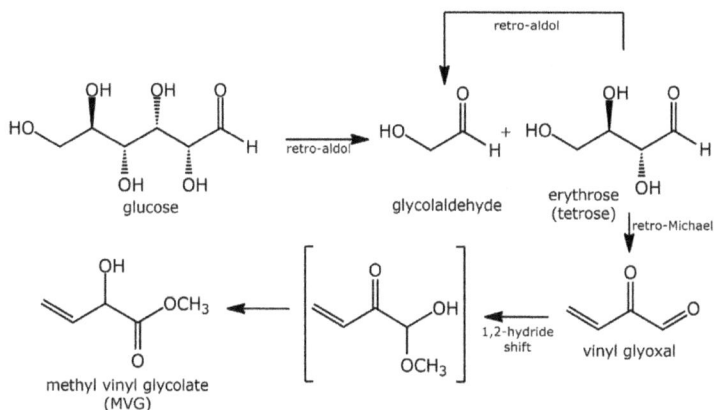

Figure 3.134: Synthetic route to bio-based methyl vinyl glycolate (MVG).

Polyglycolic acid (PGA) has a T_g between 35 and 40 °C, and its T_m is reported to be in the range of 225–230 °C. PGA also exhibits an elevated degree of crystallinity, around 45–55%.

Similar to the production of PLA, PGA can be obtained by the direct polyconden-sation of glycolic acid (GA) or via the ring opening polymerization of glycolide, the cyclic dimer of glycolic acid. The direct *polycondensation* of glycolic acid is not the most efficient route because it yields low molecular weight PGA. The most common route to produce high molecular weight PGA is via a ring-opening polymerization of glycolide. Glycolide can be prepared by heating low molecular weight PGA under re-duced pressure and collecting the glycolide by distillation. Ring-opening polymeriza-tion of glycolide is catalyzed by using catalysts such a antimony (e. g., antimony trioxide) or tin compounds (e. g., tin(II) 2-ethylhexanoate).

Due to the hydrolytic instability of PGA, only limited applications are known, but some applications have been found in the biomedical field. The hydrolytic instability is caused by the sensitive ester linkage in the PGA polymer backbone which the polymer brings back to its monomer glycolic acid. Degradation starts within the amorphous re-gions, and in the second stage the crystalline regions are susceptible to hydrolytic at-tack. Upon collapse of the crystalline regions, the polymer chain dissolves.

When exposed to physiological conditions, PGA is degraded by random hydroly-sis, and apparently it is also broken down by certain enzymes, especially those with esterase activity. The degradation product, glycolic acid, is nontoxic. The main appli-cations of PGA, and its copolymers (polylactic acid-*co*-glycolic acid) or polyglycolic acid-*co*-caprolactone (Figure 3.135), have been found in the biomedical field, such as degradable surgical sutures.

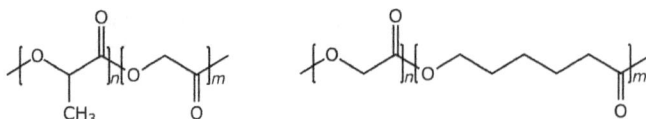

Figure 3.135: Chemicals structures of poly(lactic acid-*co*-glycolic acid) (left) and poly(glycolic acid-*co*-caprolactone) (right).

3.3.26 Poly(butylene-*co*-adipate-*co*-terephthalate) (PBAT)

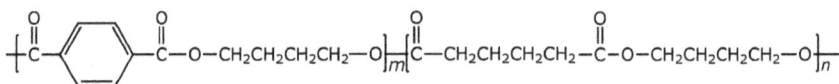

3.3.26.1 Fossil-based poly(butylene-*co*-adipate-*co*-terephthalate) (PBAT)

Poly(butylene-*co*-adipate-*co*-terephthalate) (PBAT) is a random copolyester of AA, 1,4-BDO, and TA or from DMT. AA and 1,4-BDO are polymerized to create their polyester (plus water). DMT and 1,4-BDO are also reacted to form their polyester (plus methanol). This polyester is then added to the butylene AA polyester by using tetrabutoxytitanium (TBT) as a transesterification catalyst. The result is a copolymer of the two previously prepared polymers.

Note that the chemical structure of PBAT is shown as a block copolymer due to the common synthetic method of first-synthesizing two copolymer blocks and then combining them. However, it is important to note that the actual structure of the polymer is a random copolymer of the blocks shown.

PBAT is produced commercially: by BASF under the name ecoflex®; in a blend with PLA called ecovio®, by Novamont as Origo-Bi; in a blend with starch called Mater-Bi, by JinHui Zhaolong as Ecoworld; in a blend with starch called Ecowill; and by Eastman Chemical as Eastar Bio. Furthermore, suppliers in China and other nations have also begun to produce PBAT.

PBAT is a biodegradable and certified compostable polymer due to the presence of butylene adipate groups. The high stability and mechanical properties come from the terephthalate portions in the main chain. PBAT is marketed commercially as a fully compostable plastic, with BASF's ecoflex® showing 90% degradation in an industrial composting plant after an 80-day test.

With ecovio®, BASF offers a certified compostable polymer which also contains variable bio-based content. It consists of the compostable BASF polymer ecoflex®, PLA, and other additives. In contrast to simple starch-based bioplastics, ecovio® is more resistant to mechanical stress and moisture. The main areas of use for ecovio® are plastic films such as organic waste bags, fruit and vegetable bags, carrier bags with dual-use (first for shopping, then for organic waste), and agricultural films. Furthermore, compostable packaging solutions such as paper-coating, shrink and cling films, as well as injection-molding and thermoformed products, can be produced with ecovio®.

In the same way, Novamont commercialized Mater-Bi®, a blend of PBAT (Origo-Bi®) with starch. Origo-Bi® is produced on a former PET production plant of Mosshi & Giosolfi (M&G) in Patricia (Italy).

3.3.26.2 Bio-based poly(butylene-*co*-adipate-*co*-terephthalate) (bio-PBAT)

Bio-based poly(butylene-*co*-adipate-*co*-terephthalate) (bio-PBAT) is not commercially available yet. However, producers of PBAT can switch to bio-based PBAT if the bio-based diacids (AA (bio-AA) and terephthalic acid (bio-TA)) and diols (bio-1,4-BDO) become available. Clearly, the development of bio-AA and bio-1,4-BDO is in an advanced state, so that partially bio-based PBAT could be produced in the near future.

3.3.27 Poly(butylene-*co*-succinate-*co*-terephthalate) (PBST)

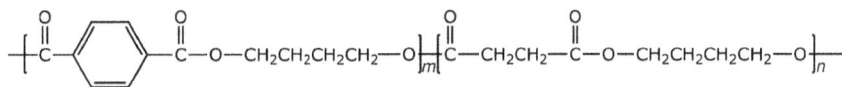

3.3.27.1 Fossil-based poly(butylene-*co*-succinate-*co*-terephthalate) (PBST)

Poly(butylene-*co*-succinate-*co*-terephthalate) (PBST) is a random copolyester of succinic acid (SA), 1,4-BDO, and TA or from DMT. Succinic acid (SA) and 1,4-BDO are polymerized to create their polyester (plus water). DMT and 1,4-BDO are also reacted to form their polyester (plus methanol). This polyester is then added to the butylene succinic acid polyester by using tetrabutoxytitanium (TBT) as a transesterification catalyst. The result is a random copolymer of the two previously prepared polymers.

Note that the chemical structure of PBST is shown as a block copolymer due to the common synthetic method of first synthesizing two copolymer blocks and then combining them. However, it is important to note that the actual structure of the polymer is a random copolymer of the blocks shown.

Fossil-based poly(butylene-*co*-succinate-*co*-terephthalate) (PBST) is commercially not available in large quantities. PBST possesses similar biodegradability and mechanical properties, compared with PBAT.

3.3.27.2 Bio-based poly(butylene-*co*-succinate-*co*-terephthalate) (bio-PBST)

In the same way as mentioned for PBAT, bio-based poly(butylene-*co*-succinate-*co*-terephthalate) (PBST) can be produced when producers switch to bio-based diacids (succinic acid (bio-SA)) and terephthalic acid (bio-TA) and diols (bio-1,4-BDO). Clearly, the development of bio-based succinic acid (bio-SA) and bio-1,4-BDO is in an advanced state, so that partially bio-based PBST (bio-PBST) could be produced.

3.3.28 Polycaprolactone (PCL)

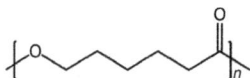

Polycaprolactone (PCL) is a *biodegradable* aliphatic polyester with a low melting point of around 60 °C and a glass-transition temperature of about – 60 °C. The most common use of polycaprolactone (PCL) is in the production of specialty polyurethanes.

Polycaprolactone is prepared by ring-opening polymerization of ε-caprolactone using a catalyst such as stannous octoate (Figure 3.136). ε-Caprolactone, or simply caprolactone, is a cyclic ester (lactone), possessing a seven-membered ring. Its name is derived from caproic acid ($CH_3CH_2CH_2CH_2CH_2COOH$). It was in the past produced on a large scale as a precursor to caprolactam. Caprolactone is treated with ammonia (NH_3) at elevated temperatures to yield the caprolactam.

Caprolactone (CL) is prepared industrially by Baeyer-Villiger oxidation of cyclohexanone with peracetic acid (CH_3COOOH), as pointed out in Figure 3.137. Cyclohexanone can be easily produced by the oxidation of cyclohexane (C_6H_{12}) in air, using a cobalt catalyst. This process co-forms also cyclohexanol, and this mixture, called "KA oil" for ketone-alcohol oil, is the main feedstock for the production of AA. Alternatively, cyclohexanone can be produced by the partial hydrogenation of phenol (C_6H_5OH). On an industrial scale, cyclohexane (C_6H_{12}) is produced by *hydrogenation* of benzene (C_6H_6) in the presence of a Raney nickel catalyst.

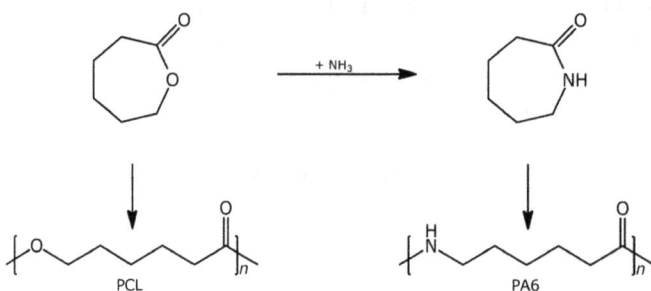

PCL   PA6

Figure 3.136: Ring opening polymerizations (ROPs) of ε-caprolactone and caprolactam.

PCL degrades slowly by hydrolysis due to its high crystallinity and hydrophobic nature. It is used in many fields, such as implantable biomaterials, biodegradable materials, and micro particles for drug-delivery systems.

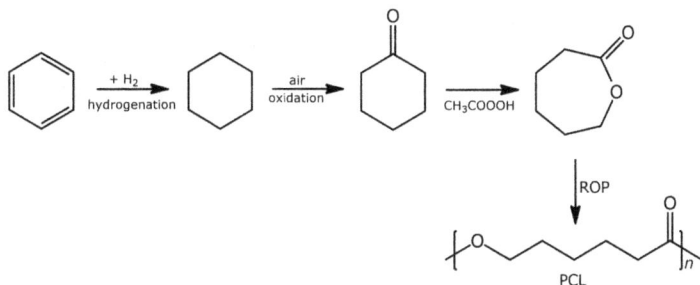

Figure 3.137: Synthesis of polycaprolactone (PCL).

PLA and polycaprolactone (PCL) copolymers have been developed to take advantage of synergistic improvement in properties offered by both polymers. For example, co-polymers have yielded more flexible materials with higher degradation rates than polycaprolactone (PCL) itself. The random copolymer (Figure 3.138) can be obtained by ring-opening polymerization of lactide and ε-caprolactone, respectively, using stannous octoate as the catalyst and low molecular weight alcohols.

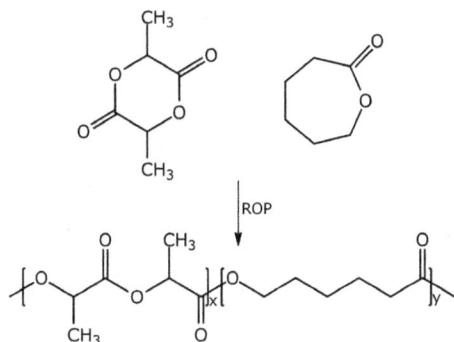

Figure 3.138: Copolymer of polylactic acid (PLA) and polycaprolactone (PCL).

3.4 Bioplastics

Multiple definitions of *bioplastic* are currently being used, with the consequence that this regularly causes some confusion. In this book, we use the definition for bioplastic as postulated by European Bioplastics. According to European Bioplastics, a plastic material is defined as a bioplastic if it is either (partly) bio-based, biodegradable, or features both properties. Note that bio-based and biodegradable are not synonymous!

Bioplastics are driving the evolution of plastics. There are two major advantages of bio-based plastic products compared to their conventional fossil-based plastic products: they save fossil resources by using renewable feedstock, and they provide the unique potential of carbon neutrality. Carbon neutrality refers to achieving net zero carbon dioxide (CO_2) emissions. It is used in the context of CO_2-releasing processes associated with transportation, energy production, agriculture, and industrial processes, including polymer-production processes. Furthermore, biodegradability is an add-on property of certain types of bioplastics. It offers additional means of recovery at the end of a product's life.

European Bioplastics is an association representing the interests of the thriving bioplastics industry in Europe. European Bioplastics believes that bioplastics are a major driver in the evolution of plastics, and that they contribute significantly to a more sustainable future. In addition, the organization supports newly developed independent and internationally certification schemes and labels for product identification and disposal, and the EU-wide implementation of bioplastics on the European market. It is important for consumers to receive transparent and correct information about bioplastics and to safeguard the public opinion of bioplastics.

There are two organizations in Europe, the Belgian certifier TÜV Austria Belgium and the German certifier DIN CERTCO, that provide *certifications* and corresponding labels on the standards. Standardization is an effort by industrial stakeholders to define generally accepted criteria and guidelines for the description of products, services, and processes. There are two different types of evaluation systems that are both commonly called standards. On the one hand, test methods describe methodological criteria and typically lay out the procedures that need to be followed. On the other hand, there are specifications that have a normative function and define a set of pass and fail criteria as the requirements that need to be met for a product or material to be compliant with the standard. While these two types are often complementary, it is the latter "specification" type of standard that ultimately defines compliance criteria. The key standardization bodies creating standards are ISO (International Organization for Standardization), CEN (European Committee for Standardisation), and ASTM (American Society for Testing and Materials). In addition, there are many national *standardization* organizations.

In November 2022, the European Commission (EC) adopted a policy framework [46] on the sourcing, labeling, and use of bio-based plastics, and the use of biodegradable and compostable plastics. This was announced in the European Green Deal and the ECs Circular Economy Action Plan [11].

This EU policy framework for bio-based, biodegradable, and compostable plastics aims to contribute to a sustainable plastic economy by:

– improving the understanding around these materials and clarify where these plastics can bring genuine environmental benefits, and under which conditions and applications;
– guiding citizens, public authorities, and businesses in their policy, purchasing or investing decisions; and
– preventing differences at national level and fragmentation of the market by promoting a shared understanding across the EU on the production and use of these plastics.

3.4.1 Bio-based plastics

The term bio-based means that a plastic (or product) is wholly or partly derived from biomass. Chemical building blocks (monomers) for the production of bioplastics can be obtained from distinguishable *biomass* sources (Figure 3.139). First-generation (1G) bioplastics are produced directly from food crops, such as corn, wheat, sugarcane, and sugar beet. The major concern is that, if these 1G food crops are used for bioplastics, food prices could rise and shortages might be experienced in some countries. Second-generation (2G) bioplastics have been developed to overcome the limitations of 1G bioplastics. Second-generation bioplastics are produced from non-food crops such as wood, organic waste, food crop waste, or specific biomass crops. Second-generation bioplastics are also aimed at being more cost competitive with existing fossil-based bioplastics. Third-generation (3G) bioplastics are based on improvements in the production of biomass. It takes advantage of specially engineered crops such as algae. The algae are cultured to act as a low-cost, high-energy, and entirely renewable feedstock.

Unlike *conventional plastics,*
which are made from
FOSSIL OIL,...

..*BIO-BASED plastics*
are derived from
RENEWABLE resources.

© European Bioplastics

Figure 3.139: Bio-based vs. fossil-based plastics [47].

Wholly bio-based or partially *bio-based plastics*, such as completely bio-PE or partially bio-based PET (bio-PET30), possess properties that are technically equivalent to their conventional fossil-based versions. However, they help to reduce a product's carbon footprint. Moreover, they can be mechanically recycled in existing recycling streams. These bio-based plastics are also called *drop-in bioplastics*. A drop-in bioplastic is a kind of "bio-similar" copy of the fossil-based plastic, but it is made from biomass instead of fossil feedstocks. The advantage is that drop-in bioplastics can be used almost immediately without any major technology or equipment investments, and the testing period is also quite limited. To put it simply: the user can replace the fossil-based PE with the bio-based drop-in PE, while using the same machinery, equipment, and process settings.

Figure 3.140 shows global capacities for bio-based or partially bio-based bioplastics. Clearly, the total production capacity amounts to 1.05 million metric tons in 2023 of bio-based or partially bio-based plastics. Compared with the global capacity of conventional fossil-based plastics (being 400 million metric tons in 2022), this only amounts to 0.27%

2023
Total: 1.05 million tonnes

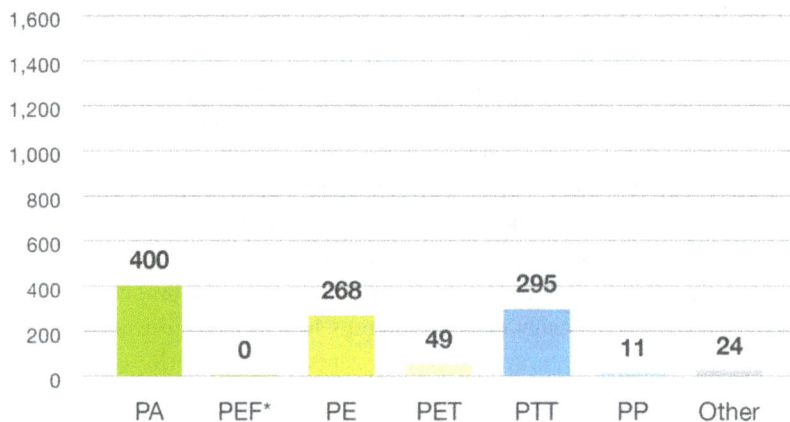

2028
Total: 2.83 million tonnes

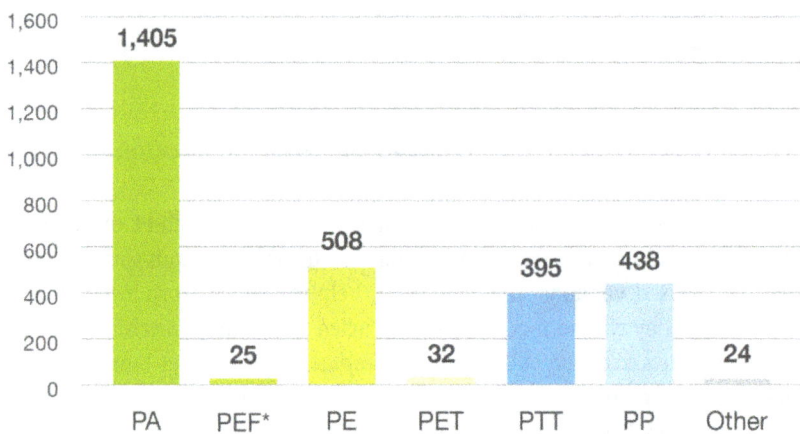

PEF is currently in development and predicted
to be available in commercial scale in 2024.

Source: European Bioplastics, nova-Institute (2023)

Figure 3.140: Global production capacities of bio-based or partially bio-based plastics 2023 vs. 2028 [47].

of the total. However, it is expected that by 2028, the global capacity for bio-based or partially bio-based plastics will increase to 2.83 million metric tons. Current major bio-based or partially bio-based plastics are PE (bio-PE, bio-based), PET (bio-PET, partially bio-based), polyamide (bio-PA, bio-based or partially bio-based), PP (bio-PP, in development stage), polyethylene furanoate (bio-PEF, in development stage), polytrimethylene terephthalate (bio-PTT, partially bio-based), among others. Note that none of these bio-plastics are biodegradable!

The ability to determine the bio-based content of a bio-based plastic is an obvious prerequisite for developing the market for bio-based plastics. Companies producing bio-based plastics can either indicate the bio-based carbon content or the bio-based mass content of their plastic product. Because both measuring methods differ in design, the typical numeric percentage values will differ, too, and must be taken into account, especially when drawing comparisons.

3.4.1.1 Bio-based carbon content

A well-established methodology to measure the *bio-based carbon content* in plastics is the ^{14}C-radiocarbon-method. The European standard NEN-EN 16640:2017 "Bio-based products – Bio-based carbon content – Determination of the bio-based carbon content of products using the radiocarbon method", describes how to measure the carbon isotope ^{14}C (radiocarbon method). This European standard corresponds with the US-standard ASTM 6866-20 "Standard Test Methods for Determining the Biobased Content of Solid, Liquid, and Gaseous Samples Using Radiocarbon Analysis". Both methods are based on the radioactive decay of ^{14}C, which can be used to estimate the age of organic materials up to roughly 60,000 years old. The ^{14}C method for estimating the bio-based carbon content is based on the fact that fossil-based materials such as crude oil or gas will contain no ^{14}C as a result of radioactive decay, whereas bio-based materials will contain new ^{14}C, taken up from the atmosphere. The ^{14}C carbon content may thus be considered as a tracer of chemicals recently synthesized from atmospheric carbon dioxide (CO_2) particularly of recently produced bio-based products. Results based on the ^{14}C methodology are expressed as a fraction of bio-based carbon on the total carbon content of the sample. Certification schemes and derived product labels based on the European and the US-standard are available – for example, by the Belgian certifier TÜV Austria Belgium or German certifier DIN CERTCO (Figure 3.141).

The bio-based carbon content for the TÜV-labels is indicated with red stars: 1 star equals a content of 20–40%; 2 stars, a content 40–60%; 3 stars, a content 60–80%; and 4 stars, above 80%. The bio-based content in bio-based plastics is also indicated in the DIN-labels.

ASTM D6866-20 was developed at the request of the US Department of Agriculture (USDA). ASTM D6866-20 is a widely used method in the bioplastics industry. Braskem, a leading Brazilian petrochemical company, is one of the many bioplastics companies that use ASTM D6866-20 to certify the bio-based content of their products. ASTM

Figure 3.141: Examples of labels to indicate the bio-based content.

D6866 was first published in 2004. There have been several versions released since. The current active version of the standard is ASTM D6866-20, effective February 2020.

Note that NEN-EN 16640:2017 expresses the carbon as a percentage of total carbon content, while the ASTM D6866-20 standard expresses the carbon as a percentage of total organic carbon (so any C-containing inorganic polymer additives are not taken into account). Figure 3.142 shows the USDA label for a material with a bio-based content of 95%, measured according to the ASTM D6866 method.

Figure 3.142: USDA bio-based label.

ISO, the International Organization for Standardization, is an independent, non-governmental international organization with a membership of 164 national standards bodies. In the past few years, ISO published standard NEN-ISO 16620, consisting of five parts. These five standards are applicable to plastic products and plastic materials, polymer resins, monomers, or additives.

- NEN-ISO 16620-1:2015 "Plastics – Biobased content – Part 1: General principles". NEN-ISO 16620-1 specifies the general principles and the calculation methods for determining the amount of bio-based content in plastic products. These calculation methods are based on the carbon mass or mass of each constituent present in the plastic products.
- NEN-ISO 16620-2:2019 "Plastics – Biobased content – Part 2: Determination of bio-based carbon content". NEN-ISO 16620-2 specifies a calculation method for the determination of the bio-based carbon content in monomers, polymers, and plastic materials and products, based on the ^{14}C-content measurement.
- NEN-ISO 16620-3:2015 "Plastics – Biobased content – Part 3: Determination of bio-based synthetic polymer content". NEN-ISO 16620-3 specifies the method of determining the amounts of bio-based part in the bio-based synthetic polymer in

plastics products. This calculation method for bio-based synthetic polymer content is based on the mass of bio-based synthetic polymer in the plastics products.
- NEN-ISO 16620-4:2016 "Plastics – Biobased content – Part 4: Determination of bio-based mass content". NEN-ISO 16620-4 specifies a method of determining the bio-based mass content in plastics products, based on the radiocarbon analysis and elemental analysis. This document is applicable to plastic products and plastic materials, polymer resins, monomers or additives that are made from bio-based or fossil-based constituents. This method is applicable, provided that the plastic product contains carbon element and that a statement giving its elemental composition and its bio-based mass content is available.
- NEN-ISO 16620-5:2017 "Plastics – Biobased content – Part 5: Declaration of bio-based carbon content, biobased synthetic polymer content and biobased mass content". NEN-ISO 16620-5 specifies the requirements for the declarations and labels of the bio-based carbon content, the bio-based synthetic polymer content, and the bio-based mass content in plastic products.

To conclude, how the measured amount of biogenic carbon is expressed as a percentage in the product, however, varies between different standards and certification schemes.

3.4.1.2 Bio-based mass content

A plastic can also be specified as bio-based by indicating its *bio-based mass content*. This method is complementary to the ^{14}C-radiocarbon-method and takes chemical elements other than the bio-based carbon into account, such as oxygen, nitrogen, and hydrogen. Consequently, measured bio-based carbon content can deviate significantly from the actual biomass content. The applied approach, for carbon to determine the bio-based content of a plastic based on isotopic measurements, cannot be used for other elements, such as oxygen, nitrogen, or hydrogen because there are no isotopes available for those elements. However, based on chemical (reaction) formula and elemental composition of bio-based and fossil-based feedstock, it is possible to calculate the bio-based content of a product taking all major elements into account. This alternative approach is specified in:
- NEN-EN 16785-1:2017 "Bio-based products – Bio-based content – Part 1: Determination of the bio-based content using the radiocarbon analysis and elemental analysis", which includes the amount of carbon, oxygen, hydrogen, and nitrogen derived from biomass in the bio-based content. The elements oxygen, hydrogen, and nitrogen in the bioplastic are considered to be derived from biomass if they are bound to carbon that has been derived from biomass. The method involves validation of a statement from the manufacturer about the composition of the product, based on evaluation of combined experimental radiocarbon and elemental analysis results.

– NEN-EN 16785:2018 "Bio-based products – Bio-based content – Part 2: Determination of the bio-based content using the material balance method", which specifies a method of determining the bio-based content in products using the material balance applied to a representative product batch in a production unit.

NEN-EN 16785 distinguishes two types of visualizations: the "bio-based content" label that shall be used on (the packaging of) the certified product or product family and the "bio-based" logo that may be used for other communication purposes (e. g., promotion and marketing). Note that the "bio-based content" label and "bio-based" logo (Figure 3.143) do not provide information on the sustainability or *environmental impact* of the product.

Figure 3.143: The "bio-based content" label, indicating the content, the product it refers and a unique registration number (left) and the "bio-based" logo (right).

Table 3.8: Gap between bio-based carbon weight fraction (in %) and the overall biomass fraction (in %).

	Bio-ethanol	Bio-ethylene	Bio-acetic acid
Molecular weight	46	28	60
Carbon atoms	2	2	2
Biomass fraction (in %)	100	100	100
Bio-based carbon weight fraction (in %)	52	85.7	40

The following example in Table 3.8 indicates the difference between the two methods described above.

3.4.2 Biodegradable plastics

Biodegradation is a chemical degradation process in which materials are metabolized into water, CO_2, and biomass (e. g., growth of the microorganism population) with the help of *microorganisms* (bacteria or fungi). So, both the biodegradability and the degradation rate of a biodegradable plastic product may be different in the soil, on the soil, in humid or dry climate, in surface water, in marine water, or in human-made systems like home composting, industrial composting, or anaerobic digestion. To claim a product's biodegradability, the ambient conditions must be specified and a timeframe for biodegradation must be set to make claims measurable and comparable. This is regulated in the applicable standards.

The property of *biodegradation* does not depend on the resource basis of a plastic but is rather directly linked to the chemical structure of the polymer and can benefit applications, in particular packaging applications. In other words, 100% bio-based plastics may be non-biodegradable, and 100% fossil-based plastics can biodegrade.

Figure 3.144 shows global production capacities for commercial *biodegradable plastics* in 2023. Clearly, the total capacity amounts to 1.14 million metric tons of biodegradable plastics. Compared with the global production capacity of conventional fossil-based plastics (400 million metric tons in 2022) this is only 0.29% of the total! However, it is expected that by 2028 the global production capacity for biodegradable plastics will increase to 4.61 million metric tons. So, in between 2022 and 2028, an additional capacity of 3.47 million metric tons will be installed worldwide, whereas the capacity increase for bio-based plastics is limited to 1.78 million metric tons. Current major biodegradable plastics are PBAT, PBS, PLA, PHAs, starch-containing polymer compounds, among others. Note that these biodegradable plastics can be bio-based, partially bio-based, or even fossil-based.

3.4.2.1 Mechanism of biodegradation

Polymers contain either a hetero-chain or a carbon backbone. The degradation mechanism of hetero-chain backbone polymers is chemical degradation via (enzyme-catalyzed) *hydrolysis*. Hydrolytic biodegradation mainly depends on the hydrolytic enzymes secreted by local microorganisms and the physico-chemical properties of the polymer. Through this process, polymer biodegradation may occur within a month. The hydrolytic biodegradation lifetime of a polymer can be controlled with the use of additives or chemicals to suit various practical applications. Hydrolytic biodegradation is the mechanism for some naturally occurring biopolymers like polysaccharides and proteins, plant source polymers such as PLA, and PBS and microbially synthesized polymers like PHAs. Many factors, such as chemical bonds, types of copolymers, thickness, water uptake, and morphology can influence the rate of hydrolytic degradation in enzyme-mediated or non-enzyme-mediated conditions.

Comparing hetero-chain and carbon chain polymers, the carbon backbone polymers biodegrade more slowly. The degradation mechanism of carbon backbone polymers is chemical degradation via oxidation or oxidative enzyme-mediated degradation.

Hydrolytic biodegradation occurs in polymers that contain hydrolysable groups, such as polysaccharides, polyesters, and polyamides, when they are exposed to moisture in biotic environmental conditions. The biodegradation of aliphatic polyesters is like the biodegradation of cellulose and chitin via enzymatic hydrolytic degradation. It has been reported that a group of esterase enzymes is responsible for the hydrolytic degradation of aliphatic polyester groups. Esterase enzymes, such as lipase, have

2023
Total: 1.14 million tonnes

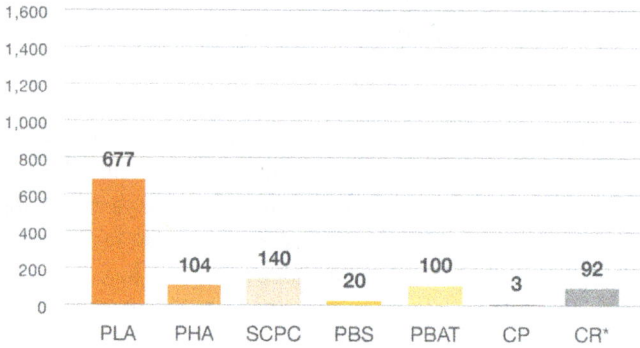

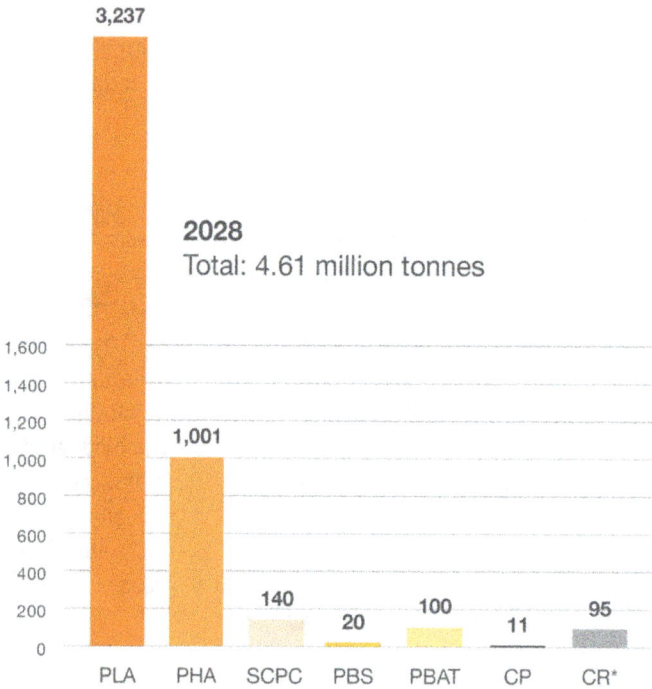

2028
Total: 4.61 million tonnes

* Regenerated cellulose films

Source: European Bioplastics, nova-Institute (2023)

Figure 3.144: Production capacities of biodegradable plastics 2023 vs. 2028 [47].

been well-studied and are more capable of hydrolyzing aliphatic polyesters than aromatic groups. Esterase enzymes are proven to hydrolyze triglycerides into fatty acid and glycerol. The main reason for this kind of degradation is that the active site of lipase has more contact with the main chain of the polymer, due to the hydrophilic nature of the aliphatic polyester, rather than the aromatic polyester, which is hydrophobic. These enzymatic reactions are heterogeneous; the hydrolytic enzymes adsorb onto the surface of the substrate polymer through the binding site of the molecules, and then the hydrolytic enzymes directly contact the ester bonds in the polymer backbone to break it down into functional group molecules. Each group of hydrolytic enzymes has a specific active binding site for substrate molecules, determined by the binding capacities of the substrate. These enzymes can hydrolyze polymers into low molecular weight compounds, which are then capable of undergoing bio-assimilation processes with the aid of naturally occurring microorganisms. It has been shown in extensive studies that bio-based polymers such as starch, cellulose, PLA, PHAs, chitin, protein, and fossil-based polyesters like polycaprolacton (PCL) and poly(butylene-co-adipate-co-terephthalate) (PBAT) readily undergo biodegradation under composting conditions.

Hydrolytic degradation breaks the polymer into fragments small enough to pass through the cell membrane of the active microorganism. These fragments enter the microorganism and are metabolized. In the case of aerobic biodegradation, this process yields a full decomposition of the polymer into carbon dioxide (CO_2) and water. In anaerobic biodegradation, carbon dioxide, water and methane are produced from the polymer. This process is called mineralization. It is the final step in biodegradation, and this means that the active microorganisms utilize carbon sources from the polymeric carbon as a nutrient.

3.4.2.2 Industrial composting

Compostable plastics are plastics that break down at composting conditions. Industrial composting conditions require elevated temperature (55–60 °C) combined with a high relative humidity and the presence of oxygen, and they are in fact the most optimal compared to other every day biodegradation conditions: in soil, surface water, or marine water. Compliance with European standard NEN-EN 13432:2000 "Packaging – Requirements for packaging recoverable through composting and biodegradation – Test scheme and evaluation criteria for the final acceptance of packaging" is considered a good measure for *industrial compostability* of biodegradable packaging materials.

Standard NEN-EN 13432:2000 requires that the following four characteristics are tested in a laboratory:
- Biodegradability, namely fragmentation and loss of visibility in the final compost – this is measured in a pilot composting test in which specimens of the test material are composted with biowaste for three months. After this time, the mass of test material residues must amount to less than 10% of the original mass. Other

methods should be used to measure the biodegradability of the packaging materials. Packaging materials are mixed with biowaste and spontaneously composted for 12 weeks in realistic composting conditions. At the end of the composting cycle, the disintegration is measured by sieving of the compost and the calculation of a mass balance. The influence of the tested sample on the quality of the compost can be studied by using the compost obtained at the end of the composting process for further measurements such as chemical analyses and ecotoxicity tests. Additionally, this method can be used for visual perception and photographic documentation of the disintegration of packaging materials and for evaluating the effect of their addition on the composting process.

– Biodegradability, namely the capability of the compostable material to be converted into carbon dioxide (CO_2) under the action of microorganisms. The standard contains a mandatory threshold of at least 90% biodegradation that must be reached in less than six months.
– Absence of negative effects on the composting process.
– Amount of heavy metals must be below given maximum values, and the final compost must not be affected negatively (no reduction of agronomic value and no ecotoxicological effects on plant growth).

Similar requirements are in place in the European standard NEN-EN 14995:2007 "Plastics – Evaluation of compostability – Test scheme and specifications" for non-packaging plastic items.

The same holds for ISO 18606:2013 "Packaging and the environment – Organic Recycling" and ISO 17088:2012 "Specifications for compostable plastics". ISO 18606:2013 specifies procedures and requirements for packaging that is suitable for organic recycling. Packaging is considered as recoverable by organic recycling only if all the individual components meet the requirements. Therefore, packaging is not considered recoverable by organic recycling if only some of the components meet the requirements laid down in this International Standard. However, if the components can be easily, physically separated before disposal, then the physically separated components can be individually considered for organic recycling. ISO 17088:2012 specifies procedures and requirements for the identification and labelling of plastics, and products made from plastics, that are suitable for recovery through aerobic composting.

European standard NEN-EN 13432:2000 requires at least 90% disintegration after 180 days, 90% biodegradation (CO_2 evolvement) in six months, and includes tests on ecotoxicity and heavy metal content. Note that complete *biodegradation* of a plastic product has occurred when 90% or more of the original material has been converted to CO_2. The remaining share is converted into water and biomass, which no longer contains any plastic. NEN-EN 13432:2000 is the standard for biodegradable packaging designed for treatment in industrial composting facilities and anaerobic digestion.

Labels for industrially compostable products, for example, the Seedling Logo, OK Compost, and DIN-Geprüft Industrial Compostable are presented in Figure 3.145.

Figure 3.145: Labels for industrially compostable products.

3.4.2.3 Home composting

At home composting conditions, temperature is lower and less constant compared to industrial composting conditions due to the smaller amount of compostable material and the lack of an industrial set-up. As a result of the lower temperature, the degradation rate of a material is (much) slower compared to industrial composting, depending on the type of material.

There is currently no international standard specifying the conditions for home composting of biodegradable plastics. However, there are several national standards, such as the Australian standard AS 5810:2010 "Biodegradable plastics – biodegradable plastics suitable for home composting". Belgian certifier TÜV Austria Belgium had developed the OK compost home certification scheme "Bio products – degradation in soil", requiring at least 90% degradation in 12 months at ambient temperature. Recently, the German certifier DIN CERTNO became the first certification body in the world to add the new French standard NF T 51-800:2015 to its certification scheme. The French standard is NF T 51-800:2015 "Specifications for plastics suitable for home composting". Labels proving *home compostability* are OK compost Home and the DIN-Geprüft Home Compostable mark, as depicted in Figure 3.146.

Figure 3.146: Labels for home composting.

3.4.2.4 Soil degradability

Belgian certifier TÜV Austria Belgium developed a certification scheme "Bio products – degradation in soil" based on both European standards NEN-EN 13432:2000 "Packaging – Requirements for packaging recoverable through composting and biodegradation – Test scheme and evaluation criteria for the final acceptance of packaging" and NEN-EN 14995:2007 "Plastics – Evaluation of compostability – Test scheme and specifications" and adapted those for the degradation of bioplastics in soil. The test demands are at least 90% biodegradation in two years at ambient temperatures. The European standard NEN-EN 17033:2018 "Biodegradable mulch films for use in agriculture and horticulture – Requirements and test methods" specifies the requirements for biodegradable films, manufactured from thermoplastic materials, to be used for mulching applications in agriculture and horticulture, which are not intended to be removed. A degradation of at least 90% in two years preferably at 25 °C is required. The label OK Biodegradable Soil is certified by the Belgian certifier TÜV Austria Belgium in the case a product meets the requirement of their *certification* scheme. DIN CERTCO awards DIN-Geprüft biodegradable in soil (Figure 3.147) in accordance with CEN/TR 15822 "Biodegradable plastics in or on soil – Recovery, disposal and related environmental issues".

Figure 3.147: Labels for degradability in soil.

3.4.2.5 Marine degradability

Currently, there is no standard in providing clear pass/fail criteria for the degradation of biodegradable plastics in seawater. Several standardization projects are in progress at ISO level. Research and development is ongoing to develop harmonized standards for marine biodegradation that are needed before relevant products can be introduced to the market. Recently, TÜV Austria Belgium has developed a certification scheme based on ASTM D7081-05 "Standard Specification for non-floating biodegradable plastics in the marine environ*ment*", which demands, in a simplified way, a biodegradation of at least 90% in six months. The corresponding label is OK biodegradable Marine (Figure 3.148).

Figure 3.148: Label for marine degradation.

All standards provide various test methods to evaluate the potential biodegradability of bioplastics in various disposal environmental conditions such as soil, compost, and marine water. The biodegradation behavior of plastics is mainly associated with their chemical structure. However, both the physical and chemical properties of plastics (discussed in Chapter 2) can influence the polymer biodegradation mechanism as well. Such properties include surface area, hydrophobic and hydrophilic nature, chemical nature, molecular weight, molecular weight distribution, crystallinity, crystal structure, glass transition temperature, melting temperature, and elasticity.

3.4.2.6 Oxo-degradability

Oxo-biodegradation is biodegradation as defined by the European Committee for Standardization (CEN) in CEN/TR 15351:2006 "Plastics. Guide for vocabulary in the field of degradable and biodegradable polymers and plastic items", as degradation resulting from oxidative and cell-mediated phenomena, either simultaneously or successively. This degradation is sometimes termed oxo-degradable, but this latter term describes only the first or oxidative phase of degradation and should not be used for material which degrades by the process of oxo-biodegradation as defined by CEN.

Oxo-degradable plastics are made from conventional plastics (e. g., PE or PP) supplemented with additives (often salts of manganese or iron) in order to mimic biodegradation. They cannot be considered as bioplastics and have failed to prove proper biodegradability in any environment. The standards that are claimed to confirm the biodegradability of such products, most notably the US standard ASTM D6954-18 "Standard Guide for Exposing and Testing Plastics that Degrade in the Environment by a Combination of Oxidation and Biodegradation", do not provide pass/fail criteria, leaving these misleading claims wholly unsubstantiated.

So, it is important to distinguish between oxo-degradable plastics that fragment but do not biodegrade except over a very long time, and oxo-biodegradable plastics that degrade and then biodegrade. Degradation is a process that takes place in almost all materials.

In the same way, the UK-based company Wells Plastics developed the Reverte™ technology for PE, PP, and PET. Reverte™ is a family of oxo-biodegradable additive masterbatches that has been developed for a wide range of polymers and packaging applications.

On January 16, 2018, the EC published a report on the use of oxo-degradable plastic. The document, titled "Impact of the use of oxo-degradable plastic, including oxo-degradable plastic carrier bags, on the environment" [48] forms part of the European strategy for plastics in a circular economy, which was released the same day. In this document, titled "A European Strategy for Plastics in a Circular Economy", the EC emphasizes that the Commission has started work with the intention to restrict the use of oxo-plastics in the European Union. The EC focused on three key issues relating to oxo-degradables: the biodegradability of oxo-degradable plastics in various environments; the environmental impacts in relation to littering; and recycling. The EC found there was no conclusive evidence that, in the open environment, oxo-degradables fragmented to a sufficiently low enough molecular weight to enable biodegradation.

A more recent report from the EC, of May 2018, states that micro plastics need to be restricted, including oxo-degradable plastics.

In a background paper, titled "Oxo-biodegradable plastics and other plastics with additives for degradation", European Bioplastics distances itself from additive-mediated conventional plastics such as so-called oxo-degradable plastics. The technology of additive-mediated fragmentation entails that a conventional plastic is combined with special additives, which are purported to promote the degradation of the product. Yet, the resulting fragments remain in the environment and do not biodegrade as defined in internationally accepted industry standards such as European standard NEN-EN 13432:2000 for industrial composting. Products made with additive-technology and available on the market include film applications such as shopping bags, agricultural mulch films and, most recently, certain plastic bottles. Experts from the plastics industry, waste management, and environment protection voice serious concerns about these products not meeting their claimed environmental promises.

3.4.3 Global production of bioplastics

According to the latest market data of European Bioplastics, the total global production capacity of *bioplastics*, both (partly) bio-based and/or biodegradable, will grow from 1.813 million metric tons in 2022 to 7,432 million metric tons in 2028 (Figure 3.149). The market will continuously grow when more sophisticated bioplastics are introduced on the market and new market applications are developed.

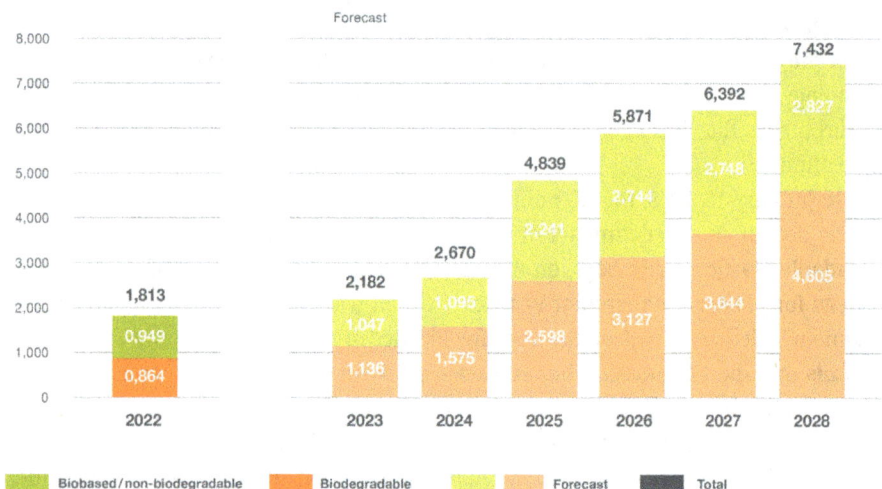

Source: European Bioplastics, nova-Institute (2023)

Figure 3.149: Global production capacities of bioplastics [47].

Innovative bioplastics, such as PLA and PHAs, are the main drivers of this growth in the field of both bio-based and/or biodegradable plastics. PHAs are an important polymer family that has been in development for a while and that now finally enters the market at commercial scale, with production capacities estimated to quadruple in the next five years. These *polyesters* are 100% bio-based and marine and soil biodegradable and feature a wide array of physical and mechanical properties depending on their chemical composition. Production capacities of PLA are also predicted to fivefold by 2028 compared to 2023. PLA is a very versatile plastic that features excellent barrier properties and is available in high-performance PLA grades that are an excellent replacement for polystyrene (PS), PP, and acrylonitrile-butadiene-styrene (ABS) in more demanding applications.

The expected market applications for bioplastics in 2028 are presented in Figure 3.150. As can be observed in the chart, both flexible (lightweight bags or pouches) and rigid packaging (cans, bottles, and jars) account for 52% of all applications. Clearly, the rigid *packaging* applications are dominated by bio-based, non-biodegradable plastics whereas flexible packaging is dominated by biodegradable plastics.

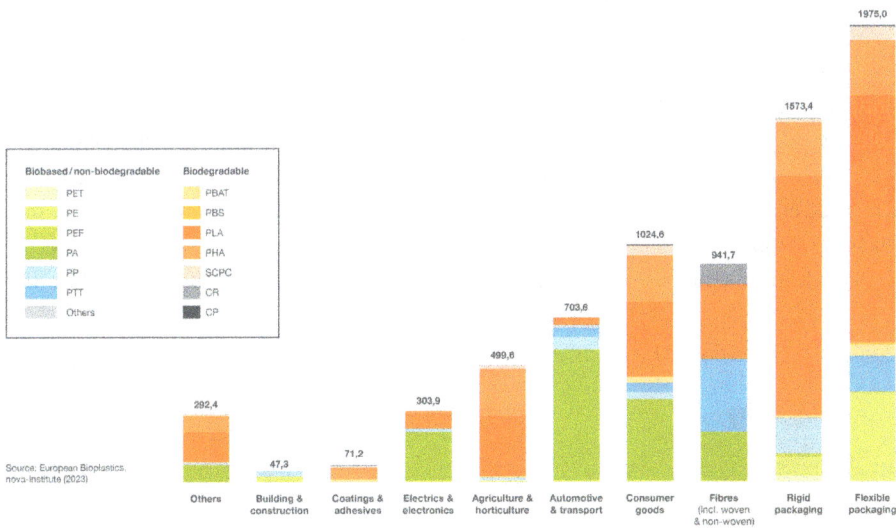

Figure 3.150: Global production capacities of bioplastics in 2028 (by market segment) [47].

In 2028, the major part of all bioplastics will be produced in Asia (71.5%) (Figure 3.151). Note that in 2023, 55.3% of all bioplastics were produced in Asia.

The transition from a fossil-based to a *bio-based economy* is essential in view of climate protection and greenhouse gas emission-reduction targets. In this way, bioplastics can make a major contribution to this transition by replacing fossil-based plastics. As indicated, bioplastics are mostly produced from carbohydrate-rich crops, such as corn, wheat, sugarcane, or sugar beet, so-called food crops or 1G feedstock.

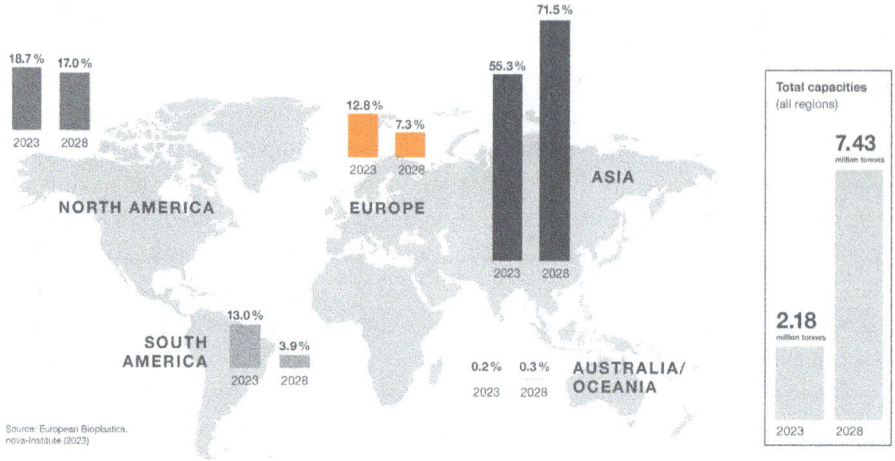

Figure 3.151: Global production capacities of bioplastics in 2028 (by region) [47].

Currently, 1G feedstock is the most efficient feedstock to produce bioplastics because it requires the least amount of land to grow on and produces the highest yields. The discussion about the use of biomass for industrial purposes is often linked to the question whether the conversion of potential food and feed to materials is ethically justifiable. The area needed to grow biomass for material use accounts for approximately 2% of global cultivated areas. Of this share, bioplastics account for about 0.02%. The sheer difference in volume shows that there is no competition between the use of biomass for food, feed, and for material use (including bioplastics). What is more, 1G feedstock bioplastics are an enabling technology that will eventually facilitate the

Land use estimation for bioplastics 2019 and 2024

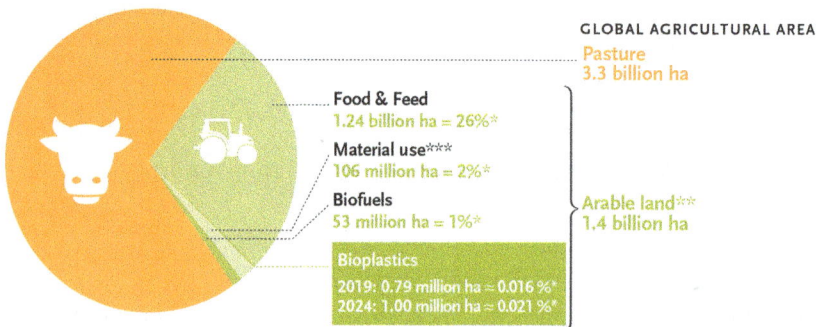

Source: European Bioplastics (2019), FAO Stats (2017), nova-Institute (2019), and Institute for Bioplastics and Biocomposites (2019). More information: **www.european-bioplastics.org**

* In relation to global agricultural area
** Including approx. 1% fallow land
*** Land-use for bioplastics is part of the 2% material use

Figure 3.152: Land use estimation for bioplastics 2019 and 2024 [47].

transition to later generations of feedstock (both 2G and 3G). Figure 3.152 gives the estimated land use for feedstock for bioplastics in 2019 and 2024.

3.5 Exercises

3.5.1 Biorefining technologies

1. Name four main features used in the IEA Biorefinery classification approach.
2. Mention two platforms which can be recognized in the development of bioplastics.
3. Describe the difference between biogas and syngas.
4. Describe the difference between cellulose and hemicellulose.
5. Which four different basic conversion processes can be distinguished?
6. Draw the IEA process flow diagram for the production of biodiesel from oil crops.

3.5.2 Biochemicals

7. Write the chemical structures and names of four top value chemicals as listed in the DOE study from 2004.
8. Name both monomers, listed in the DOE list, needed to produce PBS and PEF.
9. Write the chemical structures of both PBS and PEF.

3.5.3 Chemistry of bioplastics

10. Three different types of bioplastics can be distinguished. Describe in your own words the difference between these three types of bioplastics.
11. The Brazilian company Braskem is a supplier of bio-based polyethylene (bio-PE).
    a. Write the chemical structure of bio-PE.
    b. Which feedstock is used to produce Braskem's bio-based polyethylene?
    c. Draw the IEA process flow diagram for the production of Braskem's bio-based polyethylene.
    d. Is bio-based polyethylene bio-degradable or compostable?
    e. Discuss the recyclability of bio-based polyethylene.
12. Describe the difference between 1G and 2G feedstock for the production of bio-ethanol.
13. Olefin metathesis is an interesting process step to develop new materials. Which chemicals can be used to produce bio-based propylene (bio-PP)?
14. Inovyn recently launched so-called "bio-attributed" PVC (Biovyn™). Biovyn™ is produced from 100% renewable ethylene derived from renewable feedstock. Describe the synthesis of Biovyn™.

15. Bio-based polystyrene (bio-PS) can be prepared from bio-based 1,3-butadiene. Write the reaction scheme to produce bio-PS. Is bio-PS biodegradable or compostable?
16. Discuss a possible process route for the production of bio-based acrylic acid.
17. Polyethylene terephthalate (PET) is produced from two basic starting monomers.
    a. Give the names and chemical structures of the monomers.
    b. Write down the process routes to produce bio-based analogous of the monomers.
    c. Discuss the compostability of bio-based PET.
18. Write a process route for the production of bio-1,4-BDO.
19. DuPont's Hytrel RS™ is a partially bio-based thermoplastic polyetherester. Write the chemical structure of Hytrel RS™.
20. US-based Genomatica recently developed the GENO-CPL™ technology. Describe in detail the GENO-CPL™ technology. Explain which bio-based monomer is produced in this process.
21. Evonik recently introduced several Vestamid Terra grades onto the market. Name two examples of Vestamid Terra.
22. Write the chemical structure of PBS, and names of the two monomers needed to produce PBS.
23. Discuss both the biodegradability and recyclability of PBS.
24. TianAn Biopolymer is world's largest producer of poly(3-hyroxybutyrate-*co*-3-hydroxyvalerate) (PHBV). Write the chemical structure of PHBV.
25. Describe the industrial process route to produce polylactic acid (PLA). Why is lactide used in the process, rather than lactic acid?
26. Write the chemical structure of polyglycolic acid (PGA).

3.5.4 Bioplastics

27. Explain the difference between biodegradability and compostability.
28. Name an example of a biodegradable plastic, and a compostable plastic.
29. Discuss the term drop-in bioplastics.
30. Some plastics may use the "seedling" logo. Discuss the background story of this logo.

4 Recycling of plastics

4.1 Circular plastics

The butterfly model of the Ellen MacArthur foundation (Figure 1.5) with regard to *circular plastics* contains of 2 closed loops: a biological cycle (Chapter 3) and a cycle for technical materials. In this chapter we will focus on the technical cycle. To achieve a sustainable and responsible use of plastics, the so-called 3R model is proposed by the Ellen MacArthur Foundation [6]. This promotes *reducing*, *reusing*, and *recycling* of plastics and plastic products.

Reduce

The most important rule in the 3R-model is to *reduce*. This means that we should try to reduce the consumption of (natural) raw materials as much as possible. In practice, this comes down to not purchasing non-necessary products. Plastics are widely used in the packaging industry. This involves the packaging of fruit and meat products, to extend shelf life, and the use of plastic carrier bags. In order to reduce the use of this packaging product, politics can ban the distribution of free plastic carrier bags. In some European Member States, plastic carrier bags are now prohibited, or consumers have to pay for it. The ban is an elaboration of a European Directive (2015/270) to reduce the use of plastic bags.

Reuse

The purpose of *reuse* is to use products as long as possible and for as many purposes as possible. Reuse reduces the need to purchase new products, and consequently reduces the consumption of resources. We should try to use products and materials as long as possible. An example is to extend the product life time by repairing or by refurbishing a product.

Recycle

Recycling includes collecting products at their end of life and the reprocessing into new raw materials for new products. Besides the conservation of valuable resources, recycling saves energy as well. The extraction of raw materials from nature and the production of new materials costs more energy than the recycling of these materials. The challenge for the future will be to optimize the recycling process in order to close the loop of plastics.

A commonly used classification of plastic recycling processes is the distinction between primary, secondary, tertiary, and quaternary recycling. Figure 4.1 shows a schematic representation of the four types of recycling. All four forms of recycling can be applied to most thermoplastic plastics.

https://doi.org/10.1515/9783111201443-004

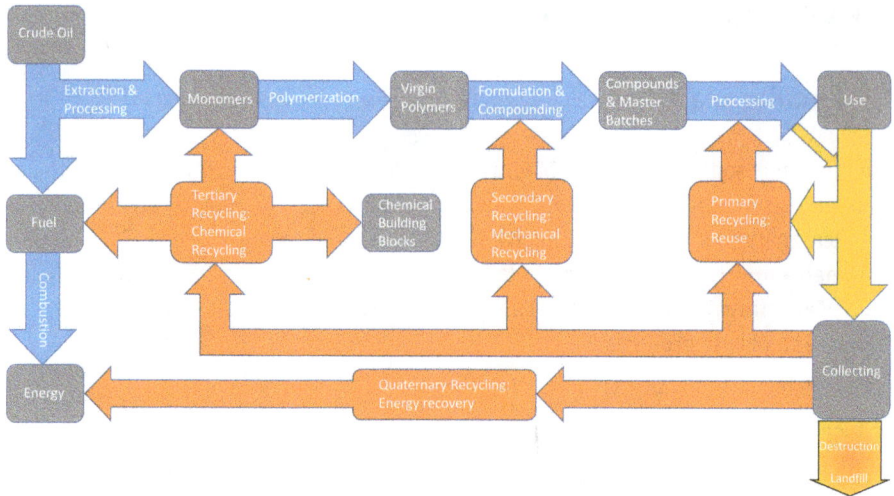

Figure 4.1: Four types of recycling: primary, secondary, tertiary, and quaternary recycling.

Primary recycling

Using uniform and not-contaminated plastic waste to manufacture plastic products by reprocessing, is named primary recycling. The recovered plastic is typically used in products with performance characteristics that are equivalent to products made from virgin plastics. Ideally, closed-loop recycling takes the recovered material and uses it in the original application. For example, plastic bottles that become new bottles, or, the grinding of crates for bottles into raw materials for production of new crates. Primary recycling is actually recycling for the same purpose. And in practice it means reuse of in-plant scrap in the primary production process.

Secondary recycling

Also known as *mechanical recycling*: using recycled plastic waste to manufacture new plastic products. The recovered plastic is typically used in products that have less demanding performance requirements than the original application. Secondary recycling often requires reformulation to meet specifications of the new product. For example, an old bottle crate can be ground into grains to make parts for a new vacuum cleaner, or, ground old bicycle tires that are used to manufacture mats.

Tertiary recycling

Tertiary recycling converts discarded plastic products into new monomers, high-value chemical building blocks or fuel feedstocks. This form of recycling is also called *chemical recycling*. Through a depolymerization process, polymer chains are broken down into raw materials.

Quaternary recycling

Actually, this form of recycling is used to convert raw materials into energy, by incinerating plastic waste and thereby generating steam and/or electricity. However, this is not a fully circular approach as referred to the above.

In this chapter, we will discuss the recycling of plastics in detail. With regard to thermoplastics, we will focus on mechanical recycling (secondary recycling), chemical recycling (tertiary recycling) and *product development*. The latter is a part of mechanical and chemical recycling, with regard to the total recycling chain.

4.2 Plastic packaging

As mentioned in Section 1.2, the largest percentage of the thermoplastics is used in the packing industry. Packaging material is however very diverse with regard to the materials used, applications, and properties. This brings along some challenges with regard to recycling. That is why a closer look at packaging material is needed, before recycling issues will be treated.

4.2.1 Polymers used in packaging

The most used plastics in *packaging* applications are [49, 50]:
- Polyethylene (PE): LDPE, LLDPE, EVAc, HDPE;
- Polypropylene (PP): hPP, cPP (homopolymer and copolymer);
- Polyethylene terephthalate (PET): a-PET, c-PET, PETG.

In *foils*, barrier polymers are used like polyamides (PAs) and ionomers. Polystyrene (PS) contains only 1–2% of the plastic packaging applications. Polyvinylchloride (PVC) is nowadays not allowed as packaging material.

Applications of PE in packaging

Different types of PE are used in the packaging industry, determined by the properties and applications (Table 4.1).

LDPE and HDPE are the most commonly used polymers in the packaging industry. Both polymers differ in properties (Table 4.2).

Figure 4.2 shows some applications of *LDPE* for different kinds of packaging [52]. Typically, LDPE is applied for every day, cheap products. Note that these applications can be used to a maximum temperature of 80 °C.

LLDPE has a higher puncture resistance than LDPE. Blended with LDPE, LLDPE is used for film applications (Figure 4.3).

Table 4.1: PE: polyethylene; HDPE; high-density polyethylene; LLDPE; linear low-density polyethylene; VLDPE: very low-density polyethylene; LDPE: low-density polyethylene; MDPE; medium-density polyethylene [51].

Type of PE	Density (g/cm³)	Melt flow index (g/10 min)
HDPE	0.941–0.965	0.2–3.0
MDPE	0.926–0.940	1.0–2.0
LDPE	0.915–0.925	0.3–2.6
LLDPE	0.915–0.925	0.1–10.0
VLDPE	0.870–0.914	0.026–0.1

Table 4.2: Comparison of properties of LDPE and HDPE [51].

Property	LDPE	HDPE
Structure	Branched chains	Linear chains
Density	0.915–0.925 g/cm³	0.941–0.965 g/cm³
Melting point	~115 °C	~135 °C
Crystallinity	Low crystallinity (50–60%)	Highly crystalline (>90%)
Flexibility	Lower crystalline, higher flexibility than HDPE	Higher crystallinity, more rigid than LDPE
Transparency	More amorphous than HDPE, good transparency	More crystalline than LDPE, less transparent
Heat resistance	Retains toughness and pliability over a wide temperature range, but density drops off dramatically above room temperature	Useful above 100 °C
Chemical properties	Robust, but exposure to light and oxygen results in loss of strength and loss of tear resistance	Chemically inert

Ethylene vinyl acetate (EVAc) copolymers are often used for *films*. The amount of vinyl acetate in the copolymer directs the glass and melting temperature and the crystallinity of the polymer. The amount varies between 10–40%. In this way the properties vary from thermoplastic to elastomeric (rubbery). EVAc can be blended with PE to improve the *sealing*, clarity, and adhering properties (Figure 4.4).

HDPE is often used for flacons, jerry cans, bottles, and bags. In Figure 4.5, some blow molded products of HDPE are illustrated.

Applications of PP in packaging

Just like HDPE, polypropylene (PP) is often used for production of flacons and jerry cans. Besides these applications PP can be used as a *hot melt adhesive*. For this appli-

Figure 4.2: Packaging applications for LDPE. Applications up to 80 °C.

Blended with LDPE for film applications
- Improved mechanical properties
- Improved puncture resistance
- Better elasticity
- Better clarity

Stretch film

Shrink film

Figure 4.3: Packaging applications for LLDPE [52].

cation only the atactic PP is used (Section 2.2.3.2). A random copolymer of PP and PE is used in transparent packaging. Because of the high stiffness and strength, isotactic and syndiotactic PP are used for *caps and closures*. By blending PP copolymers, the right properties can be obtained for specific applications in injection molded products or thermoformed packaging products.

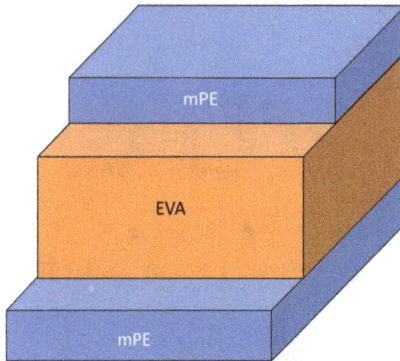

Figure 4.4: Stretch film with EVAc. The core layer of mPE is a linear metallocene polyethylene.

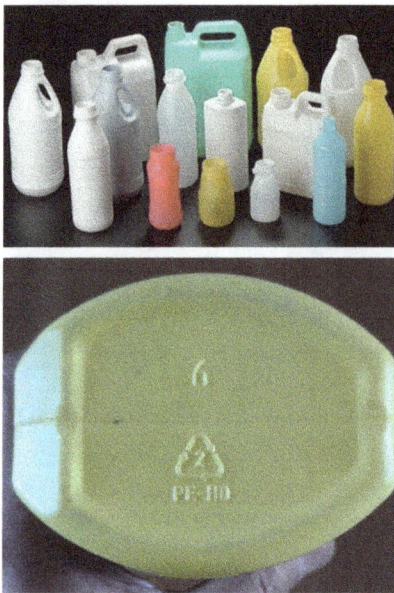

Figure 4.5: Blow molded HDPE products. Please note the differences with packaging applications for LDPE.

Applications of PET in packaging

Besides applications of *polyethylene terephthalate* (PET) in fibers and films, PET is often used in bottle and tray applications. Different grades of PET lead to different applications (Table 4.3).

In the packaging industry different types of PET can be distinguished: a-PET, c-PET, and PETG.

Amorphous PET (a-PET) is used as a thermoformed film in packaging, while crystallized PET (c-PET) is used for bottles. Amorphous PET can be used for bottles as well, when 2% isophthalic acid (IPA) is added. PETG is a modified PET, in which the monomer ethylene glycol (EG) is partly replaced by cyclohexanedimethanol (CHDM) resulting in a

Table 4.3: PET grades and their application depending on the intrinsic viscosity.

Polyester grade	Intrinsic viscosity (dL/g)
Fiber grade (textile, staple fiber)	0.40–0.70
Film grade	0.70–1.00
Bottle grade	0.70–0.78
Water and soft drink bottle grade	0.78–0.85

PET which has a much lower, or no melting point and remains transparent under all conditions, which is a big advantage in the thermoforming process. To adapt the properties of PET bottles, sometimes a SiO_x-coating is used on the inside of the bottle to prevent gas (CO_2) transportation out of the bottle.

Thermoformed PET films are used in the packaging industry for food applications. Often in combination with other techniques used in packaging. A few examples:
- self-adhesive labels with strong glues;
- inks and direct printing on PET films;
- different types of lidding film (peelable, re-sealable or permanent);
- soaker pads attached with holt melt glues (meat packaging);
- multilayer PET with PE layer for better sealing;
- sealed blister packaging.

Different challenges arise with regard to the different applications. Color changes of a-PET in the recycling process in combination with different mechanical behavior of a-PET compared to c-PET in bottle applications. In the mechanical recycling process (Section 4.3) a NIR detection system cannot distinguish between a-PET and PETG. *Multilayer films* can hardly be recycled in a mechanical way. In addition, combinations of PE and PET in freezer packaging are difficult to recycle.

4.2.2 Multilayer packaging

In packaging applications transfer of oxygen, nitrogen, carbon dioxide, or water vapor plays a very important role [52]. *Barrier properties* are important in food applications because of the following reasons:
- exchange of *moisture*: protection against drying (for instance bakery products), protection of hygroscopic products (for example milk powder);
- oxygen permeation: microbiologic spoilage (oxidation of grease, flavors, bakery products, nut products, ready-to-serve meals);

- emerging of CO_2: change of protective gas type; undesired vacuum effect (bakery products, meat products);
- exchange of aroma: loss of aroma from the product packaged; assuming a different smell (coffee, spices, washing agents).

Barrier properties of films with regard to oxygen can be expressed in the *oxygen transmission rate* (OTR). The same can be done for water, the so-called *water vapor transmission rate* (WVTR).

Mass transport through the layers can be expressed on the basis of diffusion by Fick's law. Diffusion rate depends on temperature, concentration difference however, is no driving force for transportation. In equation (4.1) a simplification is shown for a steady state situation. In this equation P = coextruded film permeation coefficient, f_n = polymer n layer thickness ratio, P_n = polymer n permeation coefficient, TM = transmission of film, and t_t = total thickness of coextruded film. For a non-steady state situation, one uses a differential equation, based on Fick's law.

$$\frac{1}{P} = \frac{f_1}{P_1} + \frac{f_2}{P_2} + \cdots + \frac{f_n}{P_n}, TM = \frac{P}{t_t}. \tag{4.1}$$

Application of EVA in packaging

Ethylene vinyl alcohol (EVA) is a copolymer used for specific barrier properties of films in packaging of food. By increasing the amount of the ethylene component in EVA, the OTR and WVTR values decrease. Polyvinyl alcohol (PVA), which is the homopolymer, is hydrophilic and soluble in water (see Section 2.3.5).

Application of PA in packaging

Polyamide (PA) is used in packaging to improve the mechanical properties. In films especially, to increase puncture resistance and strength. PA is also used in the PET bottle industry to change the barrier properties. The OTR and WVTR values of PA and PET are comparable, however, to decrease the OTR and WVTR values of PET bottles, poly(m-xylene adipamide 6) (MXD6) is used. MXD6 is nylon in which a xylene group is incorporated (Figure 4.6). A multilayer PET bottle often consists of 5 layers: PET/MXD6/PET/MXD6/PET. The storage time of these multilayer bottles is doubled, from 10 to 20 weeks.

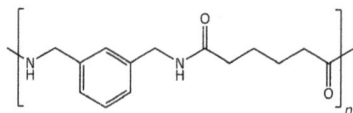

Figure 4.6: The structural formula of Nylon MXD6.

Application of ionomers in packaging

So-called *ionomers* are polyethylene polymers modified with acrylic acid. Typical ionomers are poly(ethylene-*co*-acrylic acid) ionomer (Na, Zn) and poly(ethylene-*co*-methacrylic acid) ionomer (Na, Zn) (Figure 4.7). These ionic copolymers have typically a low melting point, improved toughness, flexibility and mechanical strength. When used as films they possess a much higher clarity and gloss and provide superior hot tack, seal strength, and puncture resistance compared to unmodified PE.

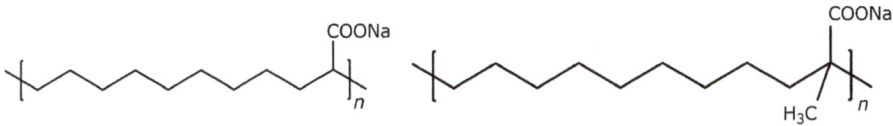

Figure 4.7: The structural formulas of poly(ethylene-*co*-acrylic acid) ionomer (Na, Zn), left, and poly(ethylene-*co*-methacrylic acid) ionomer (Na, Zn), right.

Figure 4.8 shows the design route for a multilayer (laminate) foil applied in specific food packaging. In general, the development of barrier films is based on what kind of food is packed (fish, meat, or fruit) and what kind of gas or aroma barrier is needed. Also, the moisture conditions are essential. Transportation of the product, shelf life, and appearance are important demands. Because of the fact that some layers are not compatible, an additional adhesive layer is needed.

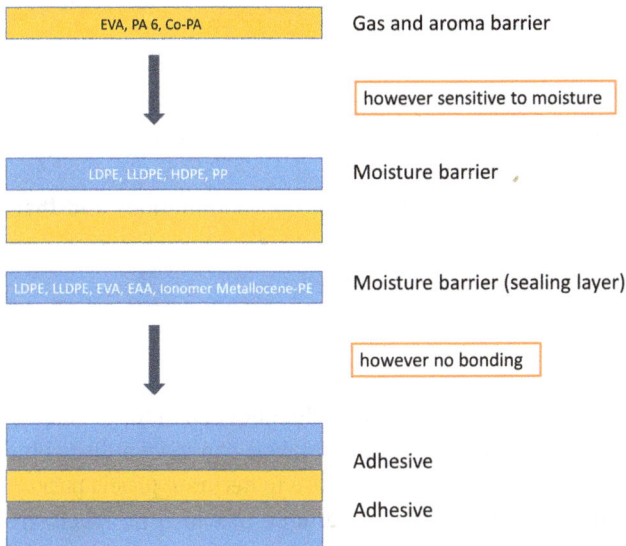

Figure 4.8: Design route for barrier films depending on the food application.

Based on these demands, barrier films are designed starting with, for example, three layers up to fourteen layers or even more (Figure 4.9). To improve adhesion between poorly adhering layers, special adhesive polymers or so-called TIE resins have been developed. These resins are typically polyethylene copolymers with polar and nonpolar repeating units. The *multilayer films* are produced by a *coextrusion process*, often stretched after extrusion to reduce film thickness. Bottles of PP or HDPE are often made out of a laminate too.

Barrier polymers are often hydrophilic and are therefore positioned in the middle of the *laminate*. The polymer on the outside is often hydrophobic, therefore an adhesive layer is needed to connect the different polymers. Sometimes a non-fogging additive is used.

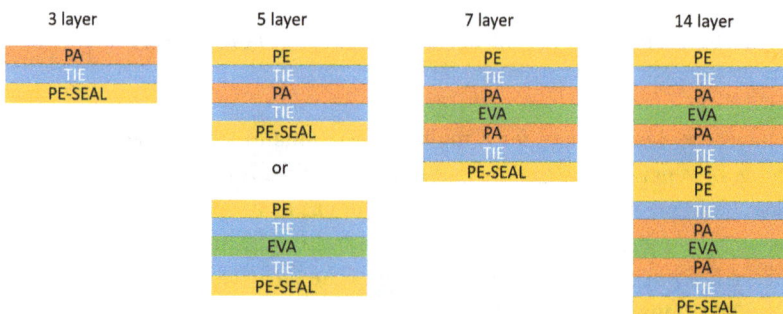

Figure 4.9: Depending on the demands of the barrier layers, multilayer films of 3 to 14 layers or even more are used in packaging. TIE is an adhesive based on a copolymer of PE.

Hydrophilic polymers (more *wettable*) are polymers containing side groups like: $-OH$, $-COO^-$, $-Al_n(OH)_m$, etc. Hydrophobic polymers (less wettable) are polymers containing side groups like: $-CH_3$, $-CH_2CH_2$, $-CF_3$, *etc.*

In Table 4.4, a diagram is shown with different layer combinations and their compatibility.

4.2.3 Decoration of packaging

With regard to the decoration of packaging, different techniques and kinds of labeling can be distinguished (Figure 4.10). In general, glue and labels can have a disturbing influence on the recycling process. However, it is possible to decorate plastic packaging without using a label. In-mold printing is a technique in which no labels are needed. Disadvantage is that the surface tension of the polymer must be increased by using a Corona treatment, to improve the adhesion of inks.

Table 4.4: ■ good adhesion; ■ marginal adhesion; ■ requires TIE layer resins.

	Acid copolymer	Ionomer	EVA	LDPE	HDPE	PP	PVDC	PA 6	EVOH	PET
PET	green	green	orange	yellow	orange	orange	orange	green	yellow	orange
EVOH	yellow	yellow	orange	orange	orange	orange	orange	orange	orange	
PA 6	orange	green	orange	orange	orange	orange	orange	orange		
PVDC	green	green	orange	green	green	green	orange			
PP	yellow	green	orange	green	green	green				
HDPE	green	green	orange	orange	orange					
LDPE	orange	green	orange	orange						
EVA	orange	green	orange							
Ionomer	orange	orange								
Acid copolymer	orange									

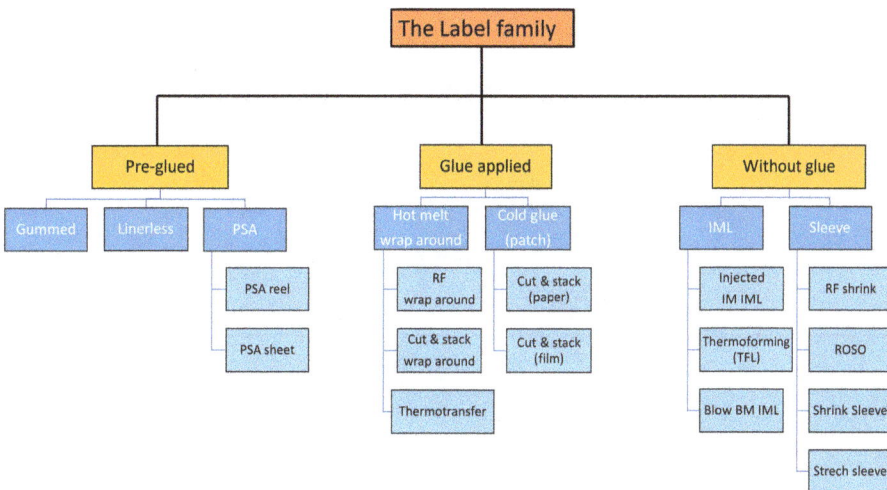

Figure 4.10: Decoration of plastic packaging. PSA = pressure sensitive application = self-adhesive labels; RF = reel fed; ROSO = roll-on shrink-on; IML = injection molded labeling.

4.3 Mechanical recycling

Mechanical recycling of plastics relates to the processing of plastic waste into secondary raw materials or products in which the chemical structure of the material does not change substantially. In principle, all thermoplastics can be mechanically recycled with little or no loss of quality. Mechanical recycling is currently the most used form of recycling and represents more than 99% of recycled quantities. However, chemical recycling becomes more feasible and will be a substantial part of recycling of plastics in the future (Section 4.4).

Waste streams that simply produce large quantities of clean plastic of one species are ideally suited for mechanical recycling. From an environmental and economic point of view, this leads to a win-win situation, in which the benefits to the environment by replacing new raw materials are generally bigger than the burden on the environment by collecting, *sorting*, transportation and recycling activities. The cost of such activities can be offset by the potential revenue from the sale of recycled products.

Plastics that cannot be sustainably recycled from an economic or environmental point of view according to the required standards, still can be valuable raw materials for other recovery solutions, such as energy production.

Although the basis of *mechanical recycling* lies in washing and sorting, other processing techniques exist. Also, in many waste processing facilities, the order of steps in the mechanical recycling chain differs. In this chapter we will deal with the different recycling steps in the most common order (Figure 4.11).

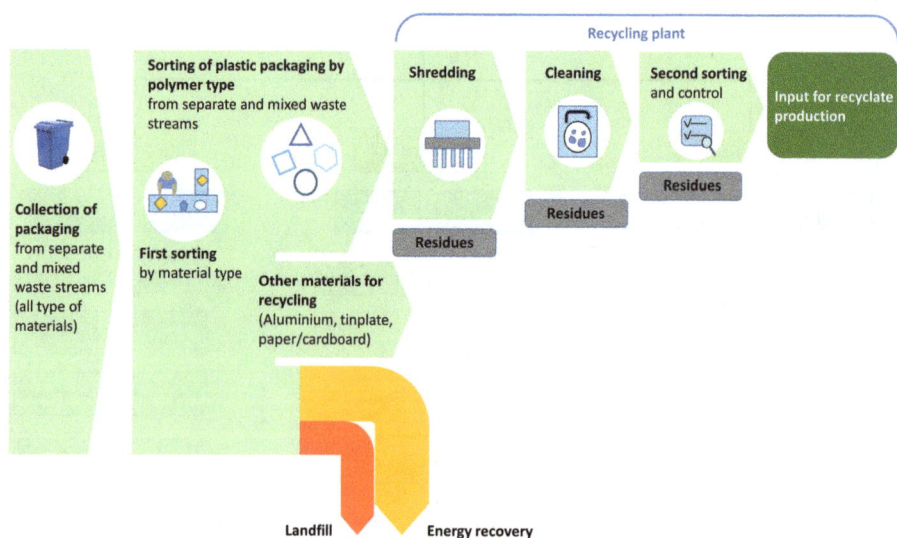

Figure 4.11: Schematic representation of the processing steps in a plastic recycling plant, starting with packaging waste.

4.3.1 First sorting

This section discusses several methods for sorting waste material [53], such as paper, plastics, organic waste, textiles, metals, wood, leather, and rubber. Waste sorting facilities receive waste from collectors and process this waste in several steps. Since the waste is a combination of materials, it requires different techniques to process and sort the waste.

There are two different concepts of sorting [54]:

- *positive sorting*: focuses on identifying and removing a desired fraction from the waste stream. This will result in a high-quality material product, however with a lower efficiency;
- *negative sorting*: focuses on identifying and removing a non-desired fraction from the waste stream. The result is quite the opposite of positive sorting; it is quite efficient, however with a lower quality of the obtained materials.

4.3.1.1 First-sorting techniques

Waste sorting facilities can adapt both methods of sorting for an optimal efficiency. To make it even better: a manual or robotics sorting approach can be added to remove contaminants and ensure quality. When waste is collected by garbage trucks and transported to a sorting plant, the waste will first have to be registered, weighed, and checked before it proceeds to the next step, which is typically classification. In this step the waste gets sorted by properties like shape, size, and density. This can be done with a trommel screen, a disc screen, or an oscillating screen. Screening is the same as filtering the waste in size, shape, and density. A few important first waste sorting techniques are summarized.

Trommel screen

The trommel screen is a long tube that resembles a sieve. The outer walls of the trommel has holes to transport the waste. The *trommel screen* is constantly rotating and is fed with waste. The holes of the trommel screen will get bigger as the waste passes through it, allowing waste of a certain size to fall through and be sorted. When the waste does not fit any of the holes it just exits the trommel screen on the other end.

Disc screen

The *disc screen* is a piece of equipment consisting of rotating discs. It separates waste through the clearance between the discs depending upon the size and the weight of the waste, while the waste moves on the rotating discs.

Oscillating screen

An *oscillating screen* is a vibrating declined bed filled with holes allowing small waste to pass through. Waste that will not fit through the holes will be transported to the other end. The oscillating screen usually has two or more beds, with decreasing hole sizes, allowing even smaller waste to be sorted at the end.

Air separation
There is another way to separate waste, which is through blowing air at undesired contaminants. The contaminants will be intercepted and sorted. Air separation is often combined with the most common spectroscopic methods (Section 4.3.3), making it an efficient and fast sorting method.

Zigzag air classifier
A zigzag *air classifier* separates waste based on weight, through an air current. Waste material is dropped into a zigzag shaped channel. The waste then gets blasted upwards by an air current. The lightweight fraction gets blown out at the top and the heavy weight fraction leaves at the bottom. This will separate the heavy weight fraction from the lightweight fraction.

Rotary air classifier
A rotary air classifier is actually a trommel screen separator with an air current that captures the lightweight fraction and separates it from the heavyweight fraction. The rotary air classifier works on the same principle as the zigzag air classifier, but within the rotary air classifier the waste gets separated with a classifier rotor through a *centrifugal force*. The lightweight fraction gets blown out by the rotary air classifier and the coarser or heavier weight fraction leaves the rotary air classifier at the bottom.

Cross-current air classifier
Waste material gets inserted through a hopper and the waste then gets blasted by an air current. This current takes away the lightweight fraction, while the heavy weight fraction drops out *via* an opening in the cross-current air classifier.

Suction hood
A suction hood sucks the lightweight fraction directly from a conveyor belt, leaving only the heavy weight fraction on the conveyor belt.

Ballistic separation
A ballistic separator is a mechanical sorting device with oscillating paddles that run along the length of the sorting deck. This steeply inclined bed with perforated plate screen will allow small waste material to fall through. Three fractions will be captured through this method: fine material, lightweight material and heavy weight material. In fact, the ballistic separator works on the same principle as the oscillating screen.

Film grabber

Waste material on a conveyor belt is led through a rotating drum with spikes. The spikes of the rotating drum will hook plastic bags and films, while other waste material will pass. Plastic bags or films usually get stuck in machinery, making it hard to sort. The *film grabber* is a good solution to this issue.

Eddy-current separation

Most common method used to separate non-magnetic metals from waste is *Eddy-current separation*. Eddy current separators (non-ferrous separators) consist of a conveyor belt system with a fast-spinning magnetic rotor at the end [55]. The rotational speed of the magnets generates an induction field, creating a rapidly changing magnetic field. Separation is based on the principle that every electrically conductive particle that is in a changing magnetic field becomes temporarily magnetic. For a short time, all the metals passing through the eddy current magnetic roller are magnetic, causing them to be 'blown away'. In this manner, a lot of alloys, non-ferrous metals, including aluminum, copper and brass can be separated from waste streams.

Eddy current separators have many applications. They can handle high capacities because the conveyor belt continuously (and automatically) separates and disposes non-ferrous metals. Important for a good separation of non-ferrous metals is an even supply of material, for example achieved *via* a vibrating gutter or transport belt. As a result, there is an even spread over the belt and a mono layer is created. This means that the layer thickness is about the same thickness as the largest particle and therefore no parts are stuck on top of each other. Application examples include:

- removal of aluminum caps for the glass recycling industry;
- removal of brass sounds and hinges in the wood recycling industry;
- removal of non-ferrous metals from snails from incineration plants;
- processing of (electronics) scrap metal or household waste;
- purifying recycled plastic flows to protect injection molding machines;
- cleaning up mining and mineral flows from, for example, broken excavator teeth;
- recovery from casting residues in metal casting industry.

Manual sorting

Manual sorting is still used nowadays, since it is efficient and cheap. Employees are placed beside a conveyor belt and manually remove contaminants from a waste stream, because sometimes the machinery will overlook certain contaminants that can be harmful for other waste streams. The manual sorting can be done on the basis of either positive sorting or negative sorting.

Robotic sorting

More and more waste sorting robots are implemented in de waste sorting plants nowadays. The robots can detect and recognize each recyclable item, from wooden parts to aluminum cans and plastic bottles, according to external characteristics like shape and color and by using a robotic arm to seize it. The robot moves recyclable items to different containers according to type. Robots have an accuracy of around 90% and can recognize items that are partly covered by other objects. They can be equipped with a visual-recognition camera and several arms.

4.3.1.2 Shredding

Shredding plays a key role in both recycling plants and production facilities [53]. *Plastic shredders* not only help with size reduction, but also with recouping waste plastic. Shredders are an indispensable component of efficient and cost-effective operations in the recycling and waste management industry. They are designed for shredding a wide variety of plastics and so, they vary from low speed to moderate speed with high torque and come in varying specifications and blade sizes (Figure 4.12).

Figure 4.12: Example of a plastic shredder.

The various applications of plastic shredders include the plastic industry, laboratories, pharmaceutical companies, manufacturing units, catering industry, food-processing facilities, bio-medical waste management plants, nursing homes, healthcare facilities, cardboard manufacturing units, and supermarkets. Shredders are often used, in these applications because of the following reasons:
- plastic shredders help recover waste during different industrial processes like molding, trimming and casting;
- shredded waste is easier to handle or reuse;

- *shredding* is an efficient *waste disposal* method, and this greatly reduces the cost of recycling for several industries;
- plastic shredders are being used to generate revenue by facilities that sell shredded plastic for use as raw material;
- they convert plastic waste into reusable raw material for manufacturing plastic containers, PET bottles and PVC products;
- plastic shredders help recoup waste disposed from commercial units;
- when used for shredding important business documents and digital storage devices that contain confidential details, plastic shredders help maintain privacy;
- by recycling and repurposing the plastic waste, companies can also reduce their carbon footprint and save the environment.

Plastic shredders are an efficient and effective solution for turning scrap materials into valuable resources no matter the size and shape of plastic. Shredders are extremely useful when working with plastic products, vinyl materials, and PVC pipes. Plastic in any unwanted form can be transformed into manageable and useful material that can be applied in (re)manufacturing. The internal mechanism of a plastic shredder typically moves in lateral, vertical or rotary direction and the speed may vary depending on the material input. Plastic shredders can be classified into the following six types based on their target application:

- hammer mills: used for shattering or pulverizing materials in a rotary drum that employs swiveling hammers;
- granulators: used for recycling plastic from production processes. Granulators also include thermoforming units for producing scrap material that is easy to handle;
- chippers: featuring high-speed rotary knives that efficiently reduce material to flakes and chips;
- grinders: using the combined power of abrasion and compression to pulverize materials and produce granular products;
- shear shredders: available in different configurations, these shredders employ rotary cutters and guillotine style knives for cutting materials according to industrial needs;
- all-purpose shredders: widely used for recycling purposes and demolition programs, all-purpose shredders are large, run at low-speed and have high torque shafts.

4.3.1.3 Washing

Although washing conditions vary for different plastics, the washing procedure often consists of two stages; a pre-washing step and a main washing step. The pre-washing occurs at ambient temperature with water to remove course solids. The main washing step is often carried out at higher temperatures around 50–80 °C. In this step, the residual consumer residue, for example food residue, is washed off along with labels and other contaminants. Detergents are often added to increase washing efficiency. For example, caustic soda can be added in order to remove glue residues. These chemicals are often applied for washing relatively more contaminated plastics such as municipal plastic waste (MPW). Other plastics such as agricultural plastic foils might not have glued labels or *pollutants* that require these extreme conditions. For such materials, the main washing step can also be carried out at ambient temperature with only water as the washing medium. In general, the washing process will be combined with a shredding step, either before or after.

During the washing procedure, parameters that play an important role are the temperature of the medium, amount of washing cycles, mass of the plastic waste, stirring speed of the washing unit, recirculation and type of *detergents* [56]. The composition of the *wastewater* is rather variable. It depends, amongst others, on the type and quantity of contamination of the plastics that are being washed. Contamination varies between different types of plastics and can also vary over time. Municipal plastic waste, for instance, can vary in composition between different residential areas, but also between different seasons. Mixed plastic waste originating from MPW, such as food packaging, is often contaminated with organic residues of food. Agricultural plastics are mostly contaminated with soil and organic residues. Both can therefore lead to higher *chemical oxygen demand* (COD) concentrations compared to plastics without such organic residues.

The washing of PET meat *packaging* delivers washing water containing *fatty acids* and *proteins* (Figure 4.13). It may be possible to generate value from a waste resource. For example, certain plastic recycling companies already produce polyhydroxyalkanoates (PHAs). However, the conditions of the wastewater should be suitable for the production of those bioplastics. The pH, temperature and biochemical oxygen demand (BOD) should all be within a suitable range in order to achieve a viable yield. In addition, the microorganisms should not be too severely affected by the presence of (eco)toxic compounds in the wastewater.

In order to produce bioplastics, one would need an acidification step to produce volatile fatty acids (VFAs), a fermentation step to convert VFAs to medium chain fatty acids, followed aerobic production of *polyhydroxyalkanoates* (PHAs), extraction and purification of PHA and several solid liquid separators in between these steps (Chapter 3). Clearly, investments are required for the installation and operation of these kind of systems.

Kind of contaminants

There are several types of *contaminants* with respect to plastic waste.
– Organic contaminants: fats, fatty acids, proteins, *sugars*;
– Biological contaminants: microbial contaminants;
– Minerals: sand and clay;
– Labels: paper, ink, and glue.

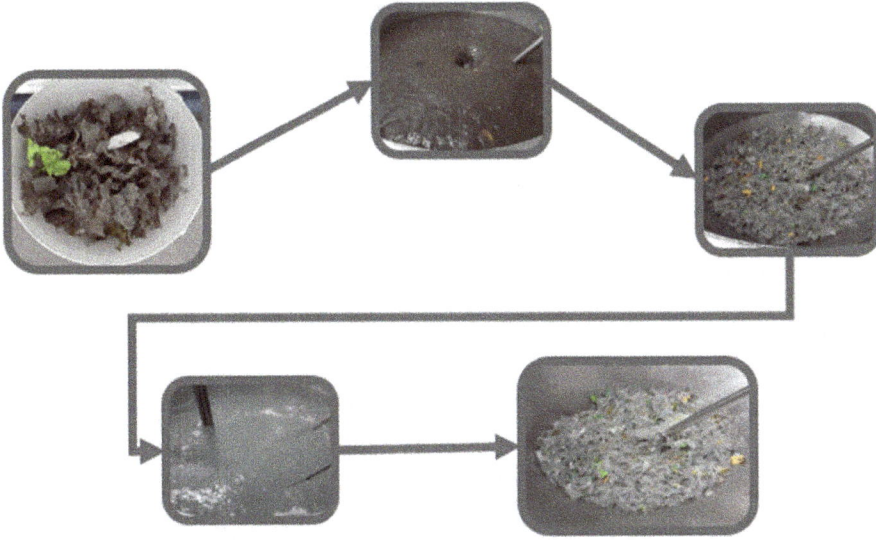

Figure 4.13: Washing procedure of PET meat packaging.

Besides those, waste streams contain additives, degradation products of additives and degraded polymers.

When treating municipal plastic waste, one of the main pollutants found in the wastewater are fibers originating from labeling materials. This paper pulp might contribute to the total COD of wastewater. Labeling materials can also contain inks and dyes which may end up in the washing wastewater (Section 4.2.3). Consequently, traces of metals and soot, in case of thermal printing, originating from these inks and dyes are found. Moreover, the detergents and caustic soda from the main washing step, and course solids from the pre-washing step can be expected in the waste wash water. For some batches of washed plastic, the presence of a single piece of contamination can also affect the composition. For instance, the presence of an ink cartridge or battery in a batch of municipal plastic waste.

Cleaning methods

Mechanical cleaning is basically separating the contaminants from the plastic by mechanical actions. This can be achieved by scrubbing, brushing, rinsing, and stirring. The contaminants are often separated by *filtration*. Solvation is possible when the contaminants can be dissolved in a medium. It depends on temperature, contact surface, and speed of mixing. Saponification can help to enhance the solvation process. Different kinds of detergents can be applied depending on the waste stream.

The mechanisms of cleaning all have their limitations, such as saturation of washing water and shape of the plastic material. During the washing procedure the washing water can be saturated with contaminants. This depends on the amount of water, chemicals added, temperature, and mechanical methods used. The same applies for diffusion and solubility processes. In addition, the efficiency of the above mentioned mechanisms depend strongly on the shape and the flexibility of the plastic particles in the waste stream. Contaminants can accumulate in voids, edges or in between foils, which influences the diffusion speed. Contaminants which are not soluble in water can only be partly removed by mechanical action.

Costs and environment

The amount of energy, water and detergents determine the efficiency of a cleaning process. The setup of the washing utility is an important factor in the efficiency of mechanical and chemical cleaning. It is obvious that temperature and time parameters play an important role as well. This topic will become more important in the future. In most cases the washing procedure takes place in a stirred tank. Turbulence will cause better and quicker cleaning of the plastic particles. However, because of the density of plastics, which is often around 1.0 kg/dm^3, particles are flowing along with the fluid. This results in a more laminar part of the flow. Evaluation of the effectiveness of the cleaning procedure can be done by determining the mechanical properties or by visual observation (eye, color analysis, microscopy) of the plastic waste.

4.3.2 Second sorting

Separating different types of plastics from one another can be performed by spectroscopic analysis techniques, density difference and size difference of the particles and marker systems. The next paragraphs will discuss a number of methods on the basis of these techniques. Note that some of the methods discussed can also be used in the first sorting step (Section 4.3.1).

4.3.2.1 Float–sink

The principle of float-sink (or drift-sink) separation is based on separating particles by density difference. The particles to be separated are brought into a liquid medium. Particles with a smaller density than the medium will start to float, while particles with a higher density will sink (Figure 4.14). In water, PP and PE can be separated from heavier plastics, such as PS, ABS and PVC. The separation between PP and PE is done by using a liquid with a density lower than that of PE (0.950 g/cm^3) and higher than that of PP (0.920 g/cm^3). As liquid phase, a mixture of water and methanol or ethyl alcohol is used. In order to change the density of one of the two plastics, a swelling agent can be used, since the difference in density of PE and PP is very small. When the PE/PP mixture is contaminated with other substances, or if pigments have been used in the plastics, pollution from the plastic residues may affect the density of the medium and thus affect separation during the process. The density of the medium

Material feed

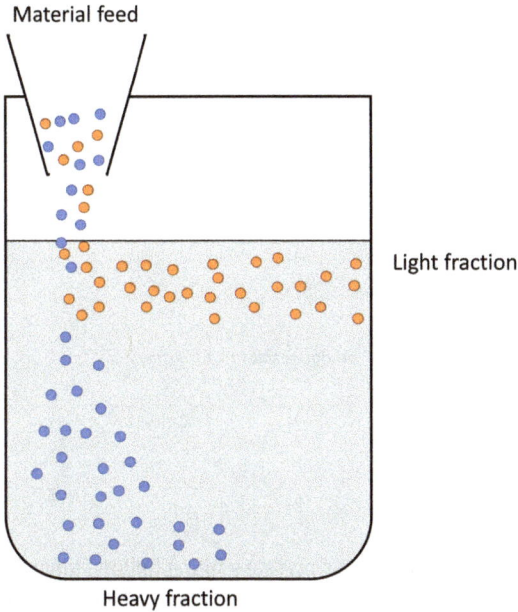

Light fraction

Heavy fraction

Figure 4.14: Float-sink process.

will vary according to its temperature as well. In addition, evaporation will change the composition of the medium, resulting in a slow float-sink process. Surface potential of the particles can vary the float-sink process and disrupt the *separation*. A lighter particle of PP can be forced to sink by clumping with heavier PE particles.

4.3.2.2 Froth flotation
Flotation uses the water-repellent properties of a number of plastic particles [57]. The plastic particles are brought into a water bath, through which air is blown. The air bubbles attach themselves to the water-repellent (hydrophobic) particles that float to the surface, while the other (hydrophilic) particles sink (Figure 4.15). This method is appropriate for separating plastics with similar densities, but different chemical nature. The water-emitting properties can be altered by adsorption of capillary active substances, by oxidation of the plastic surface, by altering surface voltage (gamma flotation) and by plasma treatment.

Most experimental flotation setups are built to separate rather clean plastic. When plastic waste from households is used, the separation by flotation does not work well. The contaminants present will disturb the flotation process.

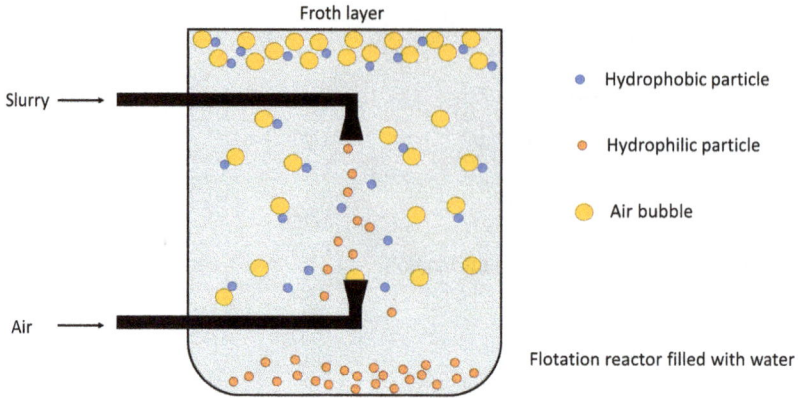

Figure 4.15: Froth flotation process.

4.3.2.3 Hydrocyclone

In a *hydrocyclone* process, a suspension is inserted under pressure into a hollow, tapered pipe (Figure 4.16) [58]. Under the influence of the centrifugal force, larger or heavier solid particles concentrate on the wall, from where they are drained (yellow arrow in the figure). The lighter or finer particles are pushed to the upper loop (blue arrow in the figure) by the liquid. This technique is very often used to improve the quality of a pre-separated batch. The use for large batches is limited because the capacity is limited.

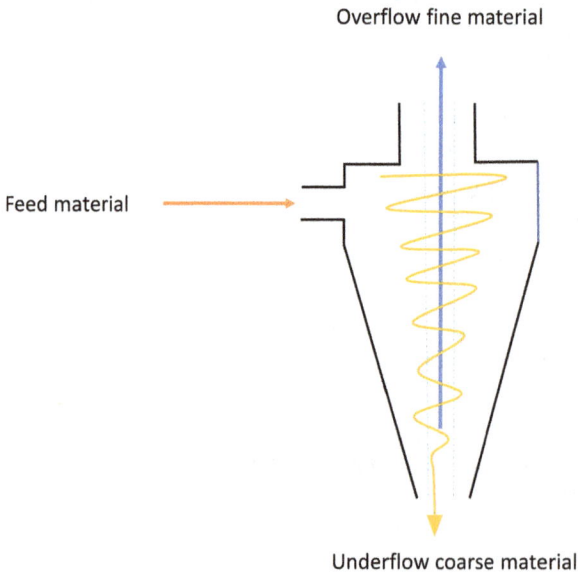

Figure 4.16: Schematic representation of sorting by a hydrocyclone [18].

4.3.2.4 Wind shifter

A *wind shifter* blows, by using a fan, an air flow through the material [59]. With this air flow, the specific weight (surface mass ratio) causes a separation between two plastics (Figure 4.17). Wind shifters can be divided into two groups. Wind shifters with an air flow either perpendicular or parallel to the material flow. The principle of perpendicular flow is to blow out light particles from the material by means of a fan that blows horizontal above the material. The principle of the parallel flow is the extraction of light particles in a vertical channel. This technique is used to separate light fractions like foils and dust.

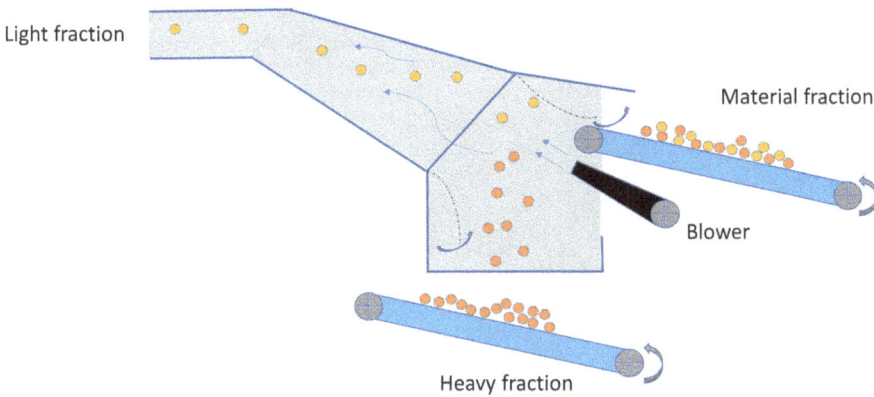

Figure 4.17: Schematic representation of sorting by a wind shifter [19].

4.3.2.5 Electrostatic separation

In case of electrostatic separation, the flow to be sorted is guided through a drum, shaker or cyclone, charging the flow electrostatically, either by exposure to an electric field or by mutual contact of the particles. This phenomenon is called tribo-electric charging. In the end, the particles fall down in an electric field. Depending on their load, the particles move to the positive or negative electrode. In this way, three or more groups can be obtained: in the case of Figure 4.18, two fairly pure groups and a mixing group that can be traced back to be separated once more.

Mixtures containing different kind of plastics can be separated, however with two types of plastics, the separation is optimal. Required size of the particles is in the range of 2–10 mm. The material to be sorted must be dried to a humidity level below 0.8% and no metal, dust, oil, fat or surfactants may be present in the flow. Moreover, the properties should not vary too much, which can cause problems for PE, since HDPE and LDPE have other electrostatic properties while PE and PP are very similar.

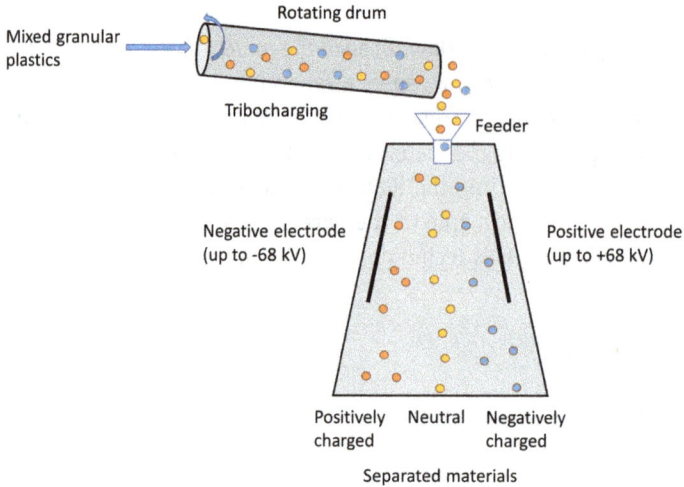

Figure 4.18: Electrostatic separation.

4.3.2.6 Magnetic-density separation

Magnetic-density separation (MDS) is a method in which polymer species can be separated by density. In this process, an artificial gravity field is created by a liquid in motion, such as water, that consists of plastic waste, and a magnetic substance, such as iron oxide. This special magnetic field results in a variable density of the flowing liquid, with the density at the top being higher than at the bottom. The polymers with different densities will sort themselves by the height of the fluid flow, where the density is equal to the polymer. With this technique, polymers with small differences in density can be separated. An example of an MDS separation system is presented schematically in Figure 4.19.

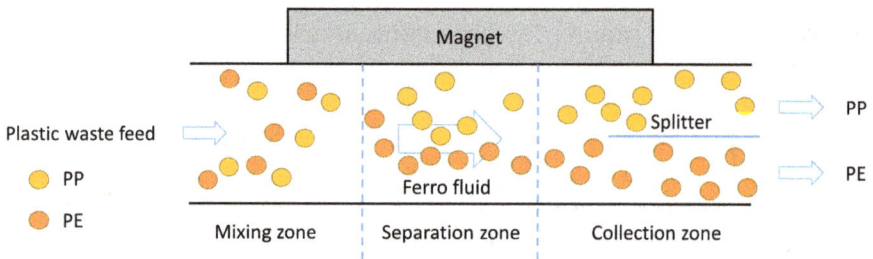

Figure 4.19: Schematic rendering of Magnetic Density Separator (MDS).

Before *plastic waste* can be separated into the MDS, the products must be shredded. These products must then be cooked in a $CaCO_3$-solution, so that air bubbles are removed. Air bubbles that remain attached to the surface of the plastic waste have a major influence on density. Therefore, wetting techniques should be used to ensure that these air bubbles do not remain present on the surface. As a drawback, these wetting methods may affect the plastic waste due to high temperatures or because of chemical interactions between plastic and the active substance.

In principal, large quantities of plastic waste can be processed with an MDS. There are processing quantities known with a plastic waste flow up to 180 kg/h. In the near future it is expected to be used on a commercial scale. At an initial mixed waste flow consisting of PP/PE in a ratio of 70:30, a pure polymer content over 95% can be achieved. Both small and larger plastic packaging can be separated within an MDS system. Another advantage of this technique is that the ferrofluid can be reused almost completely.

However, there are still a number of challenges to be solved for this separation method. In an MDS system, polymer blends, multilayers, or polymers mixed with many additives or fillers cannot be used. These factors influence the density of the polymers. As a result, these polymers end up in the wrong stream. In addition, a successful separation is highly dependent on the turbulence in the MDS system. Nevertheless, this turbulence can be properly controlled by a streamlined supply of the plastic waste and ferrous fluid.

In practice, the purity of the product separated by MDS is often not yet good enough. This is most likely caused by contaminations in the product. More research is needed to improve the quality of the MDS's final product.

4.3.2.7 Super critical fluid

By using a *super critical fluid* (SCF), like CO_2, as solvent, it is possible to separate the constituents of a multicomponent mixture by taking advantage of both the differences in volatilities of the components (distillation), and the differences in the specific interactions between the mixture components and the SCF solvent (liquid extraction). In the critical region, a substance exhibits a liquid-like density and an increased solvent capacity, which is pressure dependent. The variable solvent capacity of a SCF is the basis on which separation processes can be developed. By operating in the critical region, the pressure and temperature are used to regulate density and consequently the solvent power of a SCF. It is important however to know that the interactions between the solvent and solute molecules determine how much solute dissolves in the SCF. Thus, increasing the SCF density increases the probability that the solvent and solute molecules will interact.

4.3.3 Spectroscopic methods

There are various *spectroscopic methods* for sorting available, nowadays. These techniques are useful for the sorting of plastics, non-ferrous metals and ferrous metals. An overview of the principals of the techniques, their function, application, benefits and drawbacks will be given in the next paragraphs. All the techniques are based on different kinds of radiation with specific frequencies and wavelengths. The *electromagnetic spectrum* is the collection of all possible frequencies inside the electromagnetic spectrum. Each type of radiation of the electromagnetic spectrum has its own effect on (macro)molecules. In Figure 4.20, various radiations and their effect on molecules are illustrated [60].

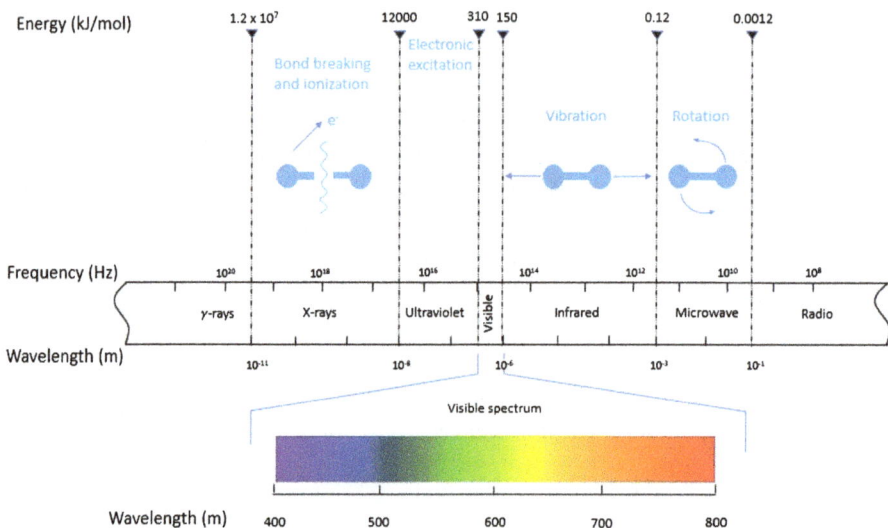

Figure 4.20: Components of electromagnetic spectrum.

Microwave radiation
When substances are irradiated with microwave radiation the molecules will start to rotate. Aside from radio waves, microwaves are the weakest radiation in the electromagnetic spectrum.

Infrared radiation
When molecules are irradiated with infrared (IR) radiation they will start to vibrate. Compared to other radiations IR is quite weak. Infrared radiation has five types of radiation: near infrared (NIR), short wavelength infrared (SWIR), mid wavelength infrared (MWIR), long wavelength infrared (LWIR), and far infrared (FIR). The shorter the wavelength the more energy it can release.

Visible light
This is the light that can be seen with the human eye. Visible light is essential for plants, because they require photosynthesis to grow. The sun is a natural source of visible

light. When our skin is exposed to light it creates vitamin D. Vitamin D strengthens bones, teeth and the immune system.

Ultraviolet light
Ultraviolet light contains a higher energy than visible light. It can carry out electronic excitation and induce chemical reactions. A long exposure to ultraviolet light can harm humans and animals.

X-ray
X-ray is the second strongest radiation in the electromagnetic spectrum. It can break and ionize bonds. It may cause cancer on the long run.

Gamma radiation
Gamma radiation is one of the strongest radiations in the electromagnetic spectrum, so strong it can penetrate through walls and thick covers. It can cause serious damage to DNA to the point of mutations, cancer and even death. Gamma radiation must be treated with utmost caution.

4.3.3.1 Near infrared
The plastic flow to be sorted passes an identification unit that uses a *Near Infrared* (NIR) identification system [61]. This detects the spectrum of the objects and thus recognizes the composition of each material. When the material does not meet the requirements set, it is blown out of the stream, by using compressed air (Figure 4.21). This separation technique is usually followed by a reduction and float-sink separation.

Black plastics cannot be separated with NIR. The materials should also not be too humid. When blowing away a piece of identified plastic, a number of neighboring pieces are usually blown along. As a consequence, these systems are more efficient if the incoming group is already relatively pure, or if the density of material on the assembly line is lower. In practice, these techniques should be used as a quality control step after an earlier separation has brought purity to around 95%. Moreover, it is better if the pieces to be analyzed do not differ too much in size. Therefore, a prior reduction step and even the formation of flakes may be necessary to obtain a desirable separation. Usually 30–350 mm size is required, sometimes for flakes 3–12 mm.

4.3.3.2 Hyperspectral imaging
Hyperspectral imaging is an innovative technique that combines the properties of digital imaging with those of spectroscopy [62]. Using this approach, it is possible to detect the spectral signature of each pixel of the acquired image in different wavelength regions (visible, near infrared and short-wave infrared) according to the characteris-

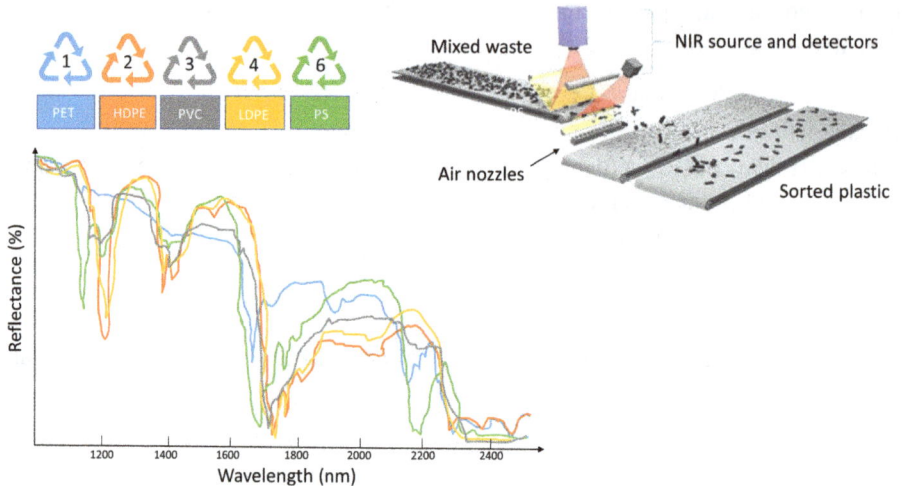

Figure 4.21: Schematic representation of NIR plastic sorting unit and NIR response (spectra) of some common plastics.

tics of the selected sensing device. A *hyperspectral image* can thus be considered as a hyper-spectral cube with two spatial dimensions and one spectral dimension. Because images made with hyperspectral cameras contain many bands, they are difficult to fully interpret by humans. Help can be offered by artificial neural networks, inspired by the human brain.

A neural network needs to be trained on data that it needs to detect. This technique in which these neural networks are used is called "deep learning" and is a subfield of machine learning, concerned with algorithms inspired by the structure and function of the brain: artificial neural networks. Neural networks can be of use in a lot of different cases, like in facial recognition, object detection, and speech recognition. Currently it is investigated if neural networks are capable of detecting and classifying different components of plastic household waste. In order to reach this goal, methods are developed that are capable of categorizing a large quantity of plastics from household waste, using a hyperspectral camera and artificial neural networks. Figure 4.22 shows preliminary results of hyperspectral imaging of common plastics.

4.3.3.3 Shortwave infrared

Most recycle centers use Shortwave Infrared (SWIR) spectroscopy for sorting plastics in the waste stream. Cost effective SWIR 1024 or 512 element line scan cameras with wavelength sensitivities ranging 1,000–2,500 nm are usually mounted onto a spectrograph. While monitoring the sorting conveyor, the spectrograph rapidly identifies the polymer type in the waste and triggers separation of the waste into its proper bin.

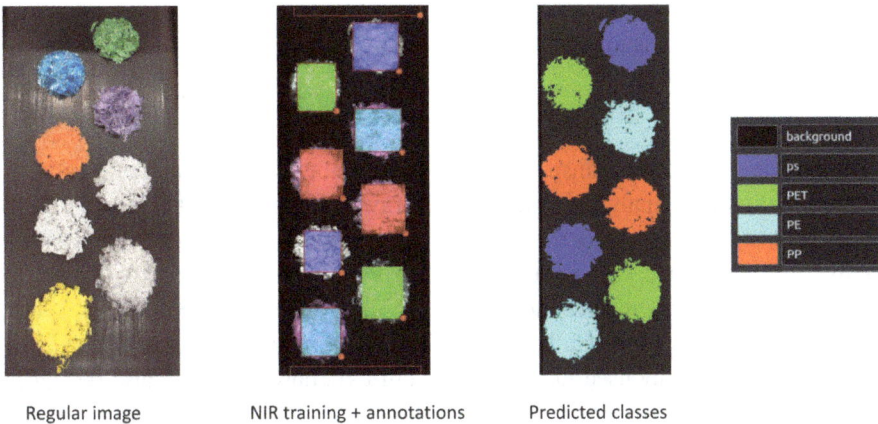

Regular image   NIR training + annotations   Predicted classes

Figure 4.22: Sorting spectrum of different kind of plastics using a hyper spectral camera [63].

NIR and SWIR are both capable of reliable identification and sorting of clear and most colored plastics. In dark materials, however, the black pigments fully absorb this type of infrared light and covers the spectral signature of the plastic itself, making proper identification challenging.

4.3.3.4 Midwave infrared

Black plastics are widely used in the automotive industry, electronics, food packages, and plastic bags; they can be found all around us. Unfortunately, instead of recycling, they are burned for energy or dumped to landfill, since there has not been efficient and reliable technology available to sort them. Hyperspectral imaging is the only technology that can identify black plastic types, when used on Midwave Infrared (MWIR) spectral range.

Certain hyperspectral cameras available on the market cover the full MWIR spectral range 3,000–5,000 nm that is required for recognizing black plastics. This allows fast and reliable sorting of not only black PS, PE, ABS and PVC, but also rubbers and *additives*, like flame retardants.

4.3.3.5 UV-VIS

UV-VIS spectroscopic data plot the intensity of transmitted or reflected light as a function of the corresponding wavelength. Due to their characteristically broad absorptive features, UV-VIS spectra are usually analyzed by automated analysis routines. In plastic recycling UV-VIS is used as an analyzing technique for:

- accurately identifying plastics for recycling to avoid contamination and potential processing problems;
- sorting of plastics;

- measuring recyclate purity;
- analyzing *moisture* content.

4.3.3.6 X-ray

In Figure 4.23, plastics are transported with an assembly line and pass an identification unit that uses an analyzing system with X-rays [58]. This detects the spectrum of the objects and thus recognizes the composition of each material. The advantage relative to NIR is that the X-rays penetrate the surface of the object and that layers of paint and other surface disturbances do not affect the identification process. When the target material is analyzed, it is blown out of the stream using compressed air. This technique is mainly used for the removal of PVC, however it can also recognize other plastics.

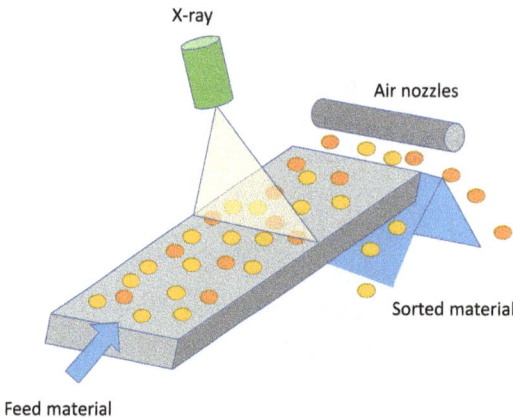

Figure 4.23: X-ray detection for sorting of plastics.

When blowing away a piece of identified plastic, a number of neighboring pieces are usually blown along. As a consequence, these systems are more efficient if the incoming group is already relatively pure, or if the density of material on the assembly line is lower. In practice, these techniques should be used as a quality control step after an earlier separation has brought purity to around 95%. Moreover, it is better if the pieces to be analyzed do not differ too much in size. Therefore, a prior reduction step and even the formation of flakes may be necessary to obtain a desirable separation.

4.3.3.7 X-ray fluorescent

X-ray fluorescent (XRF) sorting is used to determine the elemental composition of various materials [64]. Based on this analysis, objects with specific elements are detected and can be separated from the material stream. With the sorting systems, materials such as glass, ceramics, metals, minerals, plastics can be treated. In general, all solid materials containing a specific and characteristic element (or elements) above the detection limit can be identified and separated. The technology of *X-ray fluorescence* is therefore not limited to one material class or application but can be used in a wide variety of fields. Application areas for XRF sorting technology include:

- separation of leaded glass and glass ceramics;
- sorting of stainless steel, non-ferrous metals like brass, bronze, Cu, Zn, discarded metals;
- sorting of ores with different content of metals, separation of ores polluted with objectionable inclusions, like mercury;
- sorting of different minerals according to purity grade, separation of minerals polluted with objectionable inclusions;
- separation of electronic scrap coated with bromine and/or cadmium in shredded electronic scrap, enrichment of valuable metals coated electronic scrap;
- separation of any valuable metals in residues, separation of gold and silver in slag of waste incineration plants;
- the sorting system is also applicable as online quality control in the areas mentioned above.

Other than sorting different materials from plastics, distinct types of plastics cannot currently be sorted using X-ray fluorescence without the use of added *tags* in the plastics. XRF spectrometry is a non-destructive technique. Compared to other spectroscopic methods, this detection process is not affected by black *pigments*, and a clean surface is not required due to a volume detection of around 1 mm depth. However, as XRF is a spectroscopic method based on elemental analysis, the number of tracers is limited.

4.3.3.8 X-ray transmission
Sorting based on X-ray transmission (XRT) is relatively fast, capturing X-ray images within a few milliseconds. An imaging module utilizes a high intensity X-ray beam. When X-rays penetrate into the material, some of its energy gets absorbed by the material, while the rest is transmitted to a detector at the bottom. The detected radiation can be analyzed to provide information about the atomic density of the material. XRT is able to sort cast aluminum, copper and magnesium. X-ray transmission is often used in combination with an electromagnetic sensor (EMS). An example of a sorting machine that uses this analysis method is Dual Energy X-ray Transmission (DE-XRT).

4.3.3.9 Raman spectrometry
Raman spectrometry is a plastic sorting method able to sort HDPE, LDPE, PVC, PP, PS and PET [65]. Plastics are transported under the Raman spectrometer by a conveyor for analysis. An air gun is used for sorting the identified pieces of plastic, identical to the processes used in NIR and XRF. Raman spectrometry is mostly used when pure

polymer is recovered from shredded dust originating from waste household electronic appliances. Raman spectroscopy identifies the plastics based on the molecular structure. In the recycling industry where accurate sorting of massive amounts is required, the application of Raman spectroscopy has some advantages compared with IR and NIR. The following advantages make Raman measurements more robust to changes of sample characteristics:

- no reference signal is required;
- surface conditions and H_2O and CO_2 in the air or on the surface have less effect;
- the signal-to-noise ratio is easily achieved by a pumping laser.

4.3.3.10 Overview of spectroscopic methods

The *spectroscopic sorting methods* described have different advantages, disadvantages and target materials. Table 4.5 shows an overview of various techniques discussed.

Table 4.5: Overview of spectroscopic sorting techniques with their advantages and disadvantages.

Method	Target material	Advantages	Disadvantages
Near infrared (NIR) and hyperspectral imaging	ABS, PS, PP, PE, PET, PVC	Wide range of plastics can be measured rapidly	Unable to measure black and dark grey plastics
Shortwave infrared (SWIR)	Clear and colored plastics	Cost effective	Unable to measure black and dark grey plastics
Midwave infrared (MWIR)	Black plastics such as PS, PE, PP, ABS and PVC, rubbers, additives (flame retarders)	Able to measure black and dark grey plastics	Specified on black and dark grey plastic measurements
Ultraviolet/ visible spectroscopy (UV/VIS)	Tracer material added to plastics, plastic labels	Accurate measurements	Tracers needs to be added to the plastics, better alternatives possible
X-ray fluorescent (XRF)	Glass, metals, ores, minerals, electronic scrap, residues	Wide range of solid waste measurements to isolate plastics	Not able to sort different plastic types
X-ray transmission (XRT)	Aluminum, copper, magnesium	Able to sort specific metals accurately	Not able to sort different plastic types
Electromagnetic sensor (EMS)	Conductive materials (ferrous metals)	Able to quickly detect conductive materials	Not able to sort different plastic types

Table 4.5 (continued)

Method	Target material	Advantages	Disadvantages
Raman spectroscopy	HDPE, LDPE, PVC, PP, PS, PET	No reference signal is required, surface conditions have less effect	Needs pre-shredded plastics, while metals, wire, labels and other contaminations need to be removed

4.3.4 Marker systems

Separating different types of polymers is not feasible in all cases. Polymer blends and multilayer plastics are difficult to separate, limiting the overall recyclability of sorted plastics. The use of *marker systems* may, partly, solve this problem. Marker systems can be used to tag polymers with a specific chemical fingerprint, allowing the type of polymer to be detected and sorted. For example, this allows multilayers to be separated from monolayers, thus improving overall recyclability

Multiple marker systems have been developed, such as *fluorescent markers*, barcodes, or chips. The applications, impact on the recyclability, and the limitations of these marker systems will be discussed.

Fluorescent markers

A fluorescent marker can be used to mark and detect polymer species, after which the different polymer types can be separated and sorted [66]. A fluorescent molecule must be added to the polymer as a coating or additive during production. In the waste stream, the type of product can then be detected based on the wavelength it emits after irradiation of UV light. This is detected by a camera and can be compared to a database containing all the spectra of polymers available. If the spectrum matches a substance from the library, this marker can be related to a polymer type. It is therefore important that unique fluorescent markers are used for all polymer types that need to be sorted.

Furthermore, it is important that the fluorescent markers form strong emission when irradiating UV light. The strength of the emission determines the amount of fluorescent marker that is needed for successful detection. In practice, detection can already take place at concentrations below 10 ppm. However, the possible degradation also affects the strength of the emission, making it desirable to use higher concentrations.

The properties that a fluorescent marker must possess are fairly extensive:
- good compatibility with the polymers to be used;
- pigment in polymer should not affect detection;
- little to no influence on the properties of the polymer;
- little to no degradation;

- not harmful;
- location cost;
- high emissions for low concentration required.

A method has been developed whereby a fluorescent ink can be printed on packaging labels, which are detected during recycling. The method is known as PRISM (Plastic Packaging Recycling, using Intelligent Separation technologies for Materials). PRISM is already suitable for application, however, is not yet widely used. This method is extremely suitable for products with stickers, where a distinction can be made between food and non-food packaging.

In addition to the use of fluorescent stickers, fluorescent coatings can be applied. An advantage of a coating is that it can be removed during a common washing step during recycling. In a sodium hydroxide solution of 60 °C, a fluorescent coating is almost completely washed off and in this way, the separated plastic can be recycled completely. Should there be a small amount of fluorescent marker left after this washing step, the fluorescent wavelength will be changed to such an extent that this will not cause any more problems within a new life cycle of the polymer.

The stability of a fluorescent marker is extremely important for the actual application. A marker must be able to emit light throughout the life cycle of a product, so that it can be used in recycling. An example of a fluorescent marker that can be applied to food packaging is 4,4'-bis(2-benzoxazoleyl) stilbene.

The use of fluorescent markers is valuable for the identification of plastic waste, partly because of their invisibility, rapid measurability, large measuring distance and compatibility with existing recycling methods. However, there are some major drawbacks for using fluorescent markers as well. Many organic fluorescent substances are not stable under long-term UV irradiation, limiting the library of markers which can be used for different types of polymers. Other restrictions are that not all products can be stickered, partly because of their function, current production methods and marketing demands. In addition, there is a chance that a fluorescent sticker will not remain attached to the product, resulting in loss of detection and sorting efficiency during recycling. The use of a coating may be a solution for this. However, the applicability of coatings depends on the adhesiveness, and a new fabrication method will have to be developed for many products.

Barcodes and RFID chips
Besides the use of fluorescent markers, other marker systems are available, such as *barcodes* and chips. The use of barcodes for the identification of polymers, however, faces some challenges. In a common barcode, information can be stored based on the thickness ratios of black stripes. A scanner can only properly detect this 2D-information when the barcode is offered to the scanner at the right angle and distance. A barcode cannot be applied to all plastic products, because of the poor printability or because of

the distorting factor of the surface during detection. In addition, the function of a barcode is easily affected by damage. Moreover, a barcode cannot be scanned if it is wrinkled or (partially) shielded by infrared-blocking materials.

The use of normal and common barcodes, which can be found in the supermarket, will not be applicable for the identification of plastic waste on a large scale.

Magnetic micro-barcodes may be able to provide a solution for some of the issues mentioned above. They have the advantage of being invisible because of their small size (30 × 100 × 2 μm). Magnetic micro-barcodes can be magnetically oriented, because they contain five micromagnets that can have two different magnetic directions. These micromagnets can be measured after being oriented. This method is extremely suitable for microbiological applications (labeling of fatty acids or proteins in solution). However, experiments showed that it is less applicable to plastic products. Nevertheless, a magnetic barcode offers great advantages, including the transferability of the signal and the large amount of information that can be stored in a small volume.

Another identification technique is the use of identification chips. So-called RFID (Radio Frequency Identification technology) chips. The chip is a transponder, which receives a unique signal and automatically emits a unique output signal. The distance between the transponder and registration device is usually about 10 cm.

Implementing chip technology is expensive for packaging technology. For this reason, these chips are currently used at the pallet level in the plastic packaging industry, only. A new development is that RFID, as an inlay or as a sticker, is applied to the outside of the product. Both the sticker and the inlay can be removed at the washing step.

A lot of information can be processed in an RFID chip. For industry, this enables identification of the product whereabouts and study of customer behavior. The risk is, however, that this responds to the privacy of the customer.

4.3.5 Post-processing

The next step in the recycling process after sorting is post-processing [67]. Often the process proceeds with compounding of the material. In this step, depending on the application, additives can be incorporated to modify the properties of the recyclate. Pigments for coloring, fillers, fibers or *compatibilizers* can be added during the compounding step (Section 2.5.1). In this paragraph we will focus on compatibilizers since they are often used in recycling. To improve the mechanical properties of partly mixed *recyclates*, compatibilizers are added.

4.3.5.1 Compounding and compatibilizing

Most of the plastic waste recycle stream consists of polypropylene (PP) and polyethylene (PE). Often, sorting techniques cannot separate PP and PE completely. Typically, 1–5% of one of the polymers is still present in the other. PP and PE are immiscible and it is difficult to obtain good mechanical and optical properties in a polyolefin mixture. By using the right compatibilizer, it is possible to control the morphology of immiscible blends, resulting in better properties. Fortunately, there are some solutions to choose from.

Mechanical mixing

The most straightforward solution for making a good blend of PP and PE is enhanced mechanical mixing. Distributive and dispersed mixing are two important aspects. Both processes can be optimized by adding additional mixing regions in an extruder. Mechanical mixing can be achieved using a single screw extruder or a co-rotating intermeshing twin screw extruder. The use of the latter is preferred if good mechanical performance is demanded. The purchase cost and costs regarding energy usage are slightly higher for a twin screw extruder.

The miscibility of PP and PE can also be improved by adding surfactants. In Figure 4.24, ideal mixing, no mixing, and mixing using compatibilizers are illustrated.

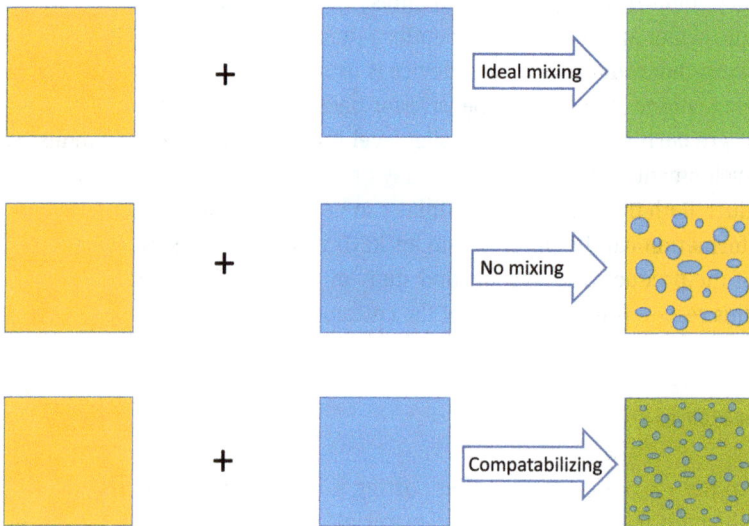

Figure 4.24: Illustration of ideal mixing, no mixing and mixing using compatibilizers.

Block and graft copolymers

If mechanical mixing is not sufficient, the mechanical properties of the blend can be enhanced by improving the compatibility of PP and PE. This can be achieved by using *block* and *graft copolymers* as compatibilizers [68].

Block and graft copolymers are frequently used as compatibilizers. Diblock copolymers and comb graft copolymers are added to lower the interfacial tension of PP and PE, resulting in a better miscibility. Diblock copolymers do this by migrating to the polymer-polymer interface. One segment of the copolymer is compatible with polymer A (like PP) and the other segment of the copolymer is compatible with polymer B (like PE) (Figure 4.25). This will reduce the interfacial tension and improve interfacial adhesion. Compatibilizers have positive effects on the mechanical properties of a blend of PP and PE. The diblock copolymer is most effective with a high number of blocks and interactions, while the comb graft copolymer is most effective in case of one long comb with multiple arms [69].

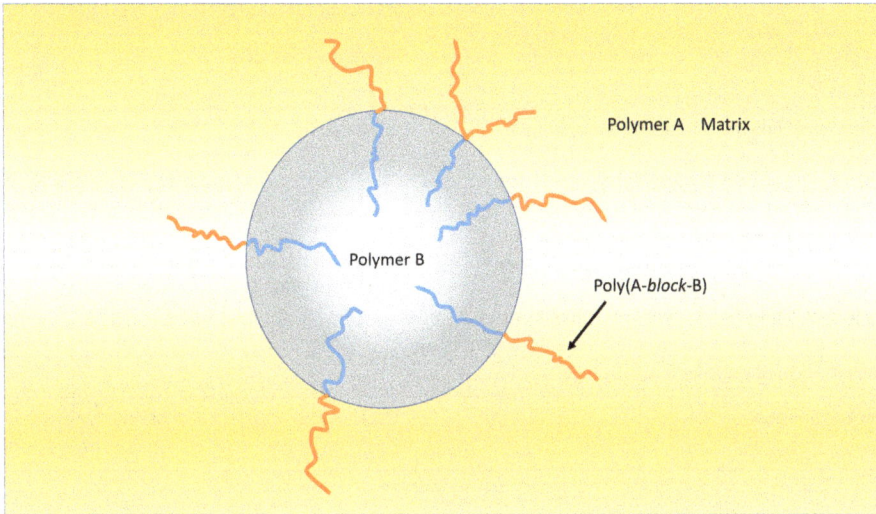

Figure 4.25: Compatibilization of polymer A and B with a diblock copolymer [68].

Reactive compatibilization

Reactive compatibilization is another option for the creation of PP and PE blends with good mechanical properties. Reactive blending or reactive processing is a method in which conditions are created to induce chemical reactions between two components. This can be done using reactive coupling agents, by esterification, by the use of low molecular weight reactive compounds and by Y-irradiation.

Reactions can take place when the components are in molten state by introduction of a third, reactive component. Another possibility is addition of a catalyst. This method will assist in the formation of a thick interphase which will result in a more stable morphology against the abusive effects of processing. By combining chemical kinetics with flow and thermal properties of the reaction ingredients and products, this technique is able to form a variety of copolymer architectures (Table 4.6) [70].

The majority of polymers used to form interchain copolymers have nucleophilic end groups (carboxylate or amine). These groups can react with electrophilic end groups such as carboxylic acids, carbodiimides, isocyanates and epoxies. Polyethylene (PE) and polypropylene (PP) can also be blended with reactive compatibilization. Reactive compatibilization using a reactive coupling agent requires malleated grafted polyolefins in order to connect the PP and PE interfaces.

Table 4.6: Chemical processes for interchain copolymer formation [70].

Type of chemical reaction	Type of obtained polymer
Chain cleavage	Block and random copolymers: AAAAABBBBB + AABBBBBAA + AABBAAABBB
End-group of polymer A reacting with end-group of polymer B	Block: AAAAABBBBB
End-group of polymer A reacting with pendant functionality of polymer B	Graft copolymer: I A – BBBBB I A – BBBBB* I
Covalent crosslinking: Pendant functionality of polymer A + pendant functionality of polymer B, or, main chain of polymer A + main chain of polymer B	Graft copolymer or crosslinked network
Ionic bond formation	Usually graft, frequently crosslinked: A B A————B

Another type of reactive compatibilization is esterification. Esterification also uses a malleated polyolefin to connect to a polymer. The other polymer needs to connect to a co-vinyl alcohol. The hydroxyl (OH) group of the co-vinyl alcohol will then react with the anhydride group of the malleated polyolefin connecting PP and PE.

Low molecular weight reactive compounds will promote *crosslinking reactions* due to radical formation and copolymerization by increasing the synergy. Peroxides will decrease the viscosity of PP due to chain scission and increase the viscosity of PE by introducing branching through crosslinking. This will reduce the viscosity mismatch between PE and PP, leading to a better miscibility.

Y-radiation is a penetrating electromagnetic radiation. Gamma (Y) rays have with the highest frequency and the lowest wavelength in the electromagnetic spectrum (Section 4.3.3). Ionizing radiations such as Y-*irradiation* provoke great changes in the structure of the polymer [71]. These radiations lead to the formation of radicals, which will lead to chain scission, crosslinking and other phenomena such as oxidative degradation in the presence of air.

The influence of the radiation is dependent on the chemical structure of the (macro) molecule. For example, rubbers have unsaturated C = C bonds which will lead to straightforward crosslinking of polymers under influence of Y-radiation. PE and PP do not have unsaturated bonds, but the radiation will still have influence on these polyolefins. PE will be able to crosslink due to abstraction of hydrogen from the second carbon forming a macroradical and due to the combination of two macroradicals. This crosslinking is possible using medium or low doses of Y-radiation and if high doses are used, PE undergoes chain scission. For PP, the presence of tertiary carbons enhances chain scission by rearrangement of macroradicals.

Other polymers like PET, PA and PLA can also by compatibilized with PE or PP. Often maleic anhydride (MA) is used as a grafting agent.

Because compatibilizing has influence on the average molar mass of the polymers, often analyzing techniques like *size exclusion chromatography* (SEC) and viscosimetry are used to determine M_w and M_n (Section 2.2.5). Other characterization techniques will be discussed in the next paragraph.

4.3.5.2 Analyses

In order to identify the chemical nature and properties of recyclates, a number of characterization techniques are typically used.

Density determination

A sample's volume is determined (using for example a pyknometer), as well as its mass. In this way, one can calculate the specific weight.

Size and shape determination

Just by the size and shape of a sample, much can be learned about its properties. By examining under the microscope, the morphology of powdery materials can be determined, giving a lot of information about the types of polymers that will be present in the sample.

Sieve test

A powder steel is placed above a series of sieves, with different mesh sizes, and by vibration the sample is sifted. In this way, the average grain size and variation in grain size can be mapped. For example, this technique is very instructive for extruding powdered recyclates, such as PVC.

Ash residue determination

The weight of a sample will be determined before and after an oven treatment. The oven temperatures are so high that only the non-volatile components (such as inorganic fillers) remain after the test.

Flame test

During the flame test, a sample is kept in a flame and its burning behavior is observed. Different plastics often show distinct fire characteristics, such as soot formation (PS), self-deafness (PPO), dripping (PET), and a green flame color (PVC). This method is therefore used very frequently to identify plastics free of charge.

Solubility test

Sample are dissolved in a number of (organic) solvents. The solubility of samples provides narrow insight into the type of polymer that is in the recyclate.

The techniques described above are relatively simple and inexpensive. However, they have the disadvantage that the results are not entirely accurate and representative. For example, the density and flame properties of a plastic may be influenced by the fillers. This is why the following more advanced techniques can also be used to map the properties of recyclates in more detail.

Fourier transform infrared spectroscopy

Fourier Transform Infrared Spectroscopy (FTIR) measures the intensity of absorption by means of infrared light (Figure 4.21). In this way, a typical spectrum is obtained for each organic compound, which provides semiquantitative information on the contents of the sample. With this technique the following observations can be done:
– the quality of the plastic processed during production;
– contaminants in the final product;
– residues of other plastics in the product;

Differential Scanning Calorimetry

This characterization technique is used to determine thermal parameters (*glass transition*, heat of melting and crystallization) of polymers (Section 2.4.3). The thermal stability of an Oxidative Induction Time (OIT) material can also be mapped. With *differential scanning calorimetry* (DSC) the following observations can be done:
– the raw material the final product;
– melting and crystallization;
– contaminations in the product;
– distinguish HDPE from LPDE and PA 6 from PA 6,6
– thermal degradation and crosslinking.

Thermogravimetric analysis

A thermogravimetric analyzer (TGA) measures the change in weight of a material according to temperature or time in a controlled atmosphere. The outcome is displayed in a thermogram. With this technique the following observations can be done:
– the amount of fillers or fiber reinforcement present in the product;
– volatiles or moisture present in a product;
– the thermal stability of a product;
– the amount of vinyl acetate in EVAc.

Melt flow rate

The value for melt flow rate (MFR) is an indication of the flowability of plastic, and a very important parameter for processing. After carrying out the measurement, the MFR value is obtained in g/10 min. Degradation of a material is usually determined by an increase in the MFR, while crosslinking reactions result in a reduction in MFR. The MFR value depends on type of application and processing technique and varies between 0 and 60 g/10 min. With this technique the following observations can be done:

– degradation of the end product;
– difference in flow behaviors.

Extrusion rheometry

Extrusion rheometry measures the viscosity of a material according to the shear rate. It is similar to MFR, but flow can be measured at much higher speeds and this is more closely linked to the actual processing conditions. With this technique the following observations can be done:

– recyclate used during processing;
– processability of the recyclate;
– extent of degradation;

Tensile properties

A number of essential mechanical properties of materials can be mapped, by pulling or pressing a specimen and measuring the movement according to the applied force. The properties that are measured include *E-modulus*, ultimate strength and *elongation at break* (Section 2.4.6). With this technique the following observations can be done:

– considering the impact of added fibers, nano clays, and other additives on the strength of the product;
– fracture that normally does not occur, simulate and compare materials.

Impact test

Through an impact test one can compare batches by their impact resistance. The impact test determines the resistance of a material against a punch or impact via a simulated test and is expressed in necessary energy (in J) per surface unit of the section (in m^2). With this technique the following observations can be done:

– determine the impact of added additives on the strength of a product;
– batches that show a different break behavior.

X-ray fluorescence spectrometry

X-ray fluorescence spectrometry (XRF) is a technique in which the composition of a sample is determined from chemical elements (Figure 4.26). Elements in a wide range can be measured with this technique. For lightweight elements, the technique is limited by the fact that the fluorescence gets lower at longer wavelengths. This radiation is difficult to detect properly because it is very easily stopped by the windows that separate the different parts of the measurement device. For heavier elements, the technique is limited by the fact that the X-rays used to induce fluorescence must have a higher energy than the characteristic radiation of the element measured. In practice, elements between sodium and uranium can be detected.

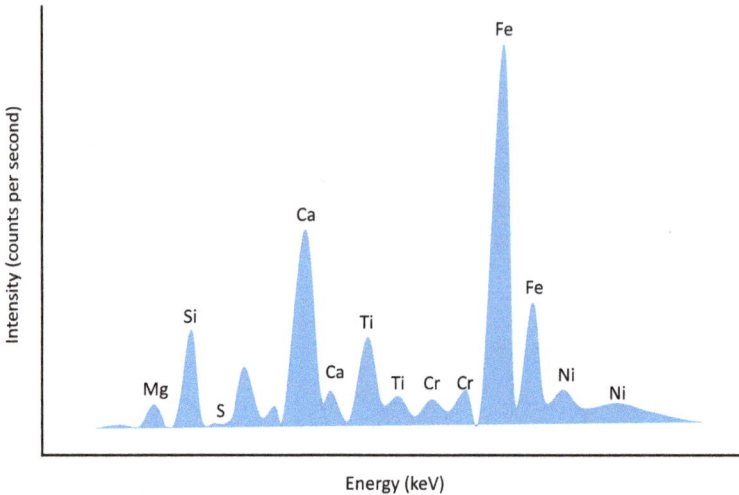

Figure 4.26: Typical X-ray fluorescence spectrum.

Inductively coupled plasma-atomic emission spectroscopy
Inductively Coupled Plasma-Atomic Emission Spectroscopy (ICP-AES), is an analytical technique that can determine the element composition of a sample. ICP-AES uses light emitted by an atom, ion or molecule when it goes from a higher energy state to a lower energy state. It is excited to a higher energy state using a plasma. Coupled with atomic emission spectroscopy, one can analyze a complete spectrum of elements. The advantage of this technique compared to XRF, is the ability to measure low concentrations with high accuracy.

4.3.6 Applications of recyclates

There are several motives for applying *recyclates* in plastic production, which together lead to a valid business case. The following factors can be considered.

Cost savings

The price of recycled plastics is in almost any case lower than the price of virgin commodities. In recent years, the quality of recyclate has increased to such an extent that it can often be used as a full-fledged and economical substitute for virgin plastics in high-quality applications.

Reducing environmental impact

Recyclate prevents the use of new feedstock and saves energy for production. This can reduce the overall *environmental impact* of plastic products. *Life cycle assessment* (LCA) studies show that recycling is in almost any case the processing method with the least environmental impact.

Reduction of scarcity of materials

The extraction of fossil fuels is expected to increase the price of plastics and reduce security of supply. By keeping plastics in a circular flow, fewer fossil fuels are needed and our dependence on crude oil is reduced.

Marketing trade-offs

Consumers are increasingly taking into account sustainability in their purchasing considerations. Recycling can be used in the marketing of products towards consumers. Add value at the product level, but also increase a company's brand value at the corporate level. Example is the *Plant Bottle*® of Coca-Cola.

Law, regulation, and green public procurement

Governments are increasingly setting targets and requirements with respect to the use of recyclates. Landfilling is prohibited and requirements on the quantities of plastic to be reused are imposed. In tenders, governments often demand the use of recycled materials.

Requirements from customers

Recyclate can already be used in many different applications. The purpose to use recyclate should always be to use the material in high-end applications. The quality not only depends on the type of application and the visual appearance, but also on the processing possibilities.

A mono-material that has been used in a high-quality application and has barely been degraded in its first life cycle can be used once more for another high-end application. High-quality products for which the material requirements are very demanding, such as applications with skin or food contact or transparent products, are still difficult to achieve with plastic recyclate. Innovations are needed, although in recent years, the *quality* and material security of recyclate has improved. As a result, recyclate can already be applied more often in premium products and for visible applications. Table 4.7 shows some applications in which recyclate has been applied in visible parts, demanding materials and premium products.

Table 4.7: Applications of various plastic recyclates.

Polymer	Applications virgin material	Applications recycled material
PP	Packaging (flacons, crates, dairy), automotive, garden furniture	(Food-grade) crates, baskets, boxes, automotive, (flower) pots, garden furniture, domestic (electric) products
HDPE	Toys, baskets, crates, food packaging (milk cartons, juice bottles, shampoo flacons), garbage bags	(Food-grade) crates, pots, (non-food) bottles, garbage bins
PET	Soft drink bottles, foil packaging, food trays	Soft drink bottles, fleece clothing, straps, housing of domestic products (also PBT originating from PET)
LDPE	Blown film, shrink film, agriculture foils, bags, toys	(Garbage) bags, shopping bags
(E)PS	Disposable cups, disposable food utensils, yoghurt trays, CD cases, Styrofoam enclosures, clear packaging	Insulation material, (light) switches, housing, CD cases, Styrofoam
PC	DVDs, (food) plates, safety glasses	Profiles, housing of electrical devices
PVC	Pipes, tubes, blister packs, window frames, floor tiles	Pipes, bins, cables, floor tiles, window frames

4.3.7 Challenges in mechanical recycling

During the recycle process of plastics some challenges are encountered with respect to the properties and business cases. In order to provide truly circular plastics, some important aspects need to be improved. In the following paragraphs these aspects will be discussed shortly.

4.3.7.1 Degradation

The *quality* of recyclates is an important issue. Of course, quality depends on the processing conditions as well as on the properties of the plastics itself. Those properties are again influenced by the use of additives. The question arises how many life cycles can a plastic endure. In general, polymers are stabilized only for processing and a first life cycle, and not for reprocessing and a second lifetime. Unfortunately, this especially accounts for packaging material made out of PP and PE.

Degradation of a polymer results from exposure to factors like heat, mechanical stress, oxygen and UV [72]. PE, used in *packaging* for example, suffers from oxidative and photo-oxidative degradation due to both processing and use. Reprocessing can cause mechanical, thermal and oxidative degradation. Degradation leads to changes in the macromolecular structure of the polymer, characterized by scission or fragmentation of the polymer chain, which generally results in a decrease of the mechanical properties. Radical chain reactions, such as formation of hydroperoxides and crosslinking, also occur during degradation and result in an altered polymer structure as well. Polyolefins often undergo this kind of degradation.

During processing and lifetime use, plastics are exposed to heat, mechanical stress, oxygen and UV as mentioned before. Depending on the polymer chain structure, these phenomena lead to more or less degradation. The effects can be summarized as follows:
– change in molecular weight, leading to a change in viscosity;
– formation of crosslinks and branched chains;
– formation of unsaturated and oxygenated compounds.

Three *degradation* mechanisms can be distinguished (Figure 4.27): mechanical, thermal and thermal oxidative degradation. In the case of mechanical degradation, C-C bonds of the polymer backbone break when shear and tensile stresses, induced by shearing and stretching during extrusion, exceed the intramolecular bonding forces. This process is dependent on the degree of polymerization: the longer the polymer chains, the higher the mechanical stresses, the more chain scission. In addition, the process is temperature dependent: for lower temperatures, the flexibility of the chain segments decreases, resulting in an increase of mechanical stresses and thus chain scission.

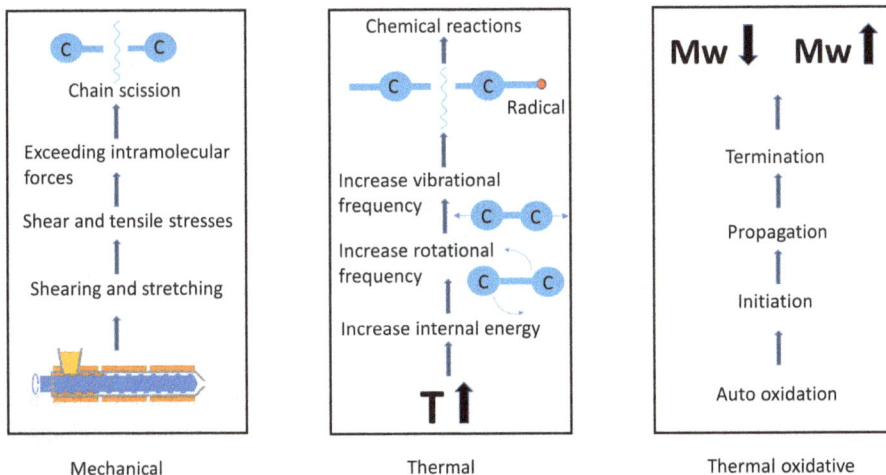

Figure 4.27: Illustration of the three degradation mechanisms observed during mechanical recycling of plastics.

Thermal degradation is induced by elevated temperatures. It results in an increase of internal energy causing different effects. First, segments of the polymer chain start to rotate at a higher frequency, causing a weakening of the intermolecular forces. Also, an increase in the vibrational energy of the bonds leads to breakage along the chain ends, resulting in radicals at the free chain ends. At last, the mobility of some additives and contaminants increases, enabling chemical reactions with energetic sites.

The last mechanism, representing thermal oxidative degradation, is caused by auto oxidation. This reaction follows the same sequence as a polymerization: initiation, propagation and termination. In the case of auto oxidation both an increase and decrease of the molecular weight can be observed.

Low density polyethylene (LDPE)

Crosslinking of the polymer chains often occurs. It is observed after many extrusion cycles that the viscosity of LDPE is increased. This crosslinking occurs due to the formation and reaction of carbon radicals.

Polypropylene (PP)

The degradation of PP typically occurs through oxidative degradation. The main effects of the degradation phenomena on the PP chain structure are decrease of the molecular weight, which results in an increase in polydispersity and formation of oxygenated functional groups. As a consequence of these structural changes, the morphology of PP can also be modified. The properties (rheological, mechanical, electrical) strongly alter as a result of the changes in structure and morphology.

Polyethylene terephthalate (PET)

The degradation mechanism of PET is chain scission, leading to reduction of molecular weight. Often the chain length is reduced significantly, however the mechanical properties are only slightly affected. Recyclates can be refreshed by virgin PET depending on the desired product properties. When degradation happens to a large extent, a post-condensation step can be performed. In post-condensation, the PET chains are connected again by a polycondensation mechanism (Section 2.3.1), usually performed in solid state.

Polystyrene (PS)

In general, after multiple reprocessing steps, an increase of (ultimate) tensile stress and a decrease of elongation at break, is observed. In other words, the PS material changes from ductile to brittle.

Polyamide (PA)
Combination of mechanical forces and heat during processing will result in various changes in the molecular structure of PAs. Chain scission will produce free radicals. Free radicals, due to their highly reactive nature, will react again with oxygen, which results in further degradation of PA.

Polycarbonate (PC)
PC can be recycled in an excellent manner. For instance, the tensile strength and E-modulus remain unchanged up to five cycles. Nevertheless, elongation at break and the toughness decrease with increasing number of cycles.

Polylactic acid (PLA)
Even after five injection molding cycles, the impact properties of PLA are still fairly acceptable, although degradation takes place to a certain extent.

4.3.7.2 Contaminants

Contaminants play a major role in the mechanical recycling of plastics. Because of the open structure of the polymer chains, contaminants can penetrate into the polymer matrix. Common *contaminants* in the PET recycling process are acids or acid-producing compounds, for example, when PET and polyvinylidene chloride (PVDC) are mixed. HCl acts as a catalyst for chain-cleavage reactions. Also, moisture can lead to cleavage by hydrolysis. Pigments and other coloring agents may result in a too brownish color in recycled plastics. Acetaldehyde, which is a degradation product of PET, can lead to health hazards in recycled products. And, misuse of plastics by consumers, in which bottles are used for storage of fuels or pesticides, results in toxic contamination of plastics.

Figure 4.28 shows some examples of different types of packaging which cause enormous problems in the plastic recycling chain due to contaminations. Blisters of medications are often made from PVC. They may contain drug residues associated with the chemical waste that is not returned to the pharmacy. Remnants of silicone in the kit syringe have a negative impact on the recyclate quality. NIR scanners in waste sorting cannot recognize black plastic objects. Silicone seals in caps affect recyclate quality. Ink residue in cartridges colors the recyclate, while actually it is not packaging and therefore belongs to chemical waste.

4.3.7.3 Multilayer packaging

Today's highly efficient solutions based on laminated plastics are often poorly recyclable, which puts *multilayer packaging* material in a difficult position (Section 4.2.2). The sheer number of different plastics available poses another problem, and though there may be a limited number of base types, there is tremendous variety to be found

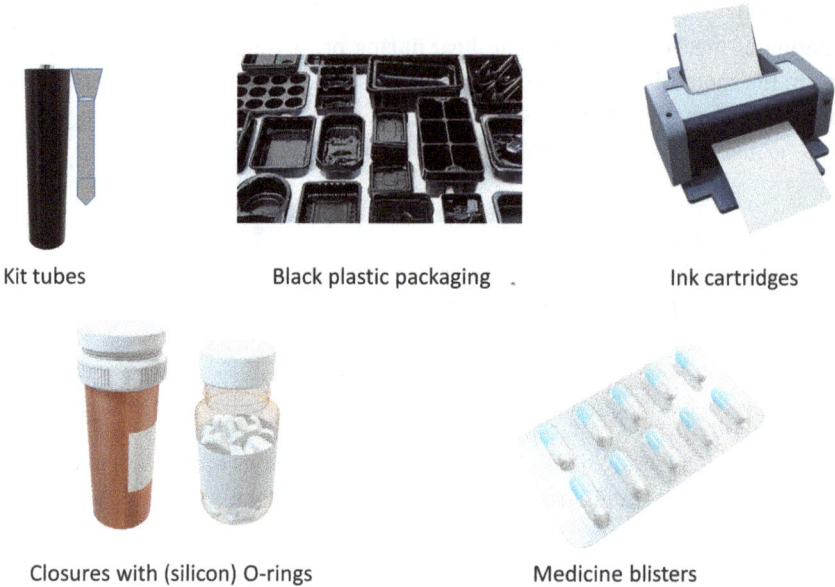

Kit tubes   Black plastic packaging   Ink cartridges

Closures with (silicon) O-rings   Medicine blisters

Figure 4.28: Examples of different packaging causing problems in the plastic recycling chain.

within them. Average polymer chain length, distribution of chain length, the use of additives and lamination, all complicate the matter of recycling materials into the same level of quality. This, however, does not mean that plastic products cannot be circular to a larger degree. A focus towards harmonizing the types of plastics can raise recycling rates, while at the same time new forms of recycling emerge, such as chemical recycling (Section 4.4), in which plastics are broken down into raw materials. In addition, in future it will be possible to separate multiple layers in packaging foils through the development of a thermo-reversible adhesive. Still, reduction of multilayer packaging takes precedence over recycling.

4.3.7.4 Color

Color often plays an important role in drawing the attention of consumers to a particular product. But color can have a practical function as well. Certain colors are used to protect light-sensitive vitamins and flavors from degradation caused by exposure to UV light. For light-sensitive products, such as long-life milk, a light barrier needs to be added to conventional PET bottles, for example by incorporating a *pigment*, typically titanium dioxide (TiO_2), or by putting a wrap or sleeve around the bottle. TiO_2 is a mineral whitener with very good coverage that is added to packaging in a variety of concentrations, either alone or in combination with other additives, such as other color pigments, carbon black, mica and silica.

In general, when coloring agents are used, it disrupts the recycling chain. A high pigment content affects the color and gloss, and with that, the reuse of recycled material. Often, color stability is not assured when using recyclates. Especially for recycled materials used for domestic appliances, this is a large drawback.

The ultimate target is a colorless or clear recycled polymer raw material, because of their higher value in the market. Different techniques for removing *colorants* and other additives from plastic waste have been developed. It can be part of a chemical recycling process, as discussed in Section 4.4.

Extraction by solvent combined with filtering or precipitation is a method often used for polyolefins. Polyolefins are very inert, thus harsh organic solvents and solvation conditions are required. Removal of carbon black by flotation can be included in this step. Precipitation and removal of pigments is a method which is still under development. The general idea is a selective precipitation of plastic components and the use of molecular sieves as adsorbent. The removal of solvent and drying of the polymer requires heat and energy. Closed-loop solvent processes are needed, because solvent residues within recyclate may cause odor among other issues.

Nowadays a lot of different colorants are available on the market. They can be divided in pigments, dyes, liquid colorants and masterbatches.

Pigments (inorganic)

Inorganic compounds are usually immiscible and insoluble, or only partially soluble, in polymers. The following groups can be distinguished:
- oxides, sulfides, carbonates, carbon black;
- oxides of iron, titanium, nickel, antimony, chrome, zinc and cobalt are stable at high temperatures and are easily dispersed into plastics;
- sulfides have usually low resistance to acids.

The size of the dispersed particles is between 10 nm–1 µm. Examples of various inorganic colorants are depicted in Figure 4.29. Special pigment are metallic flakes and pearlescent. The latter consists of thin platelets of mica, coated with TiO_2 or Fe_2O_3.

Pigments (organic)

Different kind of cyanines are used as organic pigments. Phthalocyanines are used as blue or green colorants. Phthalocyanines can cause nucleation and crystal growth in polyolefins resulting in warpage. Perylene pigments are widely used especially for melt coloration of PP and PE. Examples of various organic colorants are depicted in Figure 4.30.

In general, good inorganic or organic pigments should have a strong covering (color) power and need to be weather and chemical resistance. Also, thermal stability up to 250 °C and resistance to bleeding and migration, are important characteristics. These pigmental properties often depend on the chemical structure and size and shape of the

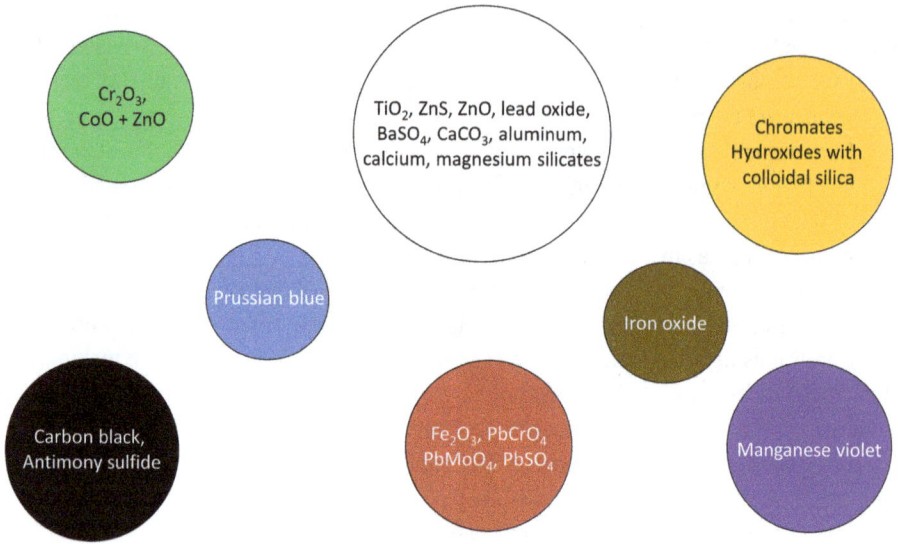

Figure 4.29: Examples of different inorganic color pigments. The colors of the circles in the figure, correspond to the colors of the pigments.

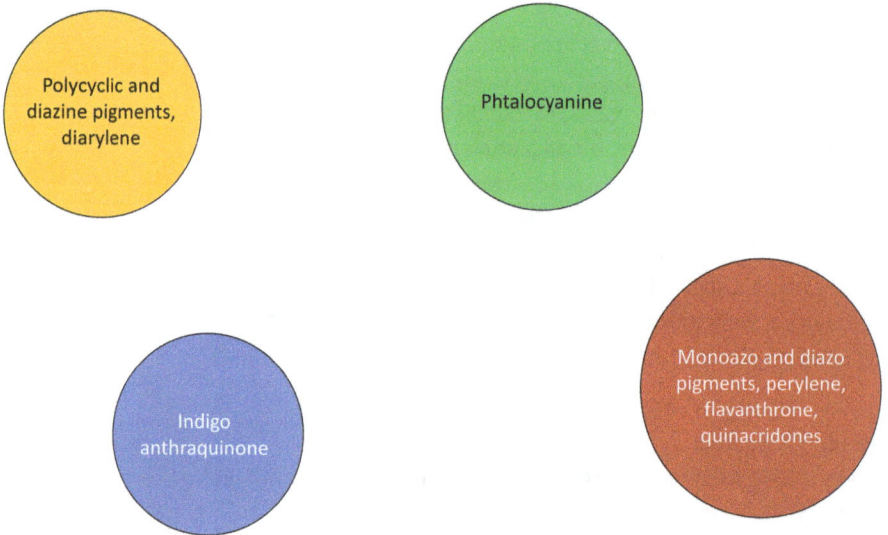

Figure 4.30: Examples of different organic color pigments. The colors of the circles in the figure, correspond to the colors of the pigments.

particles. In addition, the size distribution and crystallinity of the particles play an important role. At last their processing conditions, like temperature, determine the class of pigments that can be implemented.

Liquid colorants

A liquid master batch is a custom blend of additives and pigments dispersed in a liquid. The decision to use liquids is based on the specific requirements of the application. Due to liquid master batches, easy incorporation into the polymer matrix is achieved. Their application is often for transparent or translucent colors. Additionally, because liquid master batches are not compounded at high temperatures and pressures, they can be used when heat-sensitive pigments or additives are required. Liquid master batches can be formulated to improve the flow properties of the polymer, with fast filling times and shortened injection molding cycles, as well as improved extrusion throughput and low screw speeds. Although the usage of liquid colorants is decreasing, they are still applied extensively in transparent PET applications like bottles.

Master batches

Master batches are concentrated polymer-pigment batches with a typical composition of 40–60% of pigments, while the pigment content in a final product can vary 1–15% depending on the pigment type and color. Pigment dispersions can be stabilized by polyacrylate or ester waxes.

4.3.7.5 Odor

As a result of degradation of polymers and additives, or migration of a substance during the previous life cycle, recycled plastics can contain an unpleasant smell. This can be undesirable and greatly reduce the product value. During tests with recyclates, it can be determined whether the material meets the required demands in terms of smell. Targeted washing procedures or odorants can neutralize unpleasant odors. Several treatments based on solvent extraction and hot water and surfactant solution washes have been researched to evaluate their efficiency in reducing the smell. The use of odorants is an alternative, however most of the time the effect is just temporary. After all, it is only a replacement of an unpleasant odor for a pleasant smell.

4.4 Chemical recycling

Chemical recycling is the broad term used to describe a range of emerging technologies in the waste management industry, which allow plastics to be recycled that are difficult or uneconomic to recycle mechanically. It is a chemical process in which plastics are converted into (original) building blocks. The building blocks can be used to produce new plastics, or applied somewhere else. This allows plastic waste streams to be reused, for example in food applications. Chemical recycling processes have the potential to improve recycling rates and divert plastic waste from landfill or *incineration* [73–75].

In Section 4.3, it was explained that mechanical recycling of mixed waste streams runs into limitations. For example, multilayer packaging is not fully recyclable with mechanical recycling. Also, the quality of the output of mechanical recycling is often limited because the properties decrease after repeated recycling. A few examples of problems that typically occur during the mechanical recycling process:

- small plastic (sub)parts in source-separated material, such as bottle caps, which are not sorted out;
- plastic types in source-separated material that are not sorted out, such as PVC, PS and PLA;
- contaminated PET, PP or PE that is not detected when sorting;
- failures in the final processing of the main flows, when sorting both source-separated material and material from post-separation.

For a number of waste streams polluted with other plastics, no suitable recycling techniques are currently available. A well-known example are PET trays with relatively high levels of other plastics and contaminants [76]. Brominated EPS from the construction sector is also a severely polluted waste stream that is hard to recycle. In addition, unpleasant smell is an issue as well as color stability of recyclates in the mechanical recycling process (Section 4.3.7). Application in the food sector as packaging is usually not possible due to food safety regulations (contamination with non-food-safe substances is not excluded from mixed collection). Chemical recycling complements mechanical recycling by enabling the further extraction of value from polymer materials that have exhausted their economic potential for mechanical processing.

Nowadays, several chemical pilot plants demonstrating the viability of various chemical recycling processes are currently in operation, or in progress. Commercial plants range in size from large-scale factories with 30–200 ktons annual throughput to much smaller, modular, distributed units with a capacity of 3–10 ktons per year.

4.4.1 Chemical recycling techniques

Chemical recycling describes any technology that utilizes processes or chemical agents that directly affect the chemical nature of polymers. Those technologies fall into three distinct categories based on the position of their outputs in the *plastics supply chain* (Figure 4.31):

- Solvent-based purification
- Feedstock recycling: pyrolysis and gasification
- Depolymerization

Chemical recycling differs from mechanical recycling which uses operations to prepare waste polymers for reuse, without significantly changing the chemical structure of the material. Mechanical recycling processes a separated polymer stream, which is washed, granulated and then re-extruded to make recycled pellets that are ready for molding applications (Section 4.3). Chemical recycling, based on *depolymerization* and feedstock recycling, breaks down the macromolecular chains into shorter fractions (oligomers) or monomers using chemical, thermal or catalytic processes. Purification, on the other hand, deals with the use of solvents for removing additives from the polymers.

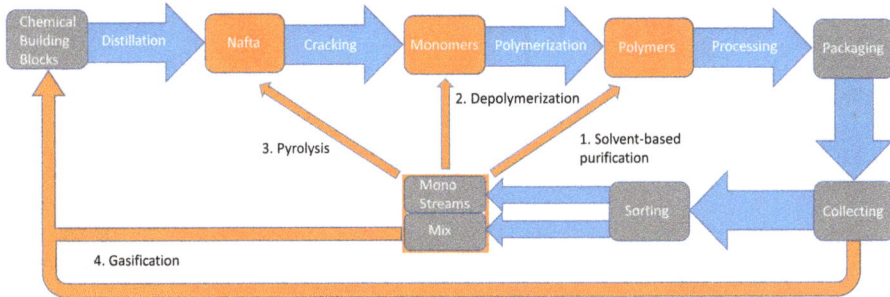

Figure 4.31: Four chemical recycling techniques and their position in the plastics supply chain.

4.4.2 Solvent-based purification

Solvent-based *purification* is a process in which plastics are dissolved in a suitable solvent or solvent mixture, after which a series of purification steps are undertaken to separate the polymer from additives and contaminants. Once the plastic feed is dissolved in the solvent(s), it can be selectively crystallized. When a solvent can dissolve either the polymer of main interest or all the other components except the target polymer, it is suitable for selective dissolution. The crucial requirement is a selective solvent. The process output is a precipitated (pure) polymer, which ideally remains unaffected by the process and can be reformulated into plastics.

At the moment, target feedstocks for solvent-based purification are PVC, PE, PP, and PS. This is a new technology and efforts are underway to scale up to a commercially viable level. Generally, waste plastics are collected as mixed polymers. Therefore, the primary challenge is the separation and recycling of waste components selectively. For PS, for instance, Fraunhofer developed a process called CreaSolv® (Figure 4.32) [77].

The non-profit organization 'PolyStyrene Loop', an initiative of a number of EPS processors, recyclers and producers, has taken the initiative to build a pilot plant in which EPS products are recycled, while the bromine-containing flame retardants are recovered. It is a physical process in which EPS is dissolved in a selective solvent and is then separated from other substances present in the feed, such as bromine.

The second type of solvent-based recycling is nonaqueous washing, typically employing solvents to have an additional effect on the pretreatment, for instance, delamination, de-inking, or increased de-odorization.

European standardization differentiates chemical and mechanical recycling by whether the process is significantly changing the chemical structure of the polymeric

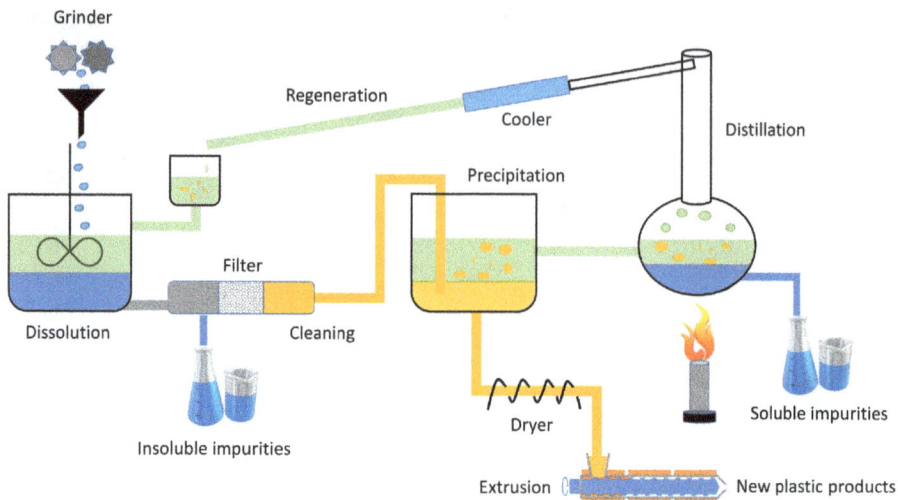

Figure 4.32: The CreaSolve® process as introduced by Fraunhofer [77].

material. Mechanical recycling does not change (on purpose) the chemical structure. Chemical recycling, however, does change the macromolecular structure. A technology leaving the polymer chain like it is therefore formally considered as mechanical recycling. Consequently, solvent-based recycling should fall under the definition of mechanical recycling. The process uses solvents, i.e., chemicals, which frequently leads to the misclassification of this technology as a chemical recycling method. Since mechanical recycling is commonly linked to the re-extrusion process, the term "physical recycling" has also been introduced to describe solvent-based and mechanical recycling because neither changes the chemical structure of the polymer waste [78].

4.4.3 Feedstock recycling

Feedstock recycling is any thermal process that converts polymers into smaller molecules, in order to form the feedstock for petrochemical-type processing. The two main processes described are *pyrolysis* and *gasification*.

4.4.3.1 Pyrolysis

In the *pyrolysis* process, plastics are broken down into a range of basic hydrocarbons by heating in the absence of oxygen, or 'cracking', sometimes referred to as thermal cracking. By utilizing a distillation process, the collected hydrocarbons can be used into products ranging from wax and oils to gas. It is possible to change the output by adjusting process time and temperature. Heavier output products can also be reintroduced into the process for additional cracking into smaller molecules. Using pyrolysis to make feedstock for polyethylene (PE) and polypropylene (PP) production could fill a large processing gap as PE and PP cannot be depolymerized directly into monomers. Furthermore, the plastics produced would consist of virgin-quality polymers and could be used in high-demanding applications, such as food packaging.

Pyrolysis production can be enhanced using catalytic degradation, where a suitable catalyst is used to promote the cracking reaction. The presence of a catalyst allows reaction temperature and time to be lowered. The process results in a much narrower product distribution and increases lightweight hydrocarbon production.

While pyrolysis-based processes can be used to recycle single-polymer plastic waste, they are particularly advantageous when it comes to dealing with contaminated and mixed-polymer waste streams. In Figure 4.33, a schematic representation of pyrolysis is shown.

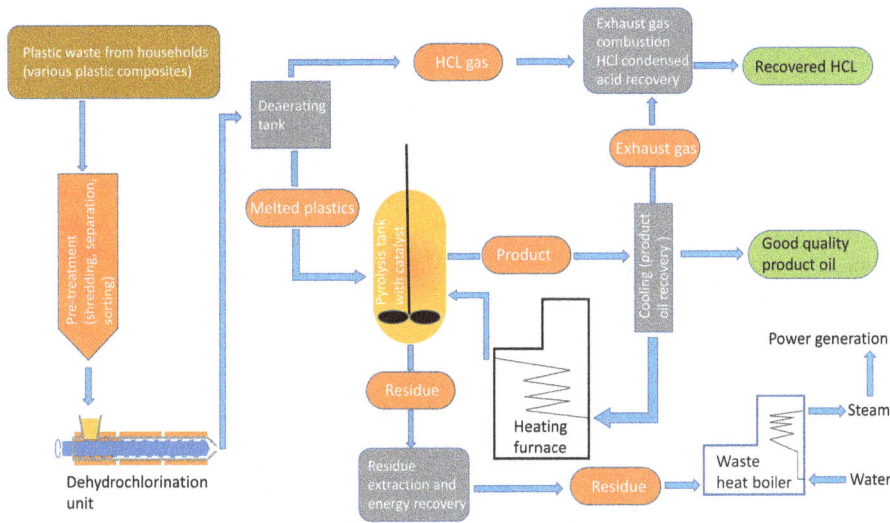

Figure 4.33: Schematic representation of a pyrolysis process.

Pyrolysis output can be processed in much the same way as oil, using conventional refining technologies to produce building blocks for polymers. Alternatively, they can be applied directly as a fuel. Historically, pyrolysis has been commercialized in applications relating to charcoal, municipal solid waste and biomass. In the waste industry, pyrolysis of mixed plastics has been in development over the last two decades, and is becoming a commercial reality with several commercial plants in operation and many more industrial-scale units expected to be commissioned over the next few years.

Pyrolysis is the heating of input material without oxygen at a temperature between 400 and 600 °C. This may include conversion by pyrolysis into oil or diesel-like fuel, syngas, salts and metals, organic acids and carbon/char. Syngas is a mixture of hydrogen (H_2), carbon monoxide (CO) and some carbon dioxide (CO_2). The ratio and quality of these products depend on the specific technology and composition of the input. First generation pyrolysis processes suffer more from shortcomings such as a high content of PET and PVC, in comparison to second generation processes. Therefore, the quality of oil or fuel is often higher in second generation pyrolysis.

Nowadays, a large number of large-scale commercial pyrolysis installations are available worldwide. Several factories specifically focus on the production of biofuels from biomass through pyrolysis. The technique is however, not (yet) widely used in the recycling of plastic waste from households. LyondellBasell worked closely together with the Karlsruhe Institute of Technology (KIT) to develop a new recycling method called MoRe-Tec (Molecular Recycling Technology). MoReTec makes use of a special catalyst in the pyrolysis process, making it faster and more energy efficient than traditional methods.

Integrated hydropyrolysis

Integrated hydropyrolysis is an advanced form of pyrolysis, producing a higher quality fuel. In contrast to conventional pyrolysis, the cracking process takes place in the presence of water. The process temperature is approximately between 300–600 °C. The integrated hydropyrolysis technique is better suited for variable heterogeneous input, and also less sensitive to oxygenated plastics (PET). As a result, hydropyrolysis is a waste processing technique that could be applied to a large part of combustible garbage flows, non-recyclable plastic waste and biomass waste. Currently, however, it is still applied in niche markets, that focus on a specific waste stream such as wood, sludge processing or roadside grass. It is expected that commercial realization will take several more years.

In general, input for integrated hydropyrolysis are separated plastics like PE, PP, PUR, PS, and residue waste streams from recycling. The output flows are oil with a low sulfur content, used as marine gas oil (MGO), *syngas* and a small proportion of salts and metals. The char produced is internally removed and used as an energy source. The ratio and the quality of products depend heavily on the input composition and the specific technology used.

4.4.3.2 Hydrothermal treatment

Hydrolysis is a reaction in which a compound is broken down by water molecules in a near-critical condition. Generally, the temperature of this *hydrothermal treatment* (HTT) is around 160–240 °C, with a corresponding pressure to keep water in the liquid state. The special properties of near-critical water makes it a good medium for dissolving organic compounds. Essential reactions of HTT are hydrolysis, dehydration, decarboxylation, and depolymerization. Hydrothermal processing has been used for recycling of waste carbon *fiber-reinforced plastics* and printed circuit boards, in a batch reactor. The ability of near-critical water to degrade the resins and plastics in the composite wastes is largely influenced by the presence of different additives and co-solvents. Hydrothermal

treatment has been proposed as a solution for the separation of mixed waste into organic and inorganic substances. Most common feedstocks for hydrolysis are:
–   plastic packaging waste (PET);
–   carbon fiber-reinforced plastics;
–   printed circuit boards;
–   polycarbonate (PC);
–   styrene-butadiene rubber (SBR);
–   polylactic acid (PLA);
–   nylon 6, nylon 6,6 (PA6, PA6,6).

This technique is used to produce synthetic *crude oil*. Separation and purification can then be handled by standard refinery operations. The technology is still at a development stage, however, with commercial operations in planning.

4.4.3.3 Gasification

Gasification is a process where mixed waste materials are heated to a very high temperature (1,000–1,500 °C) in the presence of a limited amount of oxygen, which breaks the molecules down to their simplest components to produce syngas. The syngas can then be used to produce a variety of chemicals (methanol, ammonia, hydrocarbons, acetic acid) for plastics production as well as fuel and fertilizer. Gasification is generally carried out in larger process units which are designed to achieve economies of scale. In case of pyrolysis, such units usually take in a mixed waste stream, which places less pressure on the collection and sorting system. Gasification typically requires pre-treatment to remove moisture and increase the calorific value. A very efficient gas cleaning system at elevated process temperature is needed to meet the requirements for applying syngas in chemical production. In principle, all plastics can be used as feedstock. Gasification of waste plastics leads not only to the production of syngas, but methane (CH_4) and nitrogen (N_2) as well. This output gas can be incinerated for energy or used in the production of new hydrocarbons. Gasification plants are typically built at a larger scale than pyrolysis plants.

Low temperature gasification

This specific type of gasification is performed at lower temperatures (800–1,000 °C), in the presence of oxygen. The fuel is broken down into syngas. There are many different variants of gasification, which take place in different types of installations, such as vertebrae bed gasification. Low temperature gasification is mainly applied to biomass-containing waste streams. Other installations focus on household waste, which target in particular on the production of energy.

Medium-temperature gasification

Medium temperature gasification produces a higher quality syngas with respect to lower temperature gasification. Furthermore, the process is similarly constructed. The temperature typically rises in between 900 and 1,650 °C. Although this process seems promising, it is still somewhat uncertain to what extent mixed garbage should be sorted or pre-processed before implementation.

Research is actively focusing on new technologies operating at temperatures between pyrolysis and gasification with a limited amount of oxygen. New technologies facilitate the conversion of plastic waste into either monomers or a blend of benzene, toluene, and xylene (BTX). Synova, for example, transforms biowaste and polyolefins to monomers, while BioBTX converts plastic waste into BTX. The BioBTX process is referred to as an Integrated Cascading Catalytic Pyrolysis technology. The new BTX building blocks can be used for the synthesis of polymers or other applications. Both processes can handle more polluted waste streams compared to pyrolysis and offer a higher added value of output compared to syngas. The demoplant for BioBTX is scheduled to be up and running in 2024 in Delfzijl in the Netherlands, while the demoplant for Synova is currently at the investment stage [79]. The catalytic processes described above exhibit a distinct advantage by consuming less energy due to operating at lower temperatures. However, the technology is still in development, and a potential drawback is associated with the stability of the catalyst.

4.4.4 Depolymerization

Depolymerization, sometimes referred to as *chemolysis*, is the reverse of polymerization (Section 2.3) and yields either single-monomer molecules or shorter polymer fragments known as oligomers. Those monomers are identical to those used in the preparation of polymers and because of this, the plastics prepared from depolymerized products are similar in quality to virgin materials. The main disadvantage of chemical *depolymerization* is that it can only be applied to *condensation polymers* such as PET and PAs. It cannot be used for the decomposition of most addition polymers (such as PE, PP, PVC, PS) which make up the majority of the plastic waste stream.

The production of polyester is an equilibrium reaction (Figure 4.34). When a large amount of one of the monomers is added, namely ethylene glycol (EG), the balance is disturbed and the polymer breaks down. The result of this depolymerization (glycolysis) is a liquid with such low viscosity that it can be properly purified. This way, the contaminants can be removed. The clean building blocks can be used again in a (re)polymerization process, to produce pure PET. The excess ethylene glycol (EG) is recovered and used again. Instead of ethylene glycol, other reactants can be employed (Figure 4.35), such as water (hydrolysis), methanol (methanolysis), ammonia (ammonolysis) and alcohol (alcoholysis).

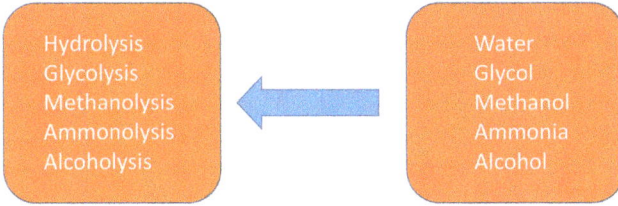

Figure 4.34: Shifting the equilibrium reaction of polycondensation.

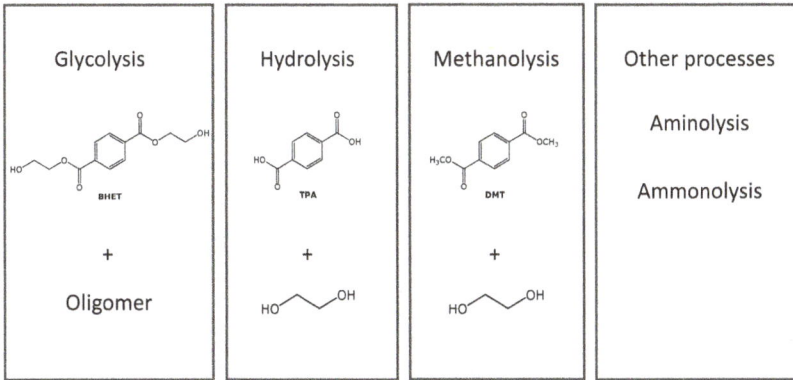

Figure 4.35: The products involved in different depolymerization processes.

Target feedstocks for depolymerization are polycondensates, which include polyesters (PET), PA, and polyurethanes (PU). Typical input material includes all-polyester carpets, multi-layer PET films, and textile products such as sports shirts.

A number of industrial plants carrying out PET depolymerization are currently in operation, based mainly on *glycolysis* and *methanolysis* treatments (Figure 4.36) Hydrolytic processes are less advanced, most of them being used at laboratory and pilot-plant scales, although several projects are being developed for commercial applications in the next few years. Ammonolysis and aminolysis-based processes are well-developed, but still less-established, treatments.

In the Netherlands the companies Cumapol, DSM-Niaga, Dufor, and Morssinkhof, in close collaboration with NHL Stenden University of Applied Sciences, developed an advanced and scalable solution for the depolymerization of PET. Their process is called CuRe Technology, used to create a fully circular low energy polyester chain. CuRe Technology aims to "cure" any type of polyester by removing the color and converting it into transparent granulate with the same properties as virgin grade polyester. A pilot plant is recently built in the city of Emmen (The Netherlands) for rapid scale-up. The Coca-Cola Company already is one of the customers of this very promising technology.

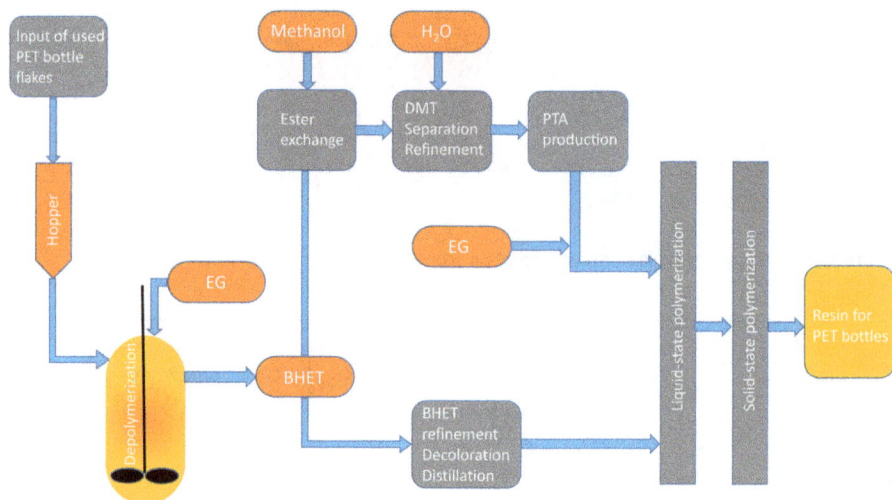

Figure 4.36: Schematic representation of the depolymerization process of PET by glycolysis.

Other companies like Aquafil (recycling of Nylon yarns), Indorama (PET bottles) Unilever in cooperation with Ioniqa (recycling of PET packaging), are implementing depolymerization techniques for the recycling of their products as well.

Enzymatic depolymerization

Polymerization methods catalyzed by enzymes are well known in polymer science. An excellent overview of biocatalysis in polymer chemistry has been provided by professor Loos [80]. However, research in the field of enzymatic depolymerization has gained more interest in recent years. *Enzymatic recycling* can be a green alternative for chemical recycling because it enables the use of mild processing conditions to regenerate the polymer's building blocks. In this technique, enzymes are utilized to catalyze the hydrolysis of chemical bonds in polymers. Ideally, enzymatic repolymerization is then applied to these monomers to complete the green cycle. While current research efforts mostly focus on PET, the enzymatic recycling of PUs and PAs is gaining momentum [81]. In the next paragraph, developments related to the enzymatic depolymerization of polyesters, PAs, and PUs will be discussed.

Recent advances in enzyme engineering and directed evolution have led to a renewed interest in biotechnology for industrial applications, and the synthesis of small molecules in the pharmaceutical industry is by now state of the art. Enzymatic polymerizations are also becoming more established recently, as they are a powerful and versatile approach that can compete with chemical and physical techniques for the production of commodity plastics, as well as for the synthesis of novel macromolecules. Examples include enzymatic monomer synthesis and enzymatic polymerizations of polyesters, PAs, and polyesteramides. Based on this, enzymatic recycling is currently emerging as a promising alternative for conventional plastic recycling because it enables the use of mild and environmentally friendly processing conditions. Especially the enzymatic recycling of PET has attracted considerable interest. Techno-economic analysis has been used to predict that enzymatic PET recycling can be cost-competitive with virgin PET manufacturing costs. Furthermore, the French company Carbios has recently launched an industrial demonstration plant that might lead to enzymatic depolymerization of PET on a large industrial scale. Since PU is after PET, the second most common plastic type that can undergo hydrolysis, it is not unthinkable to envision a similar future for enzymatic recycling of PU. While an efficient process for PU has not yet been reported due to the high complexity of the PU polymer structure, a number of enzymes have been shown to be involved in the depolymerization of PU. The Rampf company has a commercial process of depolymerization of PU. In Table 4.8, recent developments that relate to the enzymatic depolymerization of PET are listed. Almost all these developments describe the use of a cutinase for breaking down PET. Carbios tried to cover as many enzymes as possible. Only two examples were found that relate to the enzymatic depolymerization of PA or PU.

Table 4.8: Developments in the field of enzymatic depolymerization of PET.

Polymers	Types	Enzymes	Processing conditions	Company/institute
PET	Amorphous and semicrystalline PET	Cutinases together with at least one lipophilic and/or hydrophilic agent	20–80 °C, pH 4–10	Carbios
PET	Mixed PET packaging or fibers	Cutinases	Unspecified	Carbios
PET	Postconsumer PET	Cutinases, Lipases	20–100 °C, 20–300 rpm, 2–30 days	Petrobras
PET	Less than 25% crystalline PET	Unspecified	66–80 °C, ≤ 20 h	Carbios
PET	Waste PET bottles	Cutinases	67 °C, pH 8, 120 rpm	Tech University Nanjing

In Table 4.9, some examples concerning the enzymatic depolymerization of polyesters, PAs, and PUs are listed. Interestingly, Carbios has patented a multicomponent thermoplastic product, in which an enzyme capable of degrading the first thermoplastic polymer is incorporated into the second thermoplastic polymer. Potential candidates for the first thermoplastic polymer were PET, PAs (PA6 or PA6,6), PUs, and a copolymer of PET, PA6, or PA6,6, among others. This implies that Carbios is capable of enzymatically depolymerizing PAs and PUs. In addition, Nestlé has explored the use of two cutinases for enzymatic recycling of PUs in multilayer packaging material.

Table 4.9: Developments in the field of enzymatic depolymerization of polyesters, polyamides, and polyurethanes.

Polymers	Types	Enzymes	Processing conditions	Company
Polyester	Unspecified	Unnamed polypeptides	20–60 °C, pH 5–11	Carbios
PET/PLA	Semicrystalline polyesters	Cutinases, proteases, lipases, and esterases	5–65 °C	Carbios
Polyester	Unspecified polyester plastic product	Esterases	50–90 °C	Carbios
PET/PA/PU	Multicomponent plastic product	Unspecified	Reaction temperature at T_m of polymers	Carbios
PU	Multilayer packaging	Cutinases	20–50 °C, pH 6–9, 3–20 days	Nestlé

Magnetic depolymerization

The Dutch startup Ioniqa has developed a technique in which PET waste is chemically depolymerized under the influence of a magnetic liquid. During the process, BHET is produced in crystal form (Figure 4.37). BHET can then be used by existing producers as a raw material in PET production. In the process, various dyes and other contaminants are removed from the feed, leading to a high-quality material similar to the virgin resources for PET polymerization. Unlike mechanical recycling, this technique is suitable for the identified waste streams of PET trays. Although the technology is still in development, Ioniqa is working on the realization of a production facility.

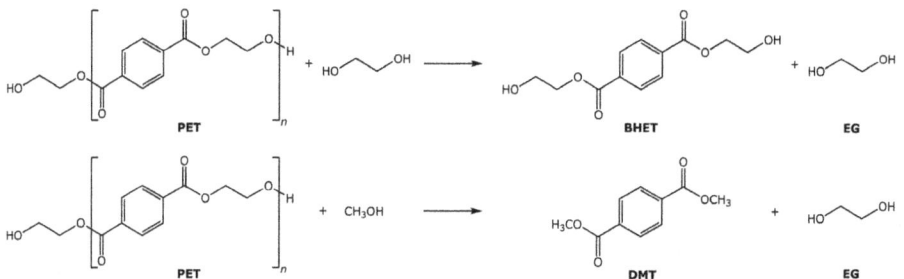

Figure 4.37: Chemical reactions involved in the glycolysis and methanolysis process.

4.5 Sustainable plastic product development

Creating a completely circular economy for every material flow may not be possible, however we move closer to a sustainable future every time we reduce the amount of waste. With regard to material flows and production, plastics can fit very well in a bio-logical cycle or a technical cycle, as explained in Chapter 1. *Biodegradable plastics* can re-enter the biosphere safely after the user phase. Conventional plastics can be designed very well for re-use to circulate without entering the biological cycle. Crucial for plastics in the *circular economy* is that the value of the materials and products is kept as high as possible for as long as possible. In other words, value creation through closed loops. Of course, not only the biological or technical aspects make a circular economy work. One has to organize closed loop activities. Important in these activities are: flow manage-ment; reversed logistics or green logistics; separating and sorting services and of course recycling services [82].

In general, the circular product design principles are based on:
– form;
– function;
– price;
– quality;
– sustainability.

The *circular product design* principles are illustrated in Figure 4.38.

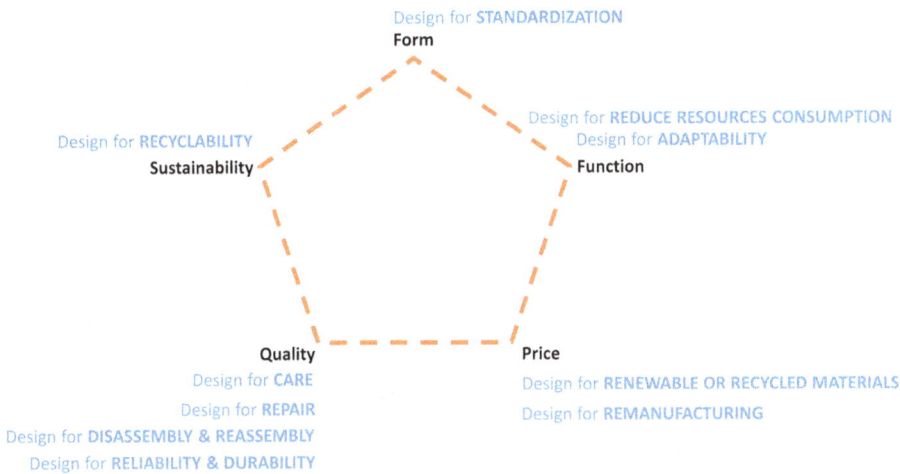

Figure 4.38: Circular product design principles.

The starting point is how to introduce plastic recyclates in products. When setting up a closed-loop system there are some challenges with regard to the circular material flows. First of all, the high upfront costs and investments, second is choice of material, and third the complexity of the supply chain. We can distinguish closed-loop recycling and open loop recycling.

- Closed-loop recycling: the material is recycled and reutilized in the same product category, for example in case of bottle-to-bottle recycling.
- Open loop recycling: the recycled material is reused as raw material for other types of products.

A wiser definition of the degree to which the loop is closed, is actually to what extent a particular organization can be held responsible for waste. In the next chapters some methods will be discussed, in which manner *recyclates* can be used in an open loop recycling system, taking into account technical aspects and organizational aspects.

4.5.1 Application of recyclates in products

The development process for products from recycled plastics will be more effective if one has a good understanding of the (im)possibilities of recyclate materials [85]. A number of themes are essential to ensure that recyclate can be used optimally when replacing virgin material or when designing a new product.

4.5.1.1 Company introduction of recyclate

The KIDV in the Netherlands is an institute which does a lot of research in the field of sustainable packaging. They developed an application for a recycle check, in which the recyclability can be determined of different packaging products. This tool can be used by companies as well as consumers. A roadmap has been developed by Philips, in the Netherlands, to facilitate the introduction of *recycling*. This roadmap mentions some specific points that can help with the application of recycled plastics. Below, the most important aspects of this roadmap will be mentioned.

Take stock of the use of plastics within the company

Create an overview of the kind and amount of plastics used within the company. This overview can be of help to determine which plastics may be worth using.

Focus on widely used polymers

Bulk plastics are widely used in products and as a result are also widely collected and recycled. To ensure a constant supply of material, it is therefore recommended to focus on one of these types of plastics.

Focus on non-visible and dark parts
Although recyclate is available in many different colors, they are still easiest to apply in dark or non-visible parts. Transparent parts, food approved products and applications with demanding mechanical requirements are still difficult to produce with recyclate.

Find and approach suppliers
Find potential suppliers of recyclate and approach them in time. In this way, their expertise can be used at an early stage. Certification systems exist with a certain quality guarantee from the supplier.

Determine the most important requirements for the product
Test the parts that will be produced with recyclate on the main properties. First, it is necessary to determine which characteristics the parts must meet and under what conditions.

Start with applications in existing products
Products that are newly marketed often have a strict release date. Existing products that can be replaced by recyclate during the life cycle often do not have this burden.

Design for recycled plastics
When experience has been gained in applying recyclate in existing products, parts can also be designed specifically for recycled plastics. This can already take into account factors such as product construction, surface masking and mold design.

4.5.1.2 Designing products with recyclate
In a new design that will contain recyclate, the properties of the recyclate must be taken into account. In the case of an existing design, a material with suitable specifications must be sought. Guidelines regarding shape and connections when designing injection mold products are available today. They often focus on aspects such as wall thicknesses, geometry, holes, ribs and loosening angles. When designing products with recyclates, the same guidelines apply in principle as when designing with virgin plastics. However, because there is a larger uncertainty factor in the properties and behavior of recyclate, the material may react differently during casting. Specific guidelines on form and connections when designing with recycled plastics are therefore difficult to compose. In general, it is wise to be careful with extreme geometries. For example, the risk of defects at sharp transitions and angles is greater compared to virgin plastics due to the uncertainty in material behavior. When existing molds are used, the settings such as temperature, pressure and cooling time can influence the outcome of the injection molding process.

Shrinkage can be stronger when applying recyclates instead of virgin plastics, especially for large molded parts. For new products specifically designed for recyclates, additional ribs can therefore be incorporated to support the material. When using existing molds, shrinkage can be reduced by applying a shorter cooling time.

4.5.1.3 Replacement of virgin plastics by recyclates

When plastic recyclate is used, in most cases this is done as a *replacement* for the same product or component that was initially produced with virgin plastic. For such an existing design a mold is already available. As a result, recyclate is required with very similar specifications and behavior as the virgin material. In practice, a certain percentage of recyclate is often mixed as a blend with virgin material. Alternatively, it can be applied in an intermediate layer of a multilayered concept.

The material properties of the recyclate should be in line with the requirements of the product fabricated with virgin plastic. The parameters representing the main requirements should be compared with the properties of the available grades of recyclate.

In case of substitution of virgin plastic to recyclate, it cannot be assumed that the process parameters of the injection molding process remain identical. Even though the main mechanical and thermal properties of the material are virtually the same, process-influencing properties, including rheological properties, may still differ. Therefore, it is important to carry out production tests with the target material, before mass production starts. In addition, the mold should be sufficiently ventilated to prevent an unpleasant smell of the final product. When a product is designed specifically for the application of recyclate, there is more freedom to operate.

In some cases, recyclate can serve as a substitute for a material other than plastics, for example wood. A new design must be created, and properties of the benchmark should be carefully considered, in order to select the suitable plastic replacement.

4.5.1.4 Visual and tactile-design considerations

When recyclate is applied in non-visible parts, the experience of a consumer will not be affected, while the cost prize can be reduced. However, depending on the requirements of the visual appearance, colored parts can be obtained from recycled plastics as well.

In products with light colors, impurities of the production material become more visible. In order to prevent visible contaminants, pure recyclate is required. Very small contaminants, for instance dark dots in the injection molding part, cannot be ruled out completely. Tests can show whether this is acceptable. Very bright colors that can be realized with virgin plastics are difficult to realize with recyclate. A product with high gloss is difficult to realize, while a color with a matte appearance is often achievable. Recyclate in dark or black colors can easily be incorporated into products. It is also the cheapest and easiest way to achieve, because it is straightforward to darken with pigments.

It is more difficult to achieve a smooth and high-gloss surface with recycled plastics with respect to virgin plastics. Actually, high gloss black can only be achieved by the use of virgin polymer which is colored black, because the light has to be able to penetrate partly. This becomes very difficult with recyclate, as it is generally no longer transparent. Recyclate will in most cases keep a mat appearance. However, silk gloss surfaces can be achieved with recyclate.

To hide visual contaminants, a texture can be applied to the surface of a plastic part. It can vary from adding a pattern to imparting a barely visible grain that hides impurities. To impart a high-quality look, the use of in-mold labels is an option. The hot polymer is sprayed against the label. By fully covering the parts under the label, these parts will not be visible. This allows requirements, with regard to the surface finishing or color, to be reduced because the part is less visible. Recyclate can then be used for the part involved.

4.5.1.5 Legislation

In a number of sectors, additional legislative requirements are required to be met by recyclate. In addition, the law and regulations also increasingly take into account a circular approach instead of a linear one. A few examples of European and American *legislation* are discussed in this paragraph.

Deutsche Gesellschaft für Kunststoffrecycling

Plastic packaging waste is sorted according to Deutsche Gesellschaft für Kunststoffrecycling (DKR) standards. These standards vary by type of plastic. One or more product specifications have been composed for each type of plastic. The material is described, the minimum purity of the material is displayed, the maximum pollution is determined, and the delivery method is specified. The main codes used in the DKR system are: DKR-310 for foils, DKR-324 for PP, DKR-328 for PET, DKR-329 for PE and DKR-350 for mixed plastics.

European Regulation No282/2008

This regulation provides for use of recycled plastics in food-contact applications. The producer must ensure that the plastic input comes from a closed and controlled cycle, so that pollution can be excluded. A deposit PET bottle is a good example of an application where a product (bottle-to-bottle) can be recycled into a new food-approved application.

European Food Safety Authority

Ultimately, the suitability of the product and the process is tested by the European Food Safety Authority (EFSA). This authority takes a look at matters relating to the collection, the effectiveness of the process of reducing pollution, and the defined use. The Morssinkhof company in the Netherlands, for example, has a EFSA *certification* for recycled PET bottles.

Food and Drug Administration
Food and Drug Administration (FDA) is the agency of the federal government of the United States, which controls the quality of food and medicines in a broad sense. It also monitors the treatment of blood, medical products and cosmetics. The FDA ensures that rules in the Public Health Service Act are complied with. The production of medicines in particular is subject to very strict rules in the United States, but they also apply to foreign companies wishing to export to the US.

Registration, evaluation, and authorization of chemicals
Registration, Evaluation and Authorization of Chemicals, or REACH, is EU regulation on the production and trade in chemical substances. It is adopted to improve the protection of human health and the environment from the risks that can be posed by chemicals, while enhancing the competitiveness of the EU chemicals industry. It also promotes alternative methods for the hazard assessment of substances in order to reduce the number of tests on animals. In principle, REACH applies to all chemical substances; not only those used in industrial processes but also in our daily lives, for example in cleaning products, paints, clothes, furniture and electrical appliances. Therefore, the regulation has an impact on most companies across the EU.

Restriction of Hazardous Substances
Restriction of Hazardous Substances (RoHS) is a guideline for limiting the use of certain hazardous substances such as lead, mercury, cadmium, chromium VI, PBB and PBDE in electrical and electronic equipment.

4.5.2 Design for recycling

During the design phase of a product, it should be taken into account that plastics remain valuable after their first life cycle. This is design for a *circular economy*. Circular economy offers a solution to the growing concerns on resource depletion and increasing waste. Design within a circular economy includes that products are designed in a way that they have a long life time and can be maintained well during use, can be reused when after their first life cycle, and can be recycled after disposal. In this way, savings are made on resources and energy, while new business models can be created.

4.5.2.1 Design within a circular economy
In 2010, the Ellen MacArthur Foundation was established to highlight the principles of the circular economy in companies around the world. Within the principles of a circular economy, the foundation formulated the following goals.

Maintaining the value in the original function
By developing a product that can be easily reused, it saves resources, labor, and energy.

Keep products in use for as long as possible in the circular economy
Products have a long life time or are reused after disposal. Ways to encourage this are to use high-quality materials, apply timeless design or ensure that material does not degrade quickly, for example by using additives against UV degradation. Another important aspect is to design products in such a way that consumers are bonding to the product. In that case, the consumer is more willing to keep the product as long as possible in use. In addition, products must be suitable for maintenance and repair.

The use of plastics in cascades
Plastics are recycled and used in lower-value applications if needed. This prevents the use of new resources and in this way the plastics can be used for a longer time. Design for recycling comes in handy in this scenario. Design for recycling does not always correspond to design for reuse, disassembly or refurbishment. Whereas design for giant screw joints is often useful, design for recycling makes it more effective to use only one material that circumvents the need for separation during the processing of a product.

Clean input material
This goes without saying and means that plastics should be clean, do not contain harmful additives, do not contain mixed or non-recyclable polymers and can be properly collected and reprocessed. As a result, it will cost less resources, energy and labor to prepare products for a new phase of use.

4.5.2.2 Design for the optimal life cycle
In an optimal *life cycle*, plastics will undergo all the steps mentioned, before they will eventually be used in the lowest-grade recovery, as a raw material for energy. The chances that a system is designed in such a way that it will actually pass through all these steps is small. In order to achieve an optimal life cycle, it requires intensive cooperation between chain parties that ensures that materials are collected and processed purely and are always used in the most valuable application.

4.5.3 Recyclability incorporating in design

Specific guidelines for various types of products and plastics have been provided in order to promote *recyclability*. An overview will be given of matters that can be applied for the purpose of recyclability.

4.5.3.1 Recyclable products

The degree of recyclability of plastics can be determined at both product and material level. If a plastic is straightforward to recycle but the product cannot be separated from other components, the material will still not be able to be recovered properly. That is why it is important to include the recyclability of the entire product. In many sectors it is known in what manner products will be processed after disposal. This can be anticipated in the design. For instance, when a product is processed in a shredder while it consists of various materials, it is useful to use fault lines. These affect the way a material will tear. In that way, it can be influenced that a fracture line will run along a screw hole to separate a plastic and a screw. On the other hand, when manual dismantling will take place during collection, it is more effective to use easily detachable connections.

4.5.3.2 Electronic domestic appliances

The recyclability of consumer electronics and household appliances can be improved by applying a number of guidelines. With regard to materials, only use materials that can be recycled and avoid the use of non-compliant coatings. In addition, limit the number of different materials and use only pure materials. With regard to connections, avoid fixed connections. Break-down by shredding or disassembly to pieces with uniform composition. Finally, recycling codes can be applied in the mold of a plastic product. This induces consumer awareness.

4.5.3.3 Material choice

The recyclability of plastics is largely related to purity and separability. These can be improved by paying attention to the following matters.

Plastic blends are mixtures of different types of plastics. Blends are generally difficult to recycle. With the exception of widely used blends, such as PC-ABS that can also be reused as a blend, this causes a lot of waste during the reprocessing of plastics. On the other hand, a high percentage of minerals in plastics can cause a change in density. When separated by specific mass, they can be mistaken for a different type of plastic (Section 4.3.2). This results in contamination of a batch, which affects the quality of the recyclate.

There are a number of organizations that have drawn up detailed guidelines per type of plastic to achieve optimal recyclability. Recycling of Used Plastics Limited (RE-COUP) is a British plastic recycling organization that regularly publishes new guidelines. Those guidelines indicate how color, barriers, additives and labels influences the recyclability and in which manner this knowledge can be incorporated into a design.

In order for a product to be in line with the circular economy, one can pursue various circular design solutions. The success of circular design is strongly related to the chain and revenue model. Depending on the market positioning and the use of the product, designers can choose a matching design approach [83, 84]. Three themes guide the design phase: a focus on long-term product use, closing the loop, and the choice of sustainable materials. These themes and the corresponding design strategies are explained below.

Focus on long-term use

By designing products for longevity, their value is extended and production of new products is prevented. It is crucial to develop products with a focus on quality, repairability, and product adhesion. After all, a high-quality product is less likely to break down and retains its value for a longer time. The balance between sound design and use is essential. For a long service life, it is not only the strength of a product that is of interest, but the user should also have the desire to use the product for a long time. Creating bonding and trust between the user and the product is therefore essential. In other words, the product must have and maintain value in the eyes of the user. When maintenance and repair of a product is convenient, it lasts longer. If fragile parts are easily replaceable, the lifespan of a product is extended.

Close the loop

To achieve a closed-loop system, the product design must be in line with processes that take place within the chain, such as reuse, refurbishing, and recycling. In case a product is discarded by the user, value can be added to the product through refurbishing. By replacing fragile and damaged parts, a product is as good as new, and ready for a new user.

By specifically considering recycling in the product design, more value is recovered from the product material at the end of its life. Different materials must be facile to separate and the design should be in line with common recycling processes. This is a strategy to achieve long-term use and system closure. Building a product from different functional modules makes repair and refurbishing convenient and cost-effective, and promotes recycling. For modularity, the connections between different components are particularly important so that the product can be assembled and disassembled several times without damaging.

Choose sustainable materials

Using recycled and safe materials in the product reduces the environmental impact. It is essential to use recycled materials to reduce the demand for primary raw materials. In other words, choose materials that do not have a negative impact on people and planet, in case you want to discard or reuse them.

Defining a design approach

Depending on the use of a product and its market positioning, a matching design approach is chosen. The design process of a short-life product has a different focus with respect to a long-life product. For example, design for recycling is very important in case of a short lifespan, because the product quickly ends up in the recycling loop, and therefore the most value can be gained in that particular phase. In case of a longer lifespan, the design focus is on maintaining the product value for as long as possible, for example by developing a high-quality product that is easy to repair. Products that are sensitive to trends or technological developments pose a challenge, because they quickly become obsolete and lose part of their value. Often, multiple design strategies are applied simultaneously so that the end product fits well with the use case, the revenue model, and the organization of the chain.

In Europe, approximately 35 million mattresses are discarded every year, significantly contributing to landfill waste due to their complex composition of glued materials. Hotel mattresses form a substantial part of this waste stream. The Auping company, in collaboration with former DSM Niaga®, developed a novel hotel mattress that contains no conventional foam or adhesives, enabling it to be recycled into high-quality new mattresses. The material sourcing is transparent, as all materials are documented in a product passport with scannable label. Using a smartphone scan, the label identifies the composition and origin of all materials used in the mattress.

Fairphone is a Dutch smartphone manufacturer dedicated to making ethical and sustainable practices the standard in the electronics industry. The company creates smartphones that prioritize environmental impact reduction, social responsibility, and product longevity. At the core of their products is a modular design approach. This means that all components, including camera, battery, screen, and plastic housing, are easily replaceable or upgradable by the user in order to extend the lifespan of the device and reduce electronic waste. Fairphone has a strong emphasis on sourcing materials like tin, tungsten, and gold from conflict-free regions. The company collaborates with manufacturers that support fair wages, safe working conditions, and labor rights, thereby reducing exploitation often associated with electronic production.

Plastic pallets are strong, easy to clean, and have a long lifespan. The market for plastic pallets is growing. However, when used with forklifts and (electric) pump trucks, they are regularly damaged. As the pallets are currently produced as a single unit, the damage is irreparable and the user has to replace the entire pallet.

Schoeller Allibert has developed a patented solution for the most common cases of damage. In the case of incorrect use of electric pump trucks, the entire top of the pallet regularly tears off the bottom. With their innovation, only a cheap plug will break instead of the entire pallet. The plugs are replaceable and, if necessary, the skids as well. Customers can order all parts separately and easily repair the pallet themselves. In this way, plastic pallets form a good alternative to wooden pallets. They have a longer lifespan of at least 10 years and can be repaired, while the material is readily recyclable and can therefore be used in the production of new plastic pallets.

4.5.4 Life-cycle assessment

Life-cycle assessment (LCA) is the factual analysis of a product's entire life cycle in terms of sustainability. Every part of a product's life cycle can have an impact on the environment in many ways. From extraction of materials from the environment, and manufacturing of the product, to the actual user phase and what happens to the product after disposal.

Figure 4.39 depicts a schematic representation of the total system of the building blocks involved in an LCA [86]. The flows in this scheme can be quantified. Basically, data will be generated describing the phases of mining or extracting natural resources, the user phase and the end of life phase. In this way, various phases of the life cycle can be compared in a quantitative way. For instance, a pair of jeans show the highest environmental impact during the user phase because it will be washed many times, wasting a lot of resources (water, detergents) and energy. Life cycle assessments (LCAs) can help in quantifying the amount of energy used, or the CO_2 *footprint* of the whole *life cycle* of a product. With LCA one can evaluate the environmental impact of a product (or service) from cradle to grave.

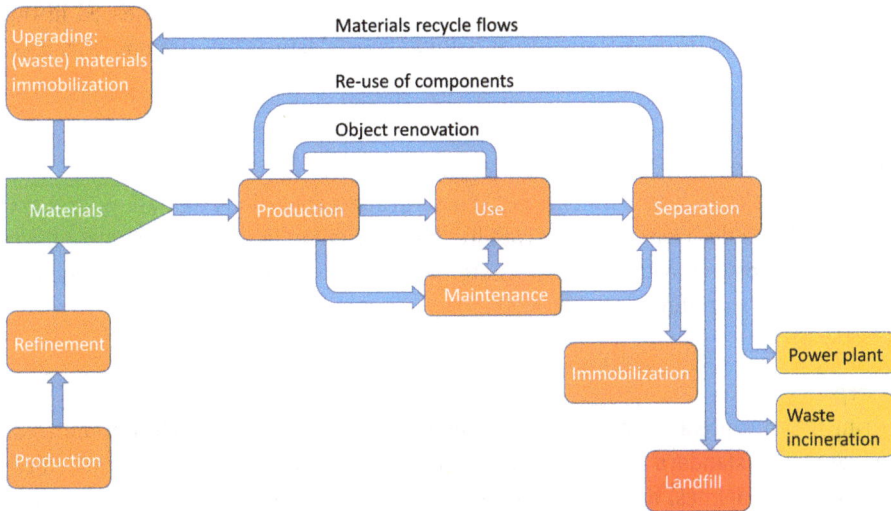

Figure 4.39: The general building blocks in LCA.

LCA can help to improve the product development process, marketing, strategic planning and policy making. In addition, consumers can learn to which extent a product is sustainable. Product designers can explore how their design choices affect the sustainability of a product. Actually, many types of LCA exist. Rule of thumb is that the more detail you want, the more input data you need and the more complete your LCA needs to be. Further, many LCA-related assessments exist, such as Environmental

Product Declarations (EPDs), studies compliant with product- or sector specific standards, single issues analyses like carbon or water footprint, circularity indicators, social LCA and long-term monitoring studies. For a lot of processes and materials, data of the indicators (eco-costs, carbon footprint, etc.) are provided in standard databases.

LCA is a standardized methodology. Standards are provided by the International Organization for Standardization (ISO). In ISO 14040 and 14044, four main phases of an LCA are described, which will be discussed below.

LCA goal and scope definition
The goal and scope definition step ensures that an LCA is performed consistently. LCA models a product, service, or system life cycle. A model is a simplification of a complex reality, so it cannot always reflect reality for 100%. Goal and scope of the LCA study is leading in making the right choices.

Inventory analysis of extractions and emissions
In the inventory analysis, one looks at all the environmental input and output associated with a product or service. This can be input for the system (extracted from the environment), for example the mining of raw materials. Or output (released in the environment), for instance emission of pollutants or plastic waste streams. Together, this leads to the complete picture.

Impact assessment
In a life cycle impact assessment (LCIA), one draws conclusions that allow to make better business decisions. One can classify the environmental impacts, evaluate them by what is most important and translate them into environmental themes such as global warming or human health.

Interpretation
During the interpretation phase, one examines that conclusions are well-substantiated. The ISO 14044 standard (Environmental management, Life cycle assessment, Requirements and guidelines) describes a number of checks to test whether conclusions are adequately supported by the data and by the procedures been used.

4.5.4.1 Incorporating LCA-models
At the moment our society is in a transition to a circular economy. With regard to circular plastics, the data obtained from LCA-models will play a crucial role. It enables the evaluation of environmental impacts corresponding to the use of alternative feedstocks for plastics in comparison to current fossil-based feedstocks. Additionally, it can quan-

tify the impact of various end of life treatments for both bio-degradable and recyclable plastics. Focusing on the required data, the following information is important.

Identification of the main recycled plastics articles

The market potential of recycled plastic articles like packaging, construction articles, automotive, including production amounts and sale prices, market shares, trends, forecasts and job creation. The potential for deployment of recycled plastics products, including information on:

- feedstock availability for polymer producers;
- technology restrictions for polymer converters;
- development status of technology;
- economic restrictions.

Life-cycle inventory

Studies on recycled polymers or specific recycled plastic articles and reviews of background reports and information are useful as input for new LCA studies. Data for relevant processes involved in recycled polymer production, and especially for polymerization and monomer production from feedstocks, is needed. Data should cover all relevant input and output flows of the process, like energy, materials, chemicals, land use, water use, emission into air, water and soil and waste. Data is requested to be representative for the average of industry production and the potential variability around this average.

Recycling incompatibilities

Data or information about potential recycling incompatibilities are needed, due to bio-based and biodegradable polymers and plastic articles, in current mechanical recycling plants for conventional plastics. What are realistic thresholds for the presence of contaminants in the mechanical recycling process of commodity polymers?

Bio-based and biodegradable polymers

In addition, data about the current presence of bio-based and biodegradable plastic articles in the plastic waste stream sent to mechanical recycling processes, is required.

Additives

Type and quantity of the most relevant *additives* used in the production of specifically recycled commodity polymers or recycled plastic articles like bottles, bags, cups and all kinds of packaging. In addition, data on the production of the additives used for specific recycled polymers and plastic articles. At last, data on the release, potential exposure and possible toxicological effects over the life cycle of additives used.

4.5.4.2 LCA for plastic packaging

LCA is a useful tool to evaluate *environmental impact* of packaging alternatives over their lifetime, from the extraction of raw material to the disposal or recycling of *packaging* at the end of life. Undertaking LCAs to compare the environmental performance of alternative materials for different packaging applications is essential if we want to take into account the environmental impact associated with the whole life cycle of packaging (mining, manufacturing process, logistics, usage and end of life route). Transport distance, sources of electricity generation, packaging shape and weight, all significantly influence the LCA results and should be considered on a case to case basis.

Results can vary significantly from one study to another, depending on key parameters and certain assumptions. As an example, the risk of producing more food waste because of the packaging design and shelf life is not always considered in LCAs, while this can have a large impact on the packaging carbon contribution. A recent study from the UK on LCAs compares various types of packaging [87]. Findings indicate that, most of the time, plastic packaging performs better than its alternatives, mainly due to its lightweight properties.

The *waste management* route in place to treat packaging at its end of life, is also shown to be a critical factor explaining variations of LCA results for packaging. Recycling always wins over virgin production on environmental indicators. Typically, recycling saves between 30% and 80% of the carbon emissions that virgin plastic processing and manufacturing generate.

Considering the fact that only about 10% of plastics are currently being recycled worldwide, there is plenty of room for improvement.

4.6 Exercises

4.6.1 Circular plastics

1. To achieve a sustainable and responsible use of plastics, the so-called 3R model can be applied.
   a. Appoint the 3Rs in the model.
   b. Four types of recycling can be distinguished. Describe, in short, the four different types.
   c. Explain why the production of plastics is still increasing worldwide.
   d. Describe the class of polymers represented by the identification code nr. 7.

4.6.2 Plastic packaging

2. Forty percent of thermoplastic polymers are used for packaging.
   a. Give the names of the most important polymers used for packaging.
   b. Name also the most common applications of these polymers.
3. Ethylene Vinyl Acetate (EVAc) is often used in packaging.
   a. Explain the function of EVAc in packaging.
   b. Give the properties of EVAc which are determined by the amount of vinyl acetate in the copolymer.
4. In the packaging industry different types of PET can be distinguished.
   a. Explain the difference between a-PET, c-PET and PETG.
   b. Discuss their different applications.
   c. Explain the different problems which arise with regard to those applications.
5. In food packaging applications, barrier properties of films are very important.
   a. Write down the reasons why barrier properties are important.
   b. Explain the abbreviations OTR and WVTR in packaging film applications.
   c. Explain why multilayer films are a problem in mechanical recycling.
   d. Explain the use of PA in packaging.
   e. Write the structural formulas of ionomers used in packaging.
   f. Name a reason for the use of adhesives in barrier films.
6. Packaging often is decorated with labels.
   a. Describe the different methods of labeling.
   b. Explain the problems with labeling regarding mechanical recycling.

4.6.3 Mechanical recycling

7. The mechanical recycling process contains many operational steps.
   a. Describe the recycling process of a processing plant from the start till the end.
   b. Explain the methods used in the first sorting step.
   c. Explain the function of shredders.
   d. Write a list of potential contaminants found in plastic waste.
   e. Explain why contaminants must be removed.
   f. Explain how washing water of plastic waste can be reused.
8. Depending on the kind of plastic waste, many sorting techniques are available nowadays.
   a. Name five mechanical sorting techniques and describe their principles.
   b. Name five spectroscopic methods used in recycling and describe their principles.
9. A promising method to sort plastics is using marker systems.
   a. Name the marker systems currently used in recycling.
   b. Discuss the pros and cons of marker systems.

10. The last step in the mechanical recycling process is post-processing.
    a. Explain the problems that can occur with PE and PP waste.
    b. Give a solution for these problems.
11. After the post-processing step, recycled material is ready for the market.
    a. Give reasons to apply recyclates in new products.
    b. Write down the applications of PP, HDPE and PET recyclates.
    c. Name three factors a company should take into account when starting to use recyclates.
12. Not all problems which arise during mechanical recycling are solved yet.
    a. Mention three remaining challenges in mechanical recycling.
    b. Explain what happens to PE, PP and PET after multiple recycling steps.
    c. Discuss what kind of packaging causes problems in the plastic recycling chain.

4.6.4 Chemical recycling

13. In addition to mechanical recycling, a new development is chemical recycling.
    a. Mention four chemical recycling techniques.
    b. Explain the applications.
    c. Name four advantages of chemical recycling compared to mechanical recycling.
    d. Discuss the disadvantages of chemical recycling.
14. Polyethylene terephthalate (PET) is a polymer which can be recycled in many ways.
    a. Describe two processes in which PET can be recycled in a chemical way.
    b. Explain the term: "shifting the equilibrium of a polycondensate".
    c. Explain why PET is recycled chemically, because PET can also be recycled mechanically.
15. Chemical recycling needs energy to run the process.
    a. Explain how efficiency can be determined for chemical recycling.
    b. Write down your opinion whether chemical recycling can replace mechanical recycling.
    c. Discuss the advantages and disadvantages of enzymatic recycling.

4.6.5 Sustainable plastic-product development

16. Starting to use recyclates in products requires some steps before introducing a product on the market.
    a. Describe the roadmap a company has to make, before introducing a recyclate in a product.
    b. Explain why sometimes recyclate is mixed with virgin material.
    c. Write down some non-plastic materials that could be replaced by recycled plastics.

17. Pigments are used for coloring of plastics.
    a. Explain why coloring is an issue in recyclates.
    b. Give an explanation why colorants have to be removed in the recycling process.
18. Besides technical issues, politics can also play an important role in recycling of plastics.
    a. Explain how politics can be of influence with regards to plastics in the circular economy.
    b. Describe what would be the best measure of politics, in your opinion, to improve plastic recycling.
19. The Ellen MacArthur Foundation described a few principles of the circular economy for companies.
    a. Write down the aim of the principles of a circular economy according to the Ellen MacArthur foundation.
    b. Give your opinion about using plastics as a raw material for energy.
20. In the development of new products, different methods can be used.
    a. Explain the method of "design for recycling".
    b. Explain the method of "design for disassembly".

Acknowledgment

We would like to thank NHL Stenden University of Applied Sciences for making it possible to write this textbook. Although the COVID-19 pandemic accelerated the writing process, without the facilitation of our university, this book would not have been written. The board and management of our university represented by Erica Schaper, Jooske Haije, and Harm-Jan Bouwers, thank you for providing full support.

In addition, we would like to acknowledge the support of our colleagues, who contributed directly or indirectly to this book. From the professorship Sustainable Polymers, we thank Corinne van Noordenne, Daan van Rooijen, Rik Brouwer, Geraldine Schnelting, Tobias van der Most, Renato Lemos Cosse, Jan Hans, Jarno Guit, and Albert Hartman. From the professorship Circular Plastics, the contributions of Marcel Crul, Sieger Pruiksma, Femke Jaarsma, Judith Ogink, Louis Lintveld, Mariska van Cronenberg, Uchechi Ubinna, Johannes van der Steen, and Siep Haakma, are much appreciated. Further we would like to thank our office managers Joyce Rotman and Djoke Bijlsma-Heeg.

Last but not least, we want to acknowledge our students who keep inspiring us. Our future generation holds the key to the successful transition towards a new (circular) plastics economy. It is because of them that we wrote this textbook.

Vincent Voet
Jan Jager
Rudy Folkersma

https://doi.org/10.1515/9783111201443-005

Bibliography

[1]   Kaur G., Uisan K., Ong K. L., Lin C. S. K. Recent trends in green and sustainable chemistry and waste valorisation: rethinking plastics in a circular economy. Current Opinion in Green and Sustainable Chemistry 2018, 9, 30–39.

[2]   Plastics Europe. The circular economy for plastics. An European overview, 2022.

[3]   Kunststof & Rubber Newsletter, February 2022, MYbusinessmedia.

[4]   Leslie H.A., van Velzen M. J. M., Brandsma S.H., Vethaak D., Garcia-Vallejo J.J., Lamoree M.H. Discovery and quantification of plastic particle pollution in human blood, Environment International, 2022.

[5]   SystemIQ. ReShaping Plastics: Pathways to a Circular, Climate Neutral Plastics System in Europe, 2022.

[6]   Ellen MacArthur Foundation and McKinsey & Company, The new plastics economy – rethinking the future of plastics, 2016.

[7]   Meadows D.H., Dennis L., Randers K., Behrens W.W. The limits to growth: A report to the Club of Rome, 1972.

[8]   Stahel W.R. The Circular Economy in the European Union: History of the Circular Economy, Springer Nature, Switzerland, 2020.

[9]   United Nations, Evaluation for agenda 2030 – providing evidence on progress and sustainability, 2017.

[10] World Economic Forum, Insight report – top 10 emerging technologies, 2019.

[11] European Commission, Circular economy action plan – for a cleaner and more competitive Europe, 2019.

[12] European Parliament. Directive on the reduction of the impact of certain plastic products on the environment, 2019.

[13] Challa G., Schouten A. J., Loos K. Polymer chemistry: an introduction. 3rd ed., University of Groningen, Groningen, The Netherlands, 2010.

[14] Aleman J., Chadwik A. V., He J., Hess M., Horie K., Jones R. G., Pratochvil P., Meisel I., Mita I., Moad G., Penczek S., Stepto R. F. T. Definitions of terms relating to the structure and processing of sols, gels, networks, and inorganic-organic hybrid materials. Pure Appl. Chem. 2007, 79, 1801–1829.

[15] Vollhardt K. P. C., Shore, E. Organic chemistry: structure and function. 4th ed., Freeman, New York, USA, 2002.

[16] Van der Vegt A. K. From polymers to plastics. 1st ed., VSSD, Delft, The Netherlands, 2006.

[17] Koltzenburg S., Masko M., Nuyken O. Polymer chemistry. 1st ed., Springer, Berlin, Germany, 2017.

[18] McMurry J. E., Fay R. C. Chemistry. 4th ed., Pearson Education, Upper Saddle River, NJ, USA, 2004.

[19] Clayden J., Greeves N., Warren S., Wothers P. Organic chemistry. 7th ed., Oxford University Press, Oxford, United Kingdom, 2002.

[20] Flory P. J. Principles of polymer chemistry. Cornell University Press, Ithaca, New York, USA, 1953.

[21] Ashter S. A. Introduction to bioplastics engineering. 1st ed., Elsevier (William Andrew), Oxford, United Kingdom, 2016.

[22] Matyjaszewski K., Sumerlin B. S., Tsarevsky N. V., Chiefari J. Controlled radical polymerization: mechanisms. 1st ed., Oxford University Press, Oxford, United Kingdom, 2016.

[23] Szwarc M., Levy M., Milkovich R. Polymerization initiated by electron transfer to monomer. A new method of formation of block polymers. J. Am. Chem. Soc. 1956, 78, 2656–2657.

[24] Ziegler K., Holtzkamp E., Breil H., Martin H. Das Mülheimer Normaldruck-Polyäthylen-Verfahren. Angew. Chem. 1955, 67, 541–547.

[25] Brintzinger H. H., Fischer D., Muelhaupt R., Rieger B., Waymouth R. M. Stereospecific olefin polymerization with chiral metallocene catalysts. Angew. Chem., Int. Ed. Engl. 1995, 34, 1143–1170.

[26] Brandrup J., Immergut E. H., Grulke E. A. Polymer handbook. 4th ed., Wiley, New York, USA, 1999.

https://doi.org/10.1515/9783111201443-006

[27] Odian G. Principles of polymerization. 3th ed., Wiley, New York, USA, 1991.
[28] Chandrasekaran S. Click reactions in organic synthesis. 1st ed. Wiley-VCH Verlag, Weinheim, Germany, 2016.
[29] Trommsdorff E., Koehle H., Lagally P. Makromol. Chem. 1948, 1, 169.
[30] Smith W. V., Ewart E. H. Kinetics of emulsion polymerization. J. Phys. Chem. 1948, 16, 592–599.
[31] Shrivastava A. Introduction to plastics engineering. 1st ed., Elsevier (William Andrew), Oxford, United Kingdom, 2018.
[32] Atkins P., de Paula J. Atkins' physical chemistry. 7th ed., Oxford University Press, Oxford, United Kingdom, 2002.
[33] Dettenmaier M., Fischer E. W., Stamm M. Calculation of small-angle neutron scattering by macromolecules in semicrystalline state. Colloid Polym. Sci. 1980, 258, 343–349.
[34] Voet V. S. D. Block copolymers based on poly(vinylidene fluoride). Ipskamp, Enschede, The Netherlands, 2015.
[35] Hamley I. W. Introduction to soft matter. Rev. ed. Wiley, Hoboken, USA, 2007.
[36] Winne J.M., Leibler L., Du Prez F.E. Dynamic covalent chemistry in polymer networks: a mechanistic perspective. Polym. Chem., 2019, 10, 6091.
[37] Fried J. R. Polymer science and technology. 3rd ed. Pearson Education, Upper Saddle River, NJ, USA, 2014.
[38] Crawford R. J. Plastics engineering. 3rd ed. Elsevier, Amsterdam, The Netherlands, 2014.
[39] Van Wijk A., Van Wijk I. 3D printing with biomaterials. 1st ed. IOS Press, Amsterdam, The Netherlands, 2015.
[40] European Bioplastics, Fact Sheet: What are bioplastics? Material types, terminology, and labels – an introduction, 2022.
[41] International Energy Agency (IEA) Bioenergy Task 42, Website Home – Task42 (ieabioenergy.com).
[42] Cherubini F. et al. Toward a common classification approach for biorefinery systems, 2009.
[43] Werpy T, and Petersen G. Top value added chemicals from biomass: volume I – results of screening for potential candidates from sugars and synthesis gas. United States: N. p., 2004. Web. doi:10.2172/15008859.
[44] Prieto M. A., Eugenio L. I. d., Galán B., Luengo J. M., Witholt B. Pseudomonas 2007, 1st ed., Springer, Berlin, Germany, 2007.
[45] Total Corbion PLA. Luminy® PLA documentation, 2020.
[46] EU policy framework on biobased, biodegradable and compostable plastics, COM(2022)682 final. 30-11-2022.
[47] European Bioplastics. 14th European Bioplastics Conference, 2019.
[48] The Impact of the Use of "Oxo-degradable" Plastic on the Environment, COM(2018)35 final 16-01-2018.
[49] Ten Klooster R., Dirken J.M., Lox F., Schilperoord A.A., Zakboek Verpakkingen. Plato Product Consultants, 2015.
[50] Thompson R., Thompson M. The manufacturing guides: graphics and packaging production. Thames & Hudson, 2012.
[51] Noorunnisa Khanam P., Al Ali AlMaadeed M., Processing and characterization of PE based composites. Adv. Man. Pol. & Comp. Sci. 2015, 1 (2), 63–79.
[52] KIDV The State of Sustainable Packaging. The Netherlands, 2020.
[53] Gobel G. Inventarisatie Scheidingstechnieken Harde Polyolefinen in PP en PE. OVAM, SORESMA, Vlaams Kunststofcentrum, Mechelen.
[54] McKinnon D., Fazakerley J., Hultermans R. Waste sorting plants extracting value from waste: an introduction. 12, 2017.
[55] Micelia H., Rossia M., Neumann R., Tavares L. M. Magnetic separation. Contaminant removal from manufactured fine aggregates by dry rare-earth magnetic separation, 2017.

[56] Osman E. Y. E. Analysis of effluent of plastic recycling processes. Report NHL Stenden and Van Hall Larenstein University of Applied Sciences. Leeuwarden, The Netherlands, 2018.

[57] Wang C. Q., Wang H., Fu J. G., Liu Y. N. Flotation separation of waste plastics for recycling – a review. Waste Manag. 41, 28–38.

[58] Gundupalli S., Hait S., Thakur A. A review on automated sorting of source-separated municipal solid waste for recycling. Waste Manag. 60, 56–74.

[59] Lee M. J., Rahimifard S. A novel separation process for recycling of post-consumer products. Elsevier, 1.4.

[60] Harris D. C. Fundamentals of spectrophotometry. Quantitative chemical analysis. 8th ed., New York: W. H. Freeman and Company, 393–398.

[61] Kosior E. D. J. Plastic packaging recycling using intelligent separation technologies for materials, Prism, Anaheim.

[62] Bonifazi G. G. C. Hyperspectral imaging applied to the waste recycling sector. 04 30, 2019.

[63] Gordon P., Riesebos B., Dijkstra M., Dijkstra K. Plastic foil classification using hyperspectral imaging and semantic segmentation convolutional neural networks. Centre of Expertise Computer Vision & Data Science, NHL Stenden University of Applied Sciences. Leeuwarden, The Netherlands, 2019.

[64] Weiss M., XRF-new applications in sensor-based-sorting using X-ray fluorescence. TK Verlag-Fachverlag für Kreislaufwirtschaft, Waste Management.

[65] Tsuchida A. H. K. Identification of shredded plastics in milliseconds using Raman spectroscopy for recycling, 2009.

[66] Maris E., Aoussat A., Naffrechoux E., Froelich D. Polymer tracer detection systems with UV fluorescence spectrometry to improve product recyclability. Miner. Eng. 2012, 29, 77–88.

[67] Kutz M. Applied plastics engineering handbook: processing and materials. Elsevier Science, Norwich, USA.

[68] Bicerano J. A practical guide to polymeric compatibilizers for polymer blends. Composites and laminates.

[69] Vervoort S., den Doelder J., Tocha E., Genoyer J., Walton K. L., Hu Y., Jeltsch K. Compatibilization of polypropylene-polyethylene blends. Polym. Eng. Sci. 2018, 58 (4), 460–465.

[70] Ajji A., Utracki L. A. Interphase and compatibilization of polymer blends. Polym. Eng. Sci. 1996, 36 (12), 1574–1585.

[71] Fel E., Khrouz L., Massardier V., Cassagnau P., Bonneviot L. Comparative study of Gamma-irradiated PP in PE polyolefins part 2: properties of PP/PE blends obtained by reactive processing with radicals obtained by high shear or Gamma-irradition. Polymer 82, 217–227.

[72] Rudolph N., Kiesel R., Aumnate C. Understanding plastics recycling: economic, ecological and technical aspects of plastic waste handling. Carl Hanser Verlag, Munich, Germany, 2017.

[73] Future of Chemical Recycling. British Plastic Federation, 2020.

[74] Horizon, The EU Research & Innovation Magazine, 2020.

[75] El Dorado of Chemical Recycling, State of Play and Policy Challenges. Zero Waste Europe, 2020.

[76] Thoden van Velzen U. Recycling options for PET trays. Wageningen Food & Biobased Research report no 1761, 2017.

[77] Schlummer M., Fell T., Mäurer A., Altnau G. The role of chemistry in plastics recycling – a comparison of physical and chemical plastics recycling. Kunststoffe International 2020, 5, 34–37.

[78] Ragaert K., Ragot C., Van Geem K.M., Kersten S., Shiran Y., De Meester S. Clarifying European terminology in plastics recycling, www.sciencedirect.nl, Current opinion in Green and Sustainable Chemistry, 2023, 44.

[79] van Eijk F., Sederel W., Groen J., van den Beuken E. Chemical recycling in a circular perspective Holland Circular Hotspot, The Netherlands Ministry of Infrastructure and Water Management, The Network Chemical Recycling of the Circular Biobased Delta, TNO, Chemport Europe, Chemelot Circular Hub, Chemistry NL and Infinity Recycling, 2023.

[80] Loos K. Biocatalysis in Polymer Chemistry; Wiley-VCH Verlag, 2010. doi:10.1002/9783527632534.

[81] Benninga J., Jager J., Voet V.S.D., Loos K.U. Enzymatic Polymerization of Polyurethanes, Chapter 5, 71–87, Sustainable Green Chemistry in Polymer Research, ACS Publicstions, 2023.

[82] CE Delft Publications, Newsletter Business Cases Circular Economy. The Netherlands, 2020.

[83] den Hollander M., Idema M., Joore P. Circular design research in the Netherlands, CLICKNL and Taskforce for Applied Research SIA, part of the Netherlands Organisation for Scientific Research (NOW), 2023.

[84] Inspiratiegids Circulair Ontwerp, Rijkswaterstaat, 2020.

[85] Rijksdienst voor Ondernemend Nederland, Partners for Innovation. NRK. Ontwerpen met Kunststof Recyclaat. The Netherlands, 2020.

[86] Vogtlander J. LCA: a practical guide for students, designers and business managers, Research Gate, 2012.

[87] Voulvoulis N., Kirkman R., Giokoumis T., Metivier P., Kyle C., Midgley V. Examining material evidence. The carbon fingerprint. Imperial College, London, UK, 2020.

Index

3D printer 75

addition polymers 13
additive manufacturing 75
additives 72, 237, 283
air classifier 222
alternating copolymer 17
alternating copolymerization 47
amorphous polymers 58, 182
anionic polymerization 39
aramids 31, 169
atactic 20

barcodes 242
barrier properties 215
bio-based carbon content 193
bio-based economy 205
bio-based mass content 195
bio-based plastics 97, 191
bio-based polyethylene 101
bio-based polypropylene 108
bioBTX 87, 118
biochemicals 95
biodegradable plastics 97, 172, 197, 271
biodegradable polymers 57, 126, 171, 188
biodegradation 197, 200
biomass 84, 91, 95, 191
bioplastics 30, 84, 97–98, 181, 189, 204
biorefining 84, 94
block copolymer 17, 48, 168, 244
block copolymer morphologies 70
blown film extrusion 74
branched macromolecules 17
building blocks 91, 96, 143

calendering 75
caps and closures 213
Carothers' equation 28
cationic polymerization 39
cellulose 87, 156
cellulose derivates 55
centrifugal force 222
certification 190, 202, 275
chain transfer 36
chain-growth polymerization 13, 35
chemical blowing agent 78
chemical crosslinks 19

chemical oxygen demand 226
chemical recycling 259
chemolysis 266
circular economy 7, 271, 276
circular plastics 209
circular product design 271
cis configuration 19, 44
closed-cell foams 78
closed-loop plastic economy 9
CO_2 footprint 281
coextrusion process 218
colorants 257
compatibilizers 243
composition drift 47
compostable 172, 181
compounder 73
compression molding 79
condensation polymers 13, 266
conducting polymers 69
contaminants 226, 255
controlled radical polymerization 38
conversion processes 92
coordination polymerization 41
copolymer composition diagram 47
copolymer engineering 50
copolymerization 46
critical micelle concentration 52
crosslinking 17
crosslinking reactions 55, 246
crude oil 2, 265
crystal growth 64
crystalline polymers 57, 182
crystallinity 57, 65, 182
crystallization 63
crystallization temperature 61
curing 32

decomposition 35
degradation 57, 253
degree of crystallization 61
degree of polymerization 20, 27, 36
dehydration 108
depolymerization 260, 266
detergents 226
Diels–Alder reaction 142
differential scanning calorimetry 61, 248
direct esterification 171

https://doi.org/10.1515/9783111201443-007

disc screen 221
disproportionation 35
drawing 77
drop-in bioplastics 191
dry spinning 77

Eddy current separator 223
elastane 170
elastomers 13
electrical insulators 69
electromagnetic spectrum 234
electrostatic charging 69
elongation at break 66, 249
E-modulus 58, 249
emulsion polymerization 53, 116
enantiomers 179
end group analysis 23
entanglements 58
environmental impact 196, 251, 284
enzymatic process 166
Enzymatic recycling 268
epoxy resins 32
European Green Deal 9
expanded polystyrene 115
extruder 73
extrusion 73

fatty acids 91, 173, 226
fermentation 85, 111, 129, 180
fiber-reinforced plastics 78, 264
fibers 13, 76
filament winding 80
fillers 72
film grabber 223
films 212
filtration 227
Flory's principle of equal reactivity 27
flotation 229
fluorescent markers 241
foams 78
foils 211
fossil-based plastics 97
fracture energy 66
free radical polymerization 35, 121
fused filament fabrication 75

gasification 86, 262, 265
gel effect 51
gel permeation chromatography 24

glass transition 60, 248
glass transition temperature 58
glycolysis 120, 267
graft copolymer 17, 48, 244
grafting from 48
grafting (on)to 48
grafting through 48
greenhouse gas 106

hand lay-up 79
HDPE 43, 100, 212
head-to-head 19
head-to-tail 19
home compostability 201
homopolymer 16
Hooke's law 58
hotmelt adhesive 212
hybrid polymers 12
hydrocyclone 230
hydrogenation 146, 159, 188
hydrogenolysis 138
hydrolysis 148, 197
hydrothermal treatment 264
hyperspectral imaging 236

ideal copolymerization 47
impact strength 68
incineration 259
industrial compostability 199
initiation 35
initiator 35
injection blow molding 74
injection molding 74
inorganic polymers 12
interfacial polymerization 31
intrinsic viscosity 24
ionomers 217
irreversible creep 62
isotactic 20

Kelvin–Voigt model 63

laminate 218
latex particle 54
LDPE 43, 100, 211
legislation 275
life cycle 277, 281
life cycle assessment 251, 281
lignin 87

lignocellulosic biomass 88, 131
linear economy 7
linear macromolecules 17
liquid crystalline phase 69
liquid crystalline polymers 69
living polymerization 40

macromolecules 1
macroplastics 6
MALDI-ToF mass spectroscopy 25
marker systems 241
Mark–Houwink equation 24
mass polymerization 51
Maxwell model 62
Mayo–Lewis equation 46
mechanical recycling 210, 220
melt spinning 76
melting temperature 60
membrane osmometry 23
metallocene catalysts 42
methanolysis 267
microorganisms 175, 196
microphase separation 70
microplastics 6
moisture 215, 238
molar mass 21
molecular weight 21
monodisperse 22
monomers 11
monosaccharide 89
morphology 57
multilayer films 215, 218, 255

natural polymers 11
Near Infrared 235
negative sorting 221
network 17
New Plastics Economy 7
Newtonian fluid 68
Newton's law of viscosity 68
nomenclature 14
non-Newtonian fluid 68
nucleation 64
number average molar mass 21
nylon rope trick 31
nylon salt 31, 159

olefin metathesis 112
oligomers 11

open-cell foams 78
organic polymers 12
oscillating screen 221
oxo-biodegradation 203
oxygen transmission rate 216

packaging 133, 205, 211, 226, 253, 284
packaging industry 4
physical blowing agent 78
physical crosslinks 19
pigments 239, 256
Plant Bottle 137, 251
plastic converter demand 3
plastic shredders 224
plastic soup 2, 6
plastic waste 232
plasticizers 72, 114
plastics 2, 13
plastics industry 2
plastics supply chain 260
platforms 85
plexiglass 127
Poisson distribution 41
pollutants 226
polyacrylic acid 123
polyacrylonitrile 44, 130
polyamides 30, 155, 161
polybutadiene rubber 44
polybutylene succinate 171
polybutylene terephthalate 146
polycarbonates 32
polycondensation 27, 185
polydispersity 22
polyesters 29, 133, 205
polyethylene 43, 98
polyethylene furanoate 143
polyethylene terephthalate 133, 214
polyhydroxyalkanoates 85, 173, 226
polyisobutylene 44
polyisoprene rubber 44
polylactic acid 179
polymer analogous reaction 54
polymerization 26
polymers 1, 11
polymethyl methacrylate 44, 127
polypropylene 44, 106
polysaccharide 89
polysiloxanes 32
polystyrene 44, 115

polytetrafluoroethylene 44
polyurethanes 34
polyvinylacetate 121
polyvinylchloride 44, 114
positive sorting 221
post-polymerization reactions 54
prepolymers 32
pressing 75
process flow diagram 84, 94
product development 211
propagation 35
proteins 226
pultrusion 79
purification 261
pyrolysis 87, 114, 118, 143, 262

quality 251–252

radical combination 35
radical polymerization 35, 127
random copolymer 16
R-configuration 20
reaction injection molding 74
reactive extrusion 74
reactivity ratio 46
recyclable 5, 277
recyclates 243, 250, 272
recycling 128, 209, 272
reduce 209
renewable feedstock 85, 92
replacement 274
Resin Identification Code 4
resin transfer molding 79
reuse 209
reversible creep 63
rheology 68
ring opening metathesis polymerization 43
ring opening polymerization 30, 42, 126,
    155, 180
rubber 13
rubbery plateau 59

Schulz–Flory distribution 29
S-configuration 20
sealing 212
semi-synthetic polymers 11
separation 229

shear thinning 68
shredding 225
silicone rubber 32
size exclusion chromatography 24, 247
solution polymerization 52, 116
sorting 220
spectroscopic methods 23, 233, 240
spherulites 64
spinneret 76
spinning 76
spray-up 79
standardization 190
starch 89, 152
steam cracking 2, 101, 157
step-growth polymerization 13, 27
stress corrosion 68
stress relaxation 63
sugars 90, 138, 148, 226
super critical fluid 233
suspension polymerization 52
Sustainable Development Goals 8
syndiotactic 20
syngas 86, 264
synthetic polymers 1, 12

tags 239
tail-to-tail 19
tautomer 54
tensile strength 66
termination 35
terpolymer 17
thermoplastic elastomers 49, 70, 168
thermoplastics 3, 13
thermosets 13
tie-molecules 64
trans configuration 19, 44
transesterification 121, 146, 171
trommel screen 221
Trommsdorff effect 51

vacuum infusion 79
viscoelasticity 59, 62
viscometry 24
viscosity 68

waste disposal 225
waste management 284

wastewater 226
water vapor transmission rate 216
weight average molar mass 22
wet spinning 77
wettable 218
wind shifter 231

X-ray fluorescence 239, 249

yield point 67
yield strength 67
Young's modulus 58, 67

Ziegler–Natta polymerization 41

www.ingramcontent.com/pod-product-compliance
Lightning Source LLC
Chambersburg PA
CBHW080937220326
41598CB00034B/5808